ACS SYMPOSIUM SERIES **493**

Macromolecular Assemblies in Polymeric Systems

Pieter Stroeve, EDITOR
University of California–Davis

Anna C. Balazs, EDITOR
University of Pittsburgh

Developed from a symposium sponsored
by the Division of Polymer Chemistry, Inc.,
at the 201st National Meeting
of the American Chemical Society,
Atlanta, Georgia,
April 14–19, 1991

American Chemical Society, Washington, DC 1992

Library of Congress Cataloging-in-Publication Data

Macromolecular assemblies in polymeric systems/ Pieter Stroeve, editor, Anna C. Balazs, editor.

   p.   cm.—(ACS Symposium Series, 0097–6156; 493).

"Developed from a symposium sponsored by the Division of Polymer Chemistry, Inc., at the 201st National Meeting of the American Chemical Society, Atlanta, Georgia, April 14–19, 1991."

Includes bibliographical references and index.

ISBN 0–8412–2427–7

1. Macromolecules—Congresses. 2. Polymers—Congresses.
I. Stroeve, Pieter, 1945– . II. Balazs, Anna C., 1953– .
III. American Chemical Society. Division of Polymer Chemistry.
IV. American Chemical Society. Meeting (201st: 1991: Atlanta, Ga.).
V. Series.

QD380.M23   1992
547.7—dc20   92–15020
                                 CIP

The paper used in this publication meets the minimum requirements of American National Standard for Information Sciences—Permanence of Paper for Printed Library Materials, ANSI Z39.48–1984. ∞

Copyright © 1992

American Chemical Society

All Rights Reserved. The appearance of the code at the bottom of the first page of each chapter in this volume indicates the copyright owner's consent that reprographic copies of the chapter may be made for personal or internal use or for the personal or internal use of specific clients. This consent is given on the condition, however, that the copier pay the stated per-copy fee through the Copyright Clearance Center, Inc., 27 Congress Street, Salem, MA 01970, for copying beyond that permitted by Sections 107 or 108 of the U.S. Copyright Law. This consent does not extend to copying or transmission by any means—graphic or electronic—for any other purpose, such as for general distribution, for advertising or promotional purposes, for creating a new collective work, for resale, or for information storage and retrieval systems. The copying fee for each chapter is indicated in the code at the bottom of the first page of the chapter.

The citation of trade names and/or names of manufacturers in this publication is not to be construed as an endorsement or as approval by ACS of the commercial products or services referenced herein; nor should the mere reference herein to any drawing, specification, chemical process, or other data be regarded as a license or as a conveyance of any right or permission to the holder, reader, or any other person or corporation, to manufacture, reproduce, use, or sell any patented invention or copyrighted work that may in any way be related thereto. Registered names, trademarks, etc., used in this publication, even without specific indication thereof, are not to be considered unprotected by law.

PRINTED IN THE UNITED STATES OF AMERICA

ACS Symposium Series

M. Joan Comstock, *Series Editor*

1992 ACS Books Advisory Board

V. Dean Adams
Tennessee Technological University

Mark Arnold
University of Iowa

David Baker
University of Tennessee

Alexis T. Bell
University of California—Berkeley

Arindam Bose
Pfizer Central Research

Robert Brady
Naval Research Laboratory

Margaret A. Cavanaugh
National Science Foundation

Dennis W. Hess
Lehigh University

Hiroshi Ito
IBM Almaden Research Center

Madeleine M. Joullie
University of Pennsylvania

Mary A. Kaiser
E. I. du Pont de Nemours and Company

Gretchen S. Kohl
Dow-Corning Corporation

Bonnie Lawlor
Institute for Scientific Information

John L. Massingill
Dow Chemical Company

Robert McGorrin
Kraft General Foods

Julius J. Menn
Plant Sciences Institute,
U.S. Department of Agriculture

Vincent Pecoraro
University of Michigan

Marshall Phillips
Delmont Laboratories

A. Truman Schwartz
Macalaster College

John R. Shapley
University of Illinois at Urbana–Champaign

Stephen A. Szabo
Conoco Inc.

Robert A. Weiss
University of Connecticut

Peter Willett
University of Sheffield (England)

Foreword

THE ACS SYMPOSIUM SERIES was founded in 1974 to provide a medium for publishing symposia quickly in book form. The format of the Series parallels that of the continuing ADVANCES IN CHEMISTRY SERIES except that, in order to save time, the papers are not typeset, but are reproduced as they are submitted by the authors in camera-ready form. Papers are reviewed under the supervision of the editors with the assistance of the Advisory Board and are selected to maintain the integrity of the symposia. Both reviews and reports of research are acceptable, because symposia may embrace both types of presentation. However, verbatim reproductions of previously published papers are not accepted.

Contents

Preface ix

1. Macromolecular Assemblies in Polymeric Systems: An Overview 1
   Pieter Stroeve and Anna C. Balazs

   MONOLAYERS AND MULTILAYER FILMS

2. Liquid-Crystalline Acetylenic Compounds: Behavior at Air–Water Interface 10
   R. C. Advincula, M. J. Roberts, X. Zhang, A. Blumstein, and R. S. Duran

3. Blends of Side-Chain Liquid-Crystalline Polymers at Air–Water Interface 20
   A. F. Thibodeaux, R. S. Duran, H. Ringsdorf, A. Schuster, A. Skoulios, P. Gramain, and W. Ford

4. Effect of Surface Pressure on Langmuir–Blodgett Polymerization of 2-Pentadecylaniline 31
   Huanchun Zhou and R. S. Duran

5. Monolayer and Thin-Film Crystallization of Isotactic Poly(methyl methacrylate) 49
   R. H. G. Brinkhuis and A. J. Schouten

6. Preparation of Multicomponent Langmuir–Blodgett Thin Films Composed of Poly(3-hexylthiophene) and 3-Octadecanoylpyrrole 64
   M. Rikukawa and M. F. Rubner

7. Langmuir–Blodgett Films of Novel Polyion Complexes of Conducting Polymers 76
   A. T. Royappa and M. F. Rubner

8. **Fluorinated, Main-Chain Chromophoric Polymer: Langmuir Layer and Fourier Transform Infrared Spectroscopic Studies** ... 83
   Pieter Stroeve, Roni Koren, L. B. Coleman, and J. D. Stenger-Smith

9. **Langmuir–Blodgett Multilayers of Fluorinated, Main-Chain Chromophoric, Optically Nonlinear Polymers** ... 94
   J. M. Hoover, R. A. Henry, G. A. Lindsay, M. P. Nadler, S. F. Nee, M. D. Seltzer, and J. D. Stenger-Smith

10. **Langmuir–Blodgett Films of Stilbazolium Chloride Polyethers: Structural Studies** ... 104
    David D. Saperstein, John F. Rabolt, J. M. Hoover, and Pieter Stroeve

11. **Static Secondary Ion Mass Spectrometric Analysis of Langmuir–Blodgett Film Multilayers: A Sampling Depth Study** ... 113
    Robert W. Johnson, Jr., Paula A. Cornelio-Clark, and Joseph A. Gardella, Jr.

12. **Langmuir–Blodgett Affinity Surfaces: Targeted Binding of Avidin to Biotin-Doped Langmuir–Blodgett Films at the Tip of an Optical Fiber Sensor** ... 122
    Shulei Zhao and W. M. Reichert

13. **Mixed Monolayers of Lecithin and Bile Acids at the Air–Aqueous Solution Interface: Effect of Temperature and Subphase pH** ... 135
    M. J. Gálvez-Ruiz and M. A. Cabrerizo-Vilchez

THREE-DIMENSIONAL SYSTEMS

14. **Photoinduced Morphological Changes in Plasmalogen Liposomes Using Visible Light** ... 154
    Valerie C. Anderson and David H. Thompson

15. **Percolation Process and Structural Study in Docusate Sodium Reverse Micelles Containing Cytochrome *c*** ... 171
    J. P. Huruguen, T. Zemb, C. Petit, and M. P. Pileni

16. **Polymerization and Phase Transitions in Deoxy Sickle Cell Hemoglobin** ... 184
    Muriel S. Prouty

17. Reduction of Phospholipid Quinones in Bilayer Membranes: Kinetics and Mechanism 202
    Charles R. Leidner, Dale H. Patterson, William M. Scheper, and Min D. Liu

18. Proteinaceous Microspheres 218
    Mark W. Grinstaff and Kenneth S. Suslick

19. Molecular Recognition in Gels, Monolayers, and Solids 227
    Kevin L. Prime, Yen-Ho Chu, Walther Schmid, Christopher T. Seto, James K. Chen, Andreas Spaltenstein, Jonathan A. Zerkowski, and George M. Whitesides

SCANNING PROBE MICROSCOPY
OF MACROMOLECULAR ASSEMBLIES

20. Scanning Probe Microscopy of Surfactant Bilayers and Monolayers 242
    J. A. N. Zasadzinski, J. T. Woodward, M. L. Longo, and B. Dixon-Northern

21. Elucidation of Macromolecular Assemblies: Use of Scanning Tunneling Microscopy for Molecular Bioengineering of Cellular Self-Assemblies in Molecular Device Design 256
    Vincent B. Pizziconi and Darren L. Page

POLYMERS AND LIQUID CRYSTALS

22. Chiral Liquid-Crystalline Copolymers for Electrooptical Applications 280
    E. Chiellini, G. Galli, A. S. Angeloni, M. Laus, and D. Caretti

23. Side-Chain Crystallinity and Thermal Transitions in Thermotropic Liquid-Crystalline Poly(γ-alkyl-α,L-glutamate)s 292
    William H. Daly, Ioan I. Negulescu, Paul S. Russo, and Drew S. Poche

24. A Novel Route to Poly(ether ketone)–Polycondensate Block Copolymers 300
    Robert J. Kumpf, Dittmar Nerger, Christopher Lantman, Harald Pielartzik, and Rolf Wehrmann

INDEXES

Author Index 315

Affiliation Index 316

Subject Index 316

Preface

THE RESEARCH AREA OF MACROMOLECULAR ASSEMBLIES is very broad. Macromolecular assemblies occur in nature but can also be fabricated by self-assembly, Langmuir–Blodgett deposition, corona-onset poling, organic molecular beam epitaxy, and many other techniques. Monolayers, bilayers, multilayers, micelles, vesicles, liquid crystals, polymer–colloid associations, and lipid tubules are examples of macromolecular assemblies. The physical and chemical properties of these systems are of great interest to scientists and engineers. Integrated optical devices of the future presumably would be constructed using these layers of organized molecules. The use of ultrathin layers would mean that chemical, biochemical, and physical sensors could be miniaturized. In the future, an understanding of the behavior of biological lipid membranes will result in the ability to devise drug delivery systems, to control the metabolism of cells, and to protect cells from the attack of a virus. The presence and behavior of macromolecules, such as polymers, in these systems is of increasing interest, and consequently the role of polymers in macromolecular assemblies is emphasized in this book.

The purpose of *Macromolecular Assemblies in Polymeric Systems* is to present the latest research results on macromolecular assemblies. To that end, in this volume there are chapters dealing with Langmuir layers and Langmuir–Blodgett films, liquid-crystalline materials, adsorption and self-assembly, biological materials, bilayers, micelles, surfactants, and polymers. Contributions in this volume come from chemists, physicists, chemical engineers, electrical engineers, biomedical engineers, polymer scientists, and biologists. Their interests range from using macromolecular assemblies to make products, e. g., microsensors and optical devices, to understanding biological phenomena. The study of the behavior of polymeric species and of their use in structures such as adsorbed monolayers, liquid crystals, and vesicles has expanded considerably in the past 10 years. The synthesis of tailor-made polymers for use in macromolecular assemblies is now of great interest. *Macromolecular Assemblies in Polymeric Systems* provides important insight into this dynamic research area and should be of interest to those who conduct interdisciplinary research.

The editors gratefully acknowledge the Division of Polymer Chemistry, Inc., and the IBM and Mobay Corporations for their financial support.

PIETER STROEVE
University of California
Davis, CA 95616–5294

ANNA C. BALAZS
University of Pittsburgh
Pittsburgh, PA 15261

April 27, 1992

Chapter 1

Macromolecular Assemblies in Polymeric Systems

An Overview

Pieter Stroeve[1] and Anna C. Balazs[2]

[1]Department of Chemical Engineering and Organized Research Program on Polymeric Ultrathin Film Systems, University of California, Davis, CA 95616
[2]Department of Materials Science and Engineering, University of Pittsburgh, Pittsburgh, PA 15261

>The areas of macromolecular assemblies in polymeric systems are reviewed and the applications for useful products or processes are discussed. The challenges in macromolecular assemblies in polymeric systems are to design and synthesize new molecules which can assemble into macromolecular structures which have useful material properties. There is a need to be able to predict the collective properties of the assembly from a knowledge of the properties of the individual molecules.

Unconventional materials such as polymers containing moieties which have optically active, magnetic, catalytic or other properties may find a variety of new applications in microelectronics, laser technology, magnetics, communications, electro-optics, bioengineering and biotechnology, catalysis and membrane separations. For example, the computer technology in the early part of the coming century will be a hybrid of microelectronics and integrated optics. Increased performance will require fabrication of macromolecular assemblies based on inexpensive materials processed at low temperatures with high yield over large areas. For optical applications, it is known that organic materials for ultrathin films have superior physical properties compared with those of inorganic materials. Organic materials have a higher damage threshold to laser irradiation than inorganic materials. Organic materials in the form of polymers offer superior mechanical properties, higher damage thresholds and improved stability compared with properties offered by low molecular weight materials. Ultrathin polymer film systems may form the basis of new logic, memory, and display elements if they can be fabricated in highly ordered form so that their structural, surface and interfacial properties can be understood and exploited.

The study of the properties of thin films has become an important scientific endeavor, particularly as more complex materials became available and are incorporated into a variety of devices. The push for micro-miniaturization in the field of electronics has stimulated research on ultrathin films. There is considerable interest in the properties of ultrathin films from less than a nanometer to about a micron thick. The study of single monolayers is important in catalysis, colloid science, biomedical science, engineering, materials science and other fields in which surface properties need to be modified. Single monolayers of a particular material can be made by a variety of techniques, but adsorption from a solution or a gas phase is probably the simplest method. The presence of a single monolayer on a substrate can have a dramatic effect on the substrate's surface properties, such as surface tension and catalytic activity. Sophisticated measurement techniques have been or are being developed to measure the surface properties when such minute quantities of material are present on the substrate.

Organized layers thicker than a monolayer are now used in microelectronics and other high technology industries. For example, ultrathin films are made in microlithographic steps, chemical vapor deposition and liquid phase epitaxy. The requirements from materials engineering considerations are that the films are uniform and have order. Crystalline ultrathin film design and growth is a major area of solid-state research. The main focus has been on inorganic materials, but recently ultrathin films of organic materials have become of considerable interest. For example, the use of ultrathin organic films has enormous potential in Q-switches, parametric oscillators and amplifiers, frequency doublers, modulators, and filters [1, 2]. Successful development of such devices has a large number of applications in telecommunications, data processing, nuclear fusion and laser irradiation. Nonlinear optical properties of inorganic materials depend on ions and attempts to process light signals faster than the order of picoseconds will cause ion vibrations which lead to energy losses and crystal damage. On the other hand, nonlinear optical properties of organics depend on the p-electron system which does not cause vibrations within the crystal structure. Polymers are the preferred materials for ultrathin films [3]. The use of polymers gives superior mechanical properties compared to organic materials of low molecular weight. The requirements for structures and molecular order within the films makes polymers the prime candidate for the design of organic ultrathin films systems for optical applications.

The construction of macromolecular assemblies can take place spontaneously such as the self-assembly of molecules adsorbed on an interface, the self-assembly into a micelle in solution or the self-assembly of polymers in solution into a polymeric liquid crystal. Macromolecular assemblies can also be fabricated by a variety of techniques, for example, Langmuir-Blodgett (L-B) deposition. With the L-B deposition technique, it is possible to construct macromolecular assemblies consisting of multilayer of molecules of precise order and composition. Depending on the molecules used in each monolayer, unusual physical properties can be imparted to the macromolecular structure.

The alignment of the rigid segments in liquid crystalline polymers constitutes another example of macromolecular self-assembly. The driving force for such molecular rearrangements arises primarily from steric forces. As a result of these forces, highly ordered fluid phases are formed, which lead to polymers with unusually high mechanical properties. Controlling this self-alignment process by external fields constitutes an important new area of research. One of the most recent developments in liquid crystalline polymers is the exploration of cholesteric mesophases. These polymers show promise as interesting optical materials, yet this new field remains largely unexplored. Further research in this area is important since processing in the liquid crystalline state can yield materials with unique optical and mechanical properties.

In biology, nature uses biomacromolecular assemblies for a wide range of functions. The assemblies can be used for technological applications or serve as models to understand molecular interactions. Proteins can self-assemble into complex structures, including reversibly assembled supramolecualar structures. Biological structures, such as cell membranes are responsible for complex functions, e.g. transport and cell recognition. Proteins and lipids can be isolated from cell membranes and their self-assembly behavior in solution or in thin film form can be studied. The Langmuir monolayer at the air-water interface and L-B films have been widely used to study the functions of macromolecular assemblies of proteins, lipids and their mixtures [4]. Also, there exists considerable interest in combining biomacromolecular assemblies and ultrathin organic film assemblies for hybrid molecular design.

There is potential for combining protein and lipid films in energy collection and transfer which is useful in the synthesis of biomolecular electronic and photonic materials. The development of biosensors, which use thin films for selective separation or molecular recognition through specific binding, is a new area of considerable interest. Another major area is the understanding of the adsorption of proteins to surfaces with respect to biocompatibility of biomaterials and biomedical devices.

The applications, future possibilities, and research needs in macromolecular assemblies have been explored by a number of publications [5-8]. It is not possible in this short introduction to give examples of the numerous areas where macromolecular assemblies are of importance. However, it is clear from recent publications and the articles in this book that there are many systems where polymers and other macromolecules play a significant if not a dominant role. The design of polymers which can form macromolecular assemblies which have unusual physical, chemical and catalytical properties is a newly emerging field.

Macromolecular Assemblies in Polymeric Systems

This book on macromolecular assemblies in polymeric systems has been divided in four sections: Monolayers and Multilayer Films; Three Dimensional Systems; Scanning Probe Microscopy of Macromolecular Assemblies; Polymers and Liquid Crystals. Contributions to the book came from authors who participated in the symposium on Macromolecular Assemblies at the ACS Meeting held in Atlanta, Georgia, April 14-19, 1991. Here we will expand on the material in this book, .

Although there were separate sessions on Langmuir layers and Langmuir-Blodgett films, liquid crystalline materials, adsorption and self-assembly, biological materials, bilayers, micelles, surfactants and polymers, in addition to a tutorial, it is clear from the papers presented at the meeting and those compiled in this book, that thin film systems are often more accessible to study than other macromolecular assemblies. Researchers may be interested ultimately in a three-dimensional assembly organized in an aqueous solution, but monolayers and multilayers are assemblies which are relatively easy to produce and can be probed by many surface analytical techniques. For example, although the interest may be in the binding of a molecule to the surface of a micelle, the binding may be studied easier by using a L-B film rather than a suspension of micelles. It is apparent that studies on Langmuir layers and L-B films have become standard procedures in many research projects on macromolecular assemblies.

There is a wealth of information that can be obtained from Langmuir layer and L-B studies. Examples, which appear in this text, are given below. The association of liquid crystalline molecules with each other or in mixtures can be studied in a Langmuir layer contained in a Langmuir trough [9,10]. The conductive properties of polymers in film form can be measured by using L-B films [11,12]. In many practical applications, L-B films already are the preferred macromolecular assembly of interest, such as in second harmonic generation (SHG) in polymeric films [13-15]. Careful experimentation with isotactic poly (methylmethacrylate) in Langmuir layer form has shown that there is present a pressure-induced crystallization process in which double helical structures of PMMA molecules are formed [16]. A new approach has been devised in which the Langmuir layer can be used to follow the kinetics of interfacial polymerization [17]. Surface specific analytical techniques can readily be applied to L-B films since the films can either be fabricated in sufficiently large areas or the L-B film can be deposited precisely on the surface of a sensor [18, 19]. For biological systems, Langmuir layer or L-B studies can yield valuable information. Researchers interested in the process have investigated the properties of mixed Langmuir layers of lecithin and bile acid in order to arrive at a more general theory of gallstone formation [20].

Three-dimensional macromolecular assemblies such as liposomes, micelles, bilayers, etc., are often found in solutions. The complex systems encountered in the biological areas include cell membranes and the structures found in the intracellular fluid. The cell membrane contains a variety of macromolecules some of which have structural elements protruding from its surface for molecular recognition. Understanding of molecular recognition can lead to new separation methods such as affinity gel electrophoresis [21]. Targeted drug delivery from liposomes to selected areas of organs in the body is another application which is actively pursued. One technique involves absorption of near or visible light by a sensitizer to effect lipid decomposition followed by the breakdown of the liposome bilayer and the release of encapsulated drugs [22]. Stability of constituents is an obvious issue to maintaining the stability of the bilayer. One study reports on the kinetics and mechanics of reduction of phospholipid quinones in bilayer membranes [23]. On the other hand, the association properties of proteins can be used to create entirely new structures, such as proteinaceous microspheres. These structures can be exploited for new technological applications [24]. The monitoring of physical changes in three-dimensional structures can be used to obtain information on the molecular interaction of the constituents in macromolecular assemblies [25, 26].

Surface analytical techniques that have been extensively used to analyze macromolecular assemblies include UV-vis, IR, and Raman spectroscopy, X-ray reflection, neutron reflection, second harmonic generation, scanning electron microscopy (SEM), scanning tunneling microscopy (STM) and atomic force microscopy (AFM). STM and AFM are particularly attractive methods since the ultimate resolution can be to the atomic scale, and therefore individual molecules can be imaged within macromolecular assemblies. Recent results have shown that AFM can be used for imaging L-B films [27]. The use of STM on L-B films has led to numerous difficulties in interpreting the data in the form of images. However, STM has met with considerable success in imaging many other macromolecular assemblies or individual macromolecules [28]. In addition of the imaging and characterization of macromolecules, these new microscopies may be useful tools for the microfabrication of molecular assembly-based devices.

In the work referred to above, polymers or biopolymers play a significant role. It is well known that polymers in solution or polymers in bulk form can have organized structures. For example, block copolymers can form specific morphologies including lamellar structures. Bulk materials made out of new block copolymers often exhibit synergistic physical properties. The synthesis of block copolymers is of considerable importance [29]. Another example is liquid crystalline polymers. Polymeric liquid crystals feature a variety of molecular architecture, stability, mechanical orientability and ease of processibility [30]. The presence of chirality in polymeric liquid crystal materials give rise to unusual ferroelectric, optical and electrooptical properties in comparison to monomeric liquid crystals. Thus, the synthesis of chiral side chains to main-chain polymers is an active area of study [30,31]. The inclusion of chromophores as side-chains to polymers or as active units in the main chain can also create optically active materials [14].

Summary

As discussed by Swalen et al. [6], the monumental challenge in macromolecular assemblies is not only to predict the collective properties of the assembly from a knowledge of the properties of the individual molecules, but also to engineer new molecules which can assemble to give structures that exhibit new and technologically useful effects. At present this is not possible. However the potential of creating new materials exists and this book gives a flavor what may be possible in the future.

Acknowledgment

This work was supported in part by a UC Davis Faculty Research Grant to P. Stroeve.

Literature Cited

1. Williams, D.J., *Nonlinear Optical Properties of Organic and Polymeric Materials*, ACS Symposium Series, No. 233, American Chemical Society, Washington, D.C. 1983.
2. Williams, D.J., *Angew Chem. Int. Ed. Engl.*, **1984**, 23, 690.
3. Garito, A.F., Singer, K.D., and Teng, C.C, in *Nonlinear Optical Properties of Organic and Polymeric Materials*, Williams, D.J. ed., ACS Symposium Series, No. 233, p. 27, American Chemical Society, Washington, DC 1983.
4. Andrade, J.F., *Thin Solid Films*, **1987**, 152, 335.
5. Kowel, S.T., Selfridge, R., Eldering, C., Matloff, N., Stroeve, P., Higgins, B.G., Srinivasan, M.P., and Coleman, L.B., *Thin Solid Films*, **1987**, 152, 377.
6. Swalen, J.D., Allara, D.L., Chandross, E.A., Garoff, S., Israelachvili, J., McCarthy, T.J., Murray, R., Pease, R.F., Rabolt, J.F., Wynne, K.J., and Yu, H., *Langmuir*, **1987**, 3, 932.
7. Stroeve, P. and Franses, E., Ed., *Molecular Engineering of Ultrathin Polymeric Films*, Elsevier Applied Science Publishers LTD, Crown House, Essex, England, 1987.
8. Ulman, A., *An Introduction to Ultrathin Organic Films*, Academic Press, San Diego, California, 1991.
9. Advincula, R.C., Robert, M.J., Zhang, X., Blumstein, A., and Duran, R.S., P. Stroeve and A.C. Balazs, Ed., in this book, 1992.
10. Thibodeaux, A.F., Duran, R.S., Ringsdorf, H., Schuster, A., Skoulios, A., Gramain, P., and Ford, W., in this book, 1992.
11. Royappa, A.T. and Rubner, M.F., in this book, 1992.
12. Rihukawa, M., and Rubner, M.F., in this book, 1992.
13. Stroeve, P., Koren, R., Coleman, L.B., and Stenger-Smith, J.D., in this book, 1992.

14. Hoover, J.M., Henry, R.A., Lindsay, G.A., Nadler, M.P., Nee, S.M., Seltzer, M.D., and Stenger-Smith, J.D., in this book, 1992.
15. Saperstein, D.D., Rabolt, J.F., Hoover, J.M., and Stroeve, P., in this book, 1992.
16. Brinkhuis, R.H.G., and Schouten, A.J., in this book, 1992.
17. Zhou, H., and Duran, R.S., in this book, 1992.
18. Johnson, R.W., Cornelio-Clark, P.A., Gardella, J.A., Jr., in this book, 1992.
19. Zhao, S., and Reichert, W.M., in this book, 1992.
20. Gálvez-Ruiz, M.J., and Gabrerizo-Vilchez, M.A., in this book, 1992.
21. Prime, K.L, Chu, Y.-H., Schmid, W., Seto, C.T., Chen, J.K., Spaltenstein, A., Zerkowski, J.A., and Whitesides, G.M., in this book, 1992.
22. Anderson, V.C. and Thompson, D.H., in this book, 1992.
23. Leidner, C.R., Patterson, D.H., Sheper, W.M., and Liu, M.D., in this book, 1992.
24. Grinstaff, M.W. and Suslick, K.S., in this book, 1992.
25. Huruguen, J.R., Zemb, T., Petit, C., and Pileni, M.P., in this book, 1992.
26. Prouty, S., in this book, 1992.
27. Zasadzinski, J.A.N., Woodward, J.T., Longo, M.L., and Dixon-Northern, B., in this book, 1992.
28. Pizziconi, V.B. and Page, D.L., in this book, 1992.
29. Kumpf, R.J., Nerger, D., Lantman, C., Pielartzik, H., and Wehrmann, R., in this book, 1992.
30. Chiellini, E., Galli, G., Angeloni, A.S., Laus, M., and Carreti, D., in this book, 1992.
31. Daly, W.H., Russo, P.S., Poche, D.S., Negulescu, E.N., Neagu, E., Neagu, R., and Kraus, M.L., in this book, 1992.

RECEIVED February 5, 1992

Monolayers and Multilayer Films

The simplest macromolecular assembly is a monolayer. Monolayers can form at practically any interface, e. g., liquid–liquid, gas–liquid, and gas–solid interfaces. The study of the monolayer at the air–water interface, often referred to as the Langmuir layer, goes back to the nineteenth century. The Langmuir trough is still the device of choice to study the interfacial behavior of the Langmuir layer by measuring surface pressure versus specific surface area isotherms. The molecular and association behavior of surface-active molecules can be deduced from the isotherms. It has become clear that the two-dimensional phase behavior of monolayers at a range of surface pressures and temperatures is quite complex, even for simple molecules.

Langmuir layers and Langmuir–Blodgett films often are systems of choice to determine the association behavior of molecules. A limitation may be that the samples are in thin-film form, i. e., mono- or multilayer. This can be a drawback if the study of the behavior of three-dimensional systems such as liposomes or micelles is desired. Nevertheless, the study of Langmuir layers and Langmuir–Blodgett films can yield valuable information on molecular orientation, phases and phase transitions, and thermodynamic parameters of two-dimensional systems.

Langmuir–Blodgett films can be fabricated over reasonably large areas so that the sample can be accessible to a variety of analytical techniques including UV-visible, IR, and Raman spectroscopies, as well as second harmonic generation, neutron reflection, scanning tunneling microscopy, and atomic force microscopy.

Chapter 2

Liquid-Crystalline Acetylenic Compounds
Behavior at Air–Water Interface

R. C. Advincula[1], M. J. Roberts[1], X. Zhang[2], A. Blumstein[2], and R. S. Duran[1]

[1]Department of Chemistry, University of Florida, Gainesville, FL 32611
[2]Department of Chemistry, University of Lowell, Lowell, MA 01854

Three substituted acetylene compounds which are known to form nematic liquid crystalline phases in the bulk, were chosen for study using the Langmuir-Blodgett technique. While being manipulated with the LB technique, greater order in the monolayer might be induced by the mesogenic groups when coupled with interactions at the water surface. Thus, there is the possibility of polymerizing a highly-oriented monomer as multilayers deposited on a substrate (Langmuir Blodgett film). In this study, each of the compounds formed unstable monolayers. However, the stability of each monolayer was improved by blending the compounds of interest with other liquid crystalline amphiphiles which are known to produce stable monolayers. Monolayers of the blends were successfully deposited on hydrophobic and hydrophilic quartz substrates.

Since the discovery of electrical conductivity in polyacetylene, this class of compounds has become the subject of intense research activity (1). The subsequent discovery of the nonlinear optical properties (2) of acetylenes has created additional interest. Polyacetylenes are characterized by large molecular hyperpolarizability as a consequence of having highly displacable electron clouds e.g. in extended π - electron systems (3). High nonlinear properties are dependent on not only the extended π-electron system but also on the overall order of the molecules. The ultimate goal of this research is the polymerization of these LC-substituted acetylenes. As a first step toward that goal the production of LB films of these materials has been achieved.

In this study, three substituted acetylene compounds (Figure 1), which are known to form nematic liquid crystalline phases in the bulk (Table I), were chosen to see if they would form Langmuir films. It is hoped that greater order will be induced by the mesogenic groups when coupled with interactions at the water surface. The interest in Langmuir film formation lies in the possibility of producing an assembly of highly-oriented monomer molecules as multilayers deposited on a substrate for subsequent polymerization.

Liquid crystals are used as display devices in which the function comes from the ability of the liquid crystalline molecules to self organize. The use of liquid crystals in making Langmuir-Blodgett films is a step towards better morphological control (4). The Langmuir-Blodgett technique offers potentially greater control over

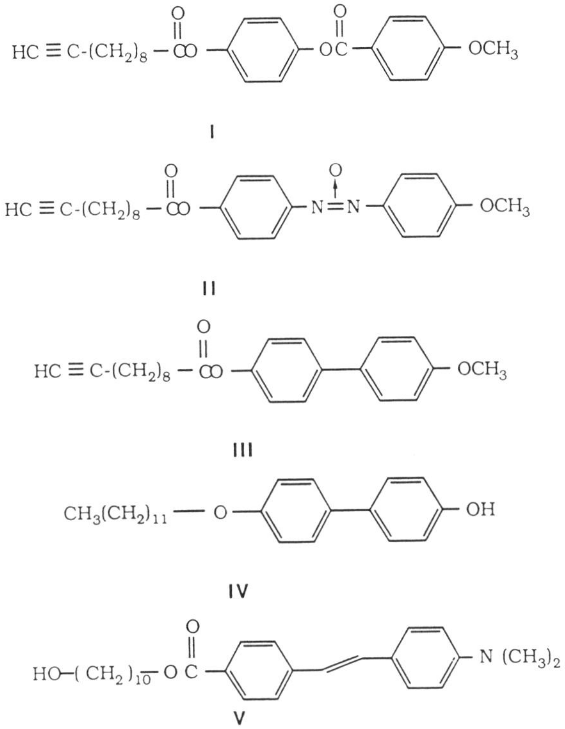

2. ADVINCULA ET AL. *Liquid-Crystalline Acetylenic Compounds*

HC≡C-(CH$_2$)$_8$-CO-O-C$_6$H$_4$-O-CO-C$_6$H$_4$-OCH$_3$

I

HC≡C-(CH$_2$)$_8$-CO-O-C$_6$H$_4$-N=N(O)-C$_6$H$_4$-OCH$_3$

II

HC≡C-(CH$_2$)$_8$-CO-O-C$_6$H$_4$-C$_6$H$_4$-OCH$_3$

III

CH$_3$(CH$_2$)$_{11}$-O-C$_6$H$_4$-C$_6$H$_4$-OH

IV

HO-(CH$_2$)$_{10}$-O-CO-C$_6$H$_4$-CH=CH-C$_6$H$_4$-N(CH$_3$)$_2$

V

Figure 1. Compounds studied.

Table I. Phase Transition of Compounds

Compound	Heating	Cooling
I	cr 38.38 n n 71.93 i	i 67.32 n n 42.19 cr
II	cr 96.23 n n 102.97 i	i 97.24 n n 75.83 cr
III	cr 76.71 i	i 65.56 n n 52.01 cr

molecular orientation and inter-molecular packing which could lead to the creation of new types of thin-film devices (4, 5).

Toward that end, this research has focussed primarily on a class of substituted acetylenes which exhibit liquid crystalline behavior in the bulk while at the same time possessing some amphiphilic character. Such compounds have characteristics which make them amenable to manipulation as monolayers using the Langmuir-Blodgett technique.

Furthermore, it has been shown that at least one important nonlinear optical property, the third-order susceptibility, is a function of the conjugation length and degree of long range order present in the material (3). In this work, the strategy was to increase inherent order of the monomers prior to polymerization (the polymerization, in essence, locks the order already present in the monolayer). The hope was that the liquid crystallinity of the monomer would be such that the acetylene group present on each molecule will be positioned as near to the ideal for polymerization as possible. The amphiphilic character of the monomer was exploited to create two-dimensional films using the Langmuir-Blodgett technique. In other words, the liquid crystallinity restricts the orientation of the monomer normal to the surface while the Langmuir-Blodgett technique constrains the monomer parallel to the surface.

Previous work with liquid crystalline monoacetylenic amphiphiles has been done most notably by three research groups. Le Moigne and coworkers have synthesized a number of liquid crystalline substituted acetylenes (including compound III) and have succeeded in polymerizing a few of these acetylenes in the bulk using Ziegler-Natta catalysts and g-irradiation in the bulk (albeit to low yield) (6). Blumstein and coworkers, who synthesized the liquid crystalline acetylenes used in this study, have found low bulk polymerization yields using g-irradiation (7). Also Ogawa has attempted to polymerize amphiphilic alkyl substituted acetylene monomers using X-ray and KrF excimer laser light (8). Although each of these workers reported low yields for polymerization of these types of monomers, we chose to work with them because the LB technique adds an additional constraint to the orientation of the monomers and thus might result in the successful polymerization of these compounds.

Experimental

The synthesis of the liquid crystalline monoacetylenic monomers (Figure 1) is described elsewhere (7). A solution of each compound, ca. 0.5 mg/ml in spectrograde chloroform (Kodak), was prepared. The monosubstituted acetylenes (I, II, and III) have eight-methylene unit tails with different mesogenic groups attached. In the bulk, the compounds have the following order in strength of liquid crystallinity: II > I > III. In addition compounds IV and V were used to increase Langmuir film stability. The behavior of the monolayers was investigated using a KSV 5000 Langmuir-Blodgett System (KSV Instruments) under a HEPA filter hood (Baker Company). The teflon LB trough was 21.6 cm long by 6.5 cm wide. A teflon barrier was used to compress the monolayers at a typical barrier speed of 20 mm/min. Water for the subphase was purified to 18 MΩ using a Milli-Q water system (Millipore). The temperature of the subphase was controlled by circulating fluid from a constant temperature bath through channels cut in the trough base. The lowest subphase temperatures were obtained by prechilling the subphase water in an ice bath. A microsyringe (Hamilton) was used for spreading chloroform solutions of the compounds at the surface prior to compression. Surface pressure was measured by one of two methods. The blends, I+V and II+V, were observed to give monolayers of such high viscosity that the Wilhelmy plate would be deflected during monolayer compression. Therefore, the surface pressure of the I+V and II+V blends was measured using a Langmuir-type floating film balance. The III+IV blend did not give such viscous monolayers and thus the surface pressure was monitored using the more convenient modified Wilhelmy balance. UV spectra of deposited films were obtained using a Perkin-Elmer UV/VIS/NIR Lambda 9 spectrophotometer.

Comparisons were made of the stability of the monolayers over time with stearic acid used as a standard for stability. First, the monolayer was compressed until it reached a given surface pressure and the barrier was stopped. The surface pressure changes were then monitored as a function of time and the slopes were compared with those of stearic acid. An alternate procedure for determining the monolayer stability was also used. In this second method, the mean molecular area of the compound or blend was monitored as a function of time while keeping the applied surface pressure constant. The applied surface pressure was kept constant by computer control of the barrier movement.

Langmuir-Blodgett films were deposited on specially prepared quartz plates. The quartz plates were cleaned using the following procedure: the plates were immersed in a detergent solution and sonicated for 10 minutes and then placed in a solution consisting of a 1:1:3 mixture containing NH_4OH, H_2O_2, and Milli-Q water respectively. The plates in the solution were then heated for 30 minutes at 80 °C. The plates were then sonicated for 10 minutes, sequentially, in each of the following solvents: $CHCl_3$, $CHCl_3$ / CH_3OH, CH_3OH. Hydrophobic quartz plates were prepared by placing a clean quartz plate in a mixture of 70 ml pure decalin, 10 ml chloroform, 20 ml carbon tetrachloride, and 2 vol % octadecyltrichlorosilane and sonicating for 2 hours.

Multilayers were deposited on 25 mm x 75 mm hydrophilic quartz plates. For dipping of the I+IV blend, the subphase temperature was controlled at ca.12°C, with a constant surface pressure of 15 mN/m. For dipping of the II+IV blend the subphase temperature was controlled at ca. 8°C, with an applied surface pressure of 15 mN/m. For dipping the III+IV blend, the subphase temperature was controlled at ca. 15°C with an applied surface pressure of 7 mN/m. The automated dipping speed was maintained at 1.0 mm/min in both directions. The dipping motion was delayed at the top of its travel for up to 10 minutes to allow time for the previously deposited layer to dry in the air. The transfer ratio and surface pressure were monitored and recorded during the dipping process.

UV spectra of the deposited III+IV blend monolayers were obtained by removing the plate from the dipping attachment, placing the quartz plate in a specially constructed sample holder, running the scan, and then returning the plate to the dipping attachment. Cross polarizers were also used to obtain the spectra. Cast films of each blend were made for comparison.

X-ray diffractograms of the deposited films were taken using a GE XRD-5 diffractometer at 35kV and 30 mA using Cu Kα radiation with a scan rate of 0.4 deg/min.

Gel permeation chromatography data were obtained using a Waters Associates GPC System with an LC-75 Perkin Elmer Spectophotometric Detector. The GPC was fitted with two Phenomenex 300 mm x 7.8 mm GPC columns in series prepacked with Phenogel 5 which has a 500 Å pore size. Polystyrene was used as a calibration standard. THF was used as the eluting solvent at a flow rate of 1 ml/min. The samples were prepared for the GPC by washing the LB layers from the substrate with a minimum amount of THF. The chromatograms of the material from the LB layers were compared to chromatograms of a THF solution of the blends.

Results and Discussion

Isotherms. Plots of surface pressure versus mean molecular area were done for all the compounds under study. For compounds I, II, and III at ambient temperature, the surface pressure onset occured at mean molecular areas (Mma) of 10 $Å^2$ and the apparent collapse pressures were very low for the compounds I and II. Even though the isotherms were reproducible, consideration of the chemical structure of the monomers suggests that this onset is too small to correspond to the formation of a true two-dimensional monolayer. Figure 2 shows isotherms recorded at various temperatures for the three monomers. It is apparent that at lower temperatures greater onset Mma (higher than 20 $Å^2$) values are obtained and the collapse pressure increased substantially. These onset Mma values are more reasonable in view of the chemical structure of the monomers.

Stability. For good deposition of the liquid crystal on a substrate, the formation of a stable monolayer is essential. A monolayer is considered stable when it displays a steady Mma at constant surface pressure over several minutes. Stability studies versus time showed that the films of compounds I, II, or III are not stable over long periods of time (>30% loss in 10 minutes) at low temperatures although they show better hysteresis as compared to higher temperatures. The order of stability is as follows III > I > II. In general, the lower temperature runs were characterized by more reasonable Mma values in the condensed phase region and more stable monolayers, but not stable enough for multilayer depositions.

In order to improve the stability of compounds I, II and III, two approaches were explored: the effect of repeated expansion and compression, and the effect of mixing other compounds which could act as "stabilizers".

After repeated expansion and compression of the films, an increase in the stability over time is observed for compounds I and II. One drawback of this approach is the apparent loss of material due to dissolution or changes in ordering (e.g. crystallization, local bilayer formation, etc.) over an extended period of time as it takes five cycles of expansion and compression to produce a reasonably stable film even at low temperatures. Another drawback is the slight decrease in Mma and collapse pressure values after expanding and compressing.

Several experiments were also performed on mixtures of the compounds. Mixing the monomers with either compound IV or V often increased the stability of the resulting monolayer film. However, since polymerization of the acetylenic compounds is desired, the amount of the "monolayer stabilizer" has to be minimized. Isotherms of blends of compound III with 5 mol % of compound IV and compound I or II with 9 mol% of compound V are shown in Figure 3. Initially, compound IV was used for both I, II and III. Concentrations of stabilizing compound IV as low as 5 mol % were adequate to stabilize monolayers containing III. On the other hand, for compounds I or II, even at 50 mol % of the stabilizing compound the resulting film was not stable and apparently no real mixing occurred. When compound V was used as a stabilizer with I and II, a minimum of only 9 mol % was necessary to produce a stable film. The combination of I or II with V produces a stable film. This behavior might be due to a greater degree of intermolecular π - orbital overlap between the aromatic rings of the mesogenic groups. In general, much higher Mma values were obtained for the condensed phases of the three compounds (I, II, III) in blends with compound IV and V. The stability of compounds I and II with compound V at constant Mma was greater than that found for pure I or II. The monolayer stability for the mixture of compounds III and IV compared to the pure compound III at constant surface pressure shows the same trend. The slope of the curve for the blend lies intermediate between the values for compounds III and IV, with the latter being the more stable monolayer because of its terminal hydroxyl group (9). A similar study using stearic acid with compounds I and II gave analogous results to blends with IV. In general, by varying the mole ratio of the mixtures and careful choice of "interactant" one can improve the monolayer stability.

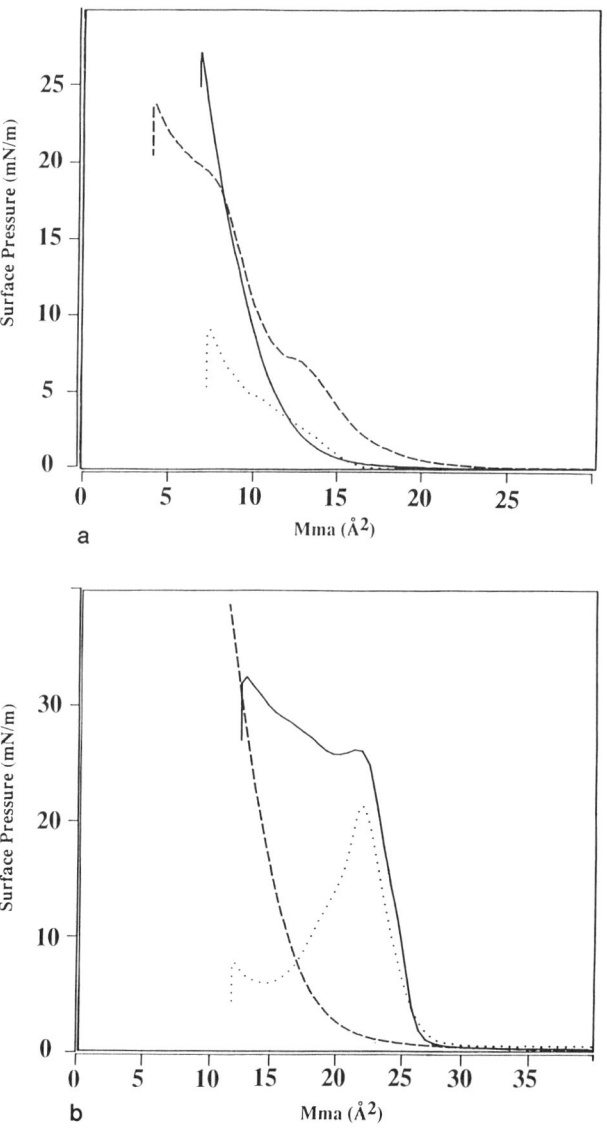

Figure 2. Isotherms recorded at ambient temperatures 2a (——— II at 26°C, --- I at 22°C, and • • • III at 22°C) and low temperatures 2b (• • • III at 16°C, ——— I at 14°C, and --- II at 6°C).

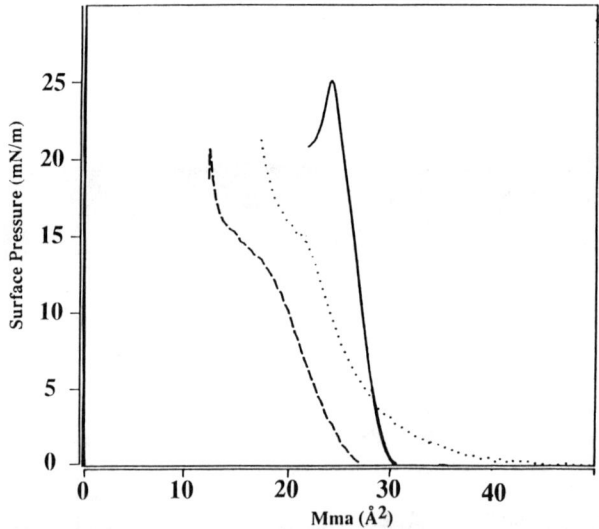

Figure 3. Isotherms of the blends at low temperature: ——— III+IV(5mol% of IV in III) at 16°C, • • • I+V(9 mol% of V in I) at 13°C, and --- II+V(9 mol% of V in II) at 13°C.

Deposition. Multilayers of stable blends of all three monomer compounds were successfully deposited on quartz substrates. Transfer of the monolayers was done at an applied surface pressure of 15 mN/m for the blends I+V and II+V since peeling off is observed at lower surface pressures and the film is not adequately stable at higher surface pressures. Blends with compounds III+IV were deposited at 7 mN/m due to similar considerations. During the depositions, the transfer ratio was recorded. The transfer ratio (TR) gives an indication of the quality of the dipping process and ideally should be unity. The III+IV monolayer was transferred to the substrate very well (TR near unity) while the substrate was moving up but that a small amount of the deposited monolayer was removed (TR approaching -0.2) when the substrate was moving down into the subphase. It is found that Z-type films were formed for the blends of all three compounds using both hydrophilic and hydrophobic quartz substrates. A total of 28 layers was deposited for the II+V blend, 16 layers for the I+V blend, and 7 layers for the III+IV blend. The blend II+V tends toward a Y-type deposition after the first 3 Z-type layers are deposited on a hydrophilic substrate. The blend I+V begins as a Y-type and tends towards Z-type deposition after the first three layers on a hydrophobic substrate.

UV Spectra The strong absorption observed in the 184 to 200 nm region is attributed to the allowed transition (E bands) and between 240 to 260 nm for the forbidden transitions of the benzene rings in the mesogenic group. A band observed at the 320 nm region for compound II is due to the azoxy chromophore conjugated to two benzene rings. Since the $p - p^*$ transition for the acetylenic group is at 173 nm, it was not observed in the spectra due to the limitations of the instrument. No significant difference was observed between the spectra of the pure compounds and the blends since only a minimal amount of the "stabilizer" was added.

X-ray Langmuir Blodgett films of I + V (16 layers) and II + V (28 layers) blends gave strong Bragg reflections in the X-ray diffractometer from which d-spacings (layer thickness) were calculated. These reflections are attributed to the stacking period of the layers. There was a correlation observed between the dipping behavior of the blends and the observed layer thickness as calculated from the diffractogram.

For blend I+V, a first order reflection was observed at an angle of 3.3 degrees. No peak corresponding to twice the d-spacing was observed at smaller angles. This corresponded to a layer thickness of 26.7 Å for the monolayers deposited at the substrate which is consistent with a unidirectional z-type deposition (head to tail packing) as observed for blend I+V. The difference from the calculated length of the molecule of approximately 30 Å could be attributed to tilting of the molecule. Other models could also be made which could account for kinks, dipole repulsions, and interdigitation along the long molecular axis.

For blend II+V, two reflections were observed in the small angle region at angles of 1.58 and 3.32 degrees and strong and medium intensities respectively. These reflections corresponded to the first and second order reflections of a stacking period of 55.86 Å. This is again consistent to the observed dipping behavior of blend II which would correspond to a Y -type head to head or tail to tail packing of the monolayers. Since the calculated thickness of a straight molecule was 30 Å, this could also correspond to a certain extent of tilting or overlap on the part of the monolayers. Tilting is plausible as the Mma for the blend of II+V at deposition was 28 Å^2 compared to an ideal packing of 22 - 25 Å^2 for compounds of this type.

Polymerization Two methods were explored to polymerize the monomer blends. The first method which involved irradiating the monolayer on the trough under air or nitrogen atmosphere with UV light (254 nm) proved unsuccessful. In the second method, Langmuir-Blodgett monomer films were exposed to a ^{60}Co g-ray source. The III + IV blend was exposed to g-irradiation dosage of 160 Mrad. The GPC data showed that the deposited III + IV blend reacted to form in 21% conversion an oligomer that was approximately 10 repeat units long, along with very much smaller portions of high molecular weight polymer. The GPC was calibrated with polystyrene standards. The percent conversion is similar to the results found for bulk polymerization (7). Without further characterization, the identity of the oligomer can not be made at this time. Using the same conditions with the II + V blend, two polymerization attempts were made with 20% and 1% respectively with an avearage number of repeat units of 35. For the I + V blend, only monomer with some degradation products was found.

Conclusion

This work has demonstrated Langmuir film and multilayer deposition properties of three functionalized acetylenic monomers. The monolayer properties of these compounds were significantly improved by blending with suitable co-surfactants. Furthermore, attempts to polymerize the compounds by g-irradiation gave similar results as that observed in the bulk. Future attempts to polymerize these monomers should focus on other ways of free radical initiation, such as vacuum UV irradaition of the LB films, laser light excitation, use of sensitizer, or perhaps macromolecular initiators. The chemical characterization is not yet complete but several conclusions may be drawn based on the work completed thus far. Compounds I through III by themselves do not form Langmuir films which will support a significant amount of surface pressure at room temperatures. However, if the temperature of the subphase is lowered below 17°C, surface pressures up to 20 mN/m were seen before collapse. Langmuir films of these pure compounds were not stable by themselves at any surface pressure. On the other hand, compounds IV and V are known to form stable Langmuir films. In addition, these compounds bear some structural characteristics in common with the other compounds. Now, it has been shown that blending these compounds leads to the production of stable Langmuir films. Following the model put forward by Ishii (10), it may be said that these "interactant" compounds affect the equilibrium behavior of the mixed monolayer, not the dynamic spreading behavior.

The stable blended monolayers were successfully transferred to quartz substrates in Z-type and Y-type depositions. The blend II+V tended toward Y-type while the blend I+V tended towards Z-type deposition. Blend III+IV also showed Z-type behavior. Work continues to increase the molecular weight and yield of the polymerization.

Acknowledgements

The authors wish to thank Dr. A. Schuster of the Ringsdorf group for providing compound IV, M. Zhao with Prof. W. Ford of Oklahoma State University for providing compound V, Dr. F. Blanchard of the Geology Department of University of Florida for allowing use of their X-ray diffractometer, and KSV Instruments for financial support. Acknowledgement is made to the Donors of The Petroleum Research Fund, administered by the American Chemical Society, for partial support of this research.

Literature Cited

(1) Aldissi, M. *Inherently Conducting Polymers*; Noyes Data Corporation, Park Ridge, NJ, **1989**.

(2) Butcher, P. N.; Cotter, D. *The Elements of Nonlinear Optics*, Cambridge University Press, New York, NY, **1990**.
(3) Le Moigne, J.; Soldera, A.; Guillon, D.; Skoulios, A. *Liquid Crystals* **1989**, vol. 6, pp. 627-639.
(4) Tieke, B. *Adv. Mater.* **1990**, vol. 2, pp. 222-231.
(5) Sakuhara, T.; Nakahara, H.; Fukuda, K. *Thin Solid Films* **1989**, vol. 159, pp. 345-351.
(6) Le Moigne, J.; Francois, B.; Guillon, D.; Hilberer, A.; Skoulios, A.; Soldera, A.; Kajzar, F. *Inst. Phys. Conf. Ser. No. 103: Section 2.4* **1989**, pp. 209-214.
(7) Zhang, X.; Ozcayir, Y.; Feng, C.; Blumstein, A. *Polymer Preprints* **1990**, vol. 31, pp. 597-598.
(8) Ogawa, K. *J. Phys. Chem.* **1989**, vol. 93, pp. 5305-5310.
(9) Tieke, B.; Lieser, G.; Wegner, G. Journal of Polymer Science: Polymer Chemistry Edition **1979**, vol.17, pp. 1631-1644.
(10) Ishii, T. *Thin Solid Films* **1989**, vol. 178, pp. 47-52.

RECEIVED September 24, 1991

Chapter 3

Blends of Side-Chain Liquid-Crystalline Polymers at Air–Water Interface

A. F. Thibodeaux[1], R. S. Duran[1], H. Ringsdorf[2], A. Schuster[2], A. Skoulios[3], P. Gramain[3], and W. Ford[4]

[1]Center for Macromolecular Science and Engineering, Department of Chemistry, University of Florida, Gainesville, FL 32611
[2]Department für Organische Chemie, Johannes Gutenburg Universität, Mainz, Germany
[3]Institut Charles Sadron, Strasbourg, France
[4]Department of Chemistry, Oklahoma State University, Stillwater, OK 74078

A previously studied side chain liquid crystalline (LC) polymer consisting of a methacrylate backbone, ethylene oxide spacer, and methoxybiphenyl mesogenic group, was found to spread to form monolayers at the air-water interface. Further investigation showed that the polymer's film properties were less than optimum for the formation of multilayers on solid substrates. To enhance the stability of the film on the water surface, the LC polymer was mixed with two well known monolayer forming compounds. First a fatty acid, stearic acid was blended with the polymer. The resulting film appeared to have properties characteristic of the average of the two pure components, indicating no attractive interaction between the compounds. The second blend material was a polymer, poly(octadecyl methacrylate). The mixed film exhibited better monolayer film properties than either of the pure polymers with respect to stability, isotherm reproducibility and temperature dependence. Isobaric analysis of the blend isotherms, stability data, and deposition results are presented and discussed.

Liquid crystalline polymers are of interest due to their self ordering capabilities, with the added advantage of greater mechanical and thermal stability over low molecular weight compounds . Our group is interested in the behavior of these compounds when they are confined to approximately two dimensions at an air/water interface. The side chain liquid crystalline polymers shown below (Figure 1) are the focus of this research and their bulk properties have been well documented (1,2,3,4,5). These polymers spread on a water surface and yield typical reproducible isotherms upon compression using a Langmuir-Blodgett (LB) trough.

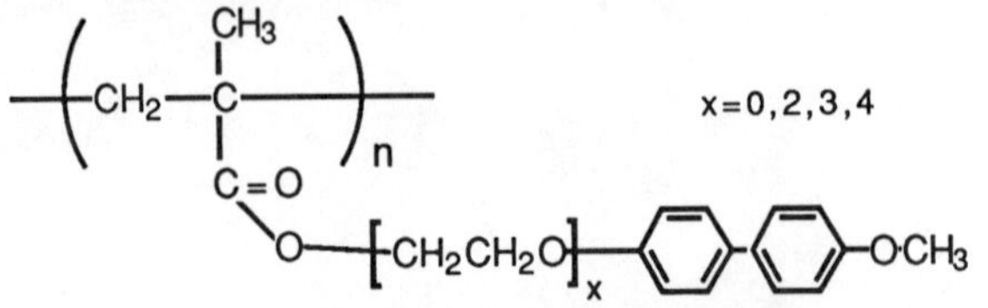

Figure 1. Structure of the side chain liquid crystalline polymers.

0097–6156/92/0493–0020$06.00/0

Monolayers can often be transferred to solid substrates to form multilayer thin films with well defined thickness and structure. This ordered deposition reduces defects while enhancing film properties. Once transferred, the film can be studied by various spectroscopic and microscopic methods. The usual vertical transfer process requires that the film be stable at a reasonable surface pressure to ensure accurate transfer ratios and ordered deposition (6). In this body of work stability of a monolayer film is defined as a negligible change in mean molecular area at a constant applied surface pressure.

Not all polymers will form stable monolayers at the air-water interface. Using a hysteresis technique (compression of the film to a predetermined surface pressure, holding for a delay time, and then re-expansion of the film to zero pressure) it was discovered that the polymer films shown in Figure 1, were not stable for long time periods in any reasonable surface pressure range. The exception was the polymer containing four ethylene oxide units (denoted as PM4) and only at large mean molecular areas and surface pressures no greater than 5 mN/m.

Previous work has shown that mixtures of film forming polymers with low molecular weight compounds and other polymers can enhance the film stability at the water surface (12-14). A low molecular weight compound (stearic acid) and an analogous polymer poly(octadecyl methacrylate) shown in Figure 2, were each blended with the polymer PM4. This paper will further discuss the behavior of these polymer mixtures in this quasi-two dimensional domain, and pose possible explanations for the results.

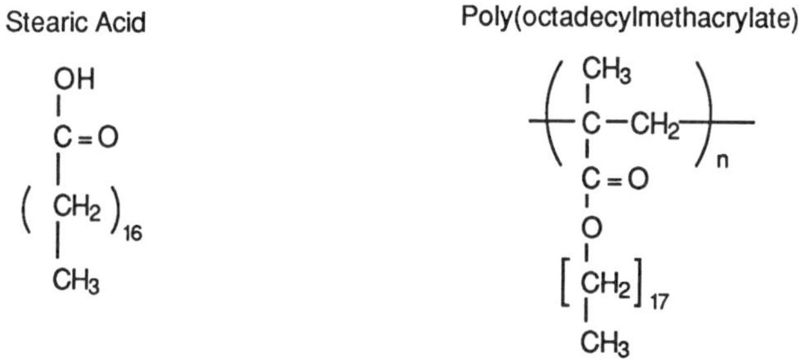

Figure 2. Compounds blended with LC polymer PM4.

Experimental

Materials. The side chain liquid crystalline polymer PM4 was polymerized free radically in THF solution (1). The tacticity of the polymer backbone was determined by ^{1}H NMR (triad analysis) and was found to be 10.7% isotactic, 34.2% heterotactic and 55.1% syndiotactic. M_w was found by SEC to be 99,800 against a calibration curve of polystyrene and M_w/M_n was determined as 2.7. The poly(octadecyl methacrylate) was polymerized radically in toluene solution, recrystallized from toluene/methanol, and had a M_w of 196,500 (M_w/M_n = 1.5) against a calibration curve of polystyrene. Tacticity of the poly(octadecyl methacrylate) was determined by 200 MHz ^{1}H NMR and was found to be approximately 25% isotactic, 50% heterotactic and 25% syndiotactic triads respectively. The stearic acid used in the blend studies was purchased from TCI Chemical Co.

Isotherms. All polymer and stearic acid solutions were made in spectra grade chloroform (Kodak) at a concentration of 0.5 to 1.0 mg/ml. Isotherms were recorded at 25°C with a compression speed of 10 mm/min (unless otherwise specified) on a KSV LB5000 Langmuir Blodgett trough using the Wilhelmy plate method of surface pressure detection. The subphase in all studies was pure water which was first deionized and then polished through a five bowl Milli-Q Plus system by Millipore® (> 18 MΩcm specific resistance). For these polymers and blends it was important to maintain the Wilhelmy plate face parallel to the length of the moving barrier and a constant distance from the end of the trough. Tests were also made with a floating barrier pressure sensor. All isotherms were run a minimum of two times, with a reproducibility error of less than 1.0 Å^2. The substrate for the multilayers was a thoroughly cleaned hydrophilic glass slide, or quartz slide when UV measurements were made. Transfer of PM4 was performed at 25°C, at a constant pressure of 5 mN/m, dipping speed of 5 mm/min and an upper delay time of 15 minutes. UV spectra were taken with a Perkin Elmer Lambda 9 instrument.

Results and Discussion

Pure Components. The film properties of PM4 were studied with respect to stability, reproducibility, deposition, and thermal dependence. The isotherms were reproducible to +/- 1 Å^2, and showed little dependence on temperature fluctuations. At first glance, there appeared to be a liquid condensed phase, a biphasic region and a solid analogous region with a collapse at approximately 35-40 mN/m (Figure 3). However, were this the case, collapse would occur at 15 Å^2, whereas the closest packing area for the methoxybiphenyl side chain in the bulk is approximately 22 Å^2. Using the hysteresis technique previously mentioned it was found that "collapse" occurs at 40 Å^2 and much lower surface pressures. The word "collapse" is used loosely with these polymers only to refer to the limiting point of film stability and monolayer thickness and does not attempt to describe the process beyond that point.

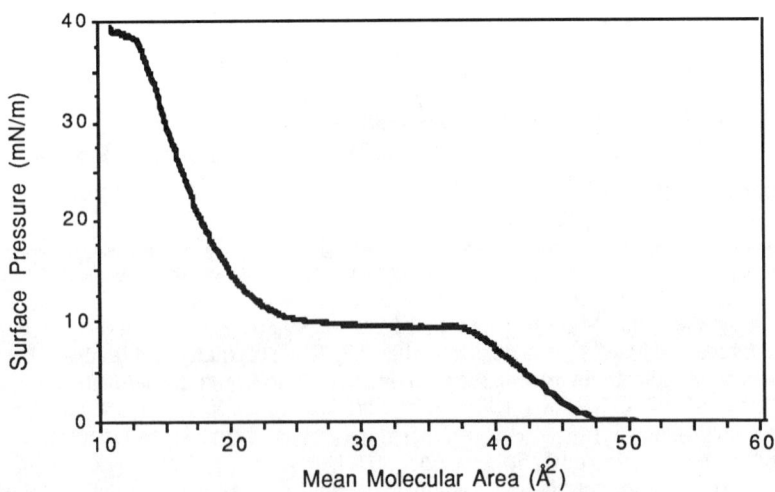

Figure 3. Surface pressure vs. mean molecular area isotherm for PM4.

Stearic acid is used as a standard monolayer forming substance whose film properties are well documented (6). It has been used, as have other long chain fatty acids, in mixing studies to improve and control film behavior. The other mixing compound is a polymeric analog of a fatty acid, poly(octadecyl methacrylate) (PODMA). It has been studied thoroughly in the bulk and at the air-water interface (8,9,10,11). Some slight discrepancies were found with respect to the deposition and isotherm shapes in this study and those referenced from previous studies. These may be due to the differences in backbone configuration of the polymer samples. The referenced studies were of isotactic or syndiotactic polymer samples, whereas the polymer sample used in this study was atactic (see Experimental).

Blends. Mixtures were made with PM4 and the above-mentioned compounds to enhance the film's properties and optimize packing. Substances with long alkyl chains were considered the best candidates for two reasons. First, they could plasticize the side chains which were rigid due to the strong interactions between mesogenic groups. Secondly, they added a more defined hydrophobic region to the film in order to "anchor" the film out of the water phase. The mixture compounds chosen were stearic acid which contains a long C-17 alkyl chain and a similar polymeric compound PODMA. Selected isotherms from these blends are shown in Figures 4a and 4b.

The above mentioned mixtures enhance film stability with respect to collapse pressure of PM4 (Figures 4a and 4b). However, a more effective way of looking at the interaction between the side chains and other groups is by comparing isobaric data of mean molecular area versus concentration (Figures 5a and 5b).

If no significant repulsive or attractive interactions were occurring between the polymer side chains and the long alkyl chains, a linear mixing relationship would be expected. Within experimental error, the mixing behavior of the PM4/stearic acid blends appears linear (Figure 5a). This indicates no net interaction between the components or phase separation. In the blend of PODMA/PM4 (Figure 5b) there is a negative deviation up to a concentration of 50 % PODMA and a positive deviation above this concentration. These deviations may indicate net attractive interactions at low concentrations of PODMA and repulsive interactions at higher concentrations.

As a contrast, results from an immiscible (in bulk and at air-water interface) polymer/polymer blend study were analyzed and isobaric concentration versus area data are shown in Figure 6. For this polymer mixture there is only a positive deviation from a linear curve indicating incompatibility. The positive deviation would not be expected if the polymer mixture were in equilibrium. In equilibrium, it would phase separate and the observed surface areas would be simply the weighted average of the pure components similar to that seen in the PM4/stearic acid case. The behavior observed in this incompatible mixture is believed to result primarily from the decreased mobility of polymer chains constrained to a free surface. Upon compression the chains do not have time to phase separate and are constrained in conformations that they would normally not assume. This technique may prove valuable as a method to quantify a "degree" of incompatibility for polymer mixtures.

For the PM4/PODMA blend a packing explanation is also reasonable. The gaps formed in the PM4 film may be large enough to accommodate only a certain number of alkyl chains before disrupting the film packing. Since the difference between the mean molecular area of collapse of the pure polymer and the optimum bulk packing area is roughly 20 $Å^2$ (the approximate packing area of one long alkyl chain) it would follow that there would be on the order of one alkyl chain for one PM4 side chain group. This would be in agreement with the observed negative deviation from linear mixing up to a 1:1 ratio of side chains. This kind of intimate packing would imply miscibility between the two components.

Isobaric data analysis of monolayer polymer blends has been previously used by Gabrielli and coworkers (12, 13, 14). They indicated that miscible monolayer results mirror the interactions of the bulk. However, it was also stated that miscibility requires that the conformations of the polymers at the air-water interface are similar to those observed in the bulk (15) It is believed that in the system PM4/PODMA, the polymer conformations in the monolayer are quite different from the bulk, most notably with respect to PODMA (1, 8, 10, 11).

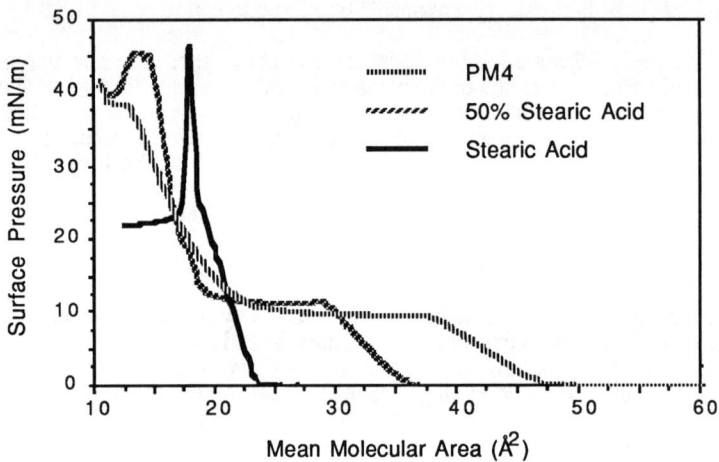

Figure 4a. Isotherms of PM4, stearic acid, and a 1:1 blend.

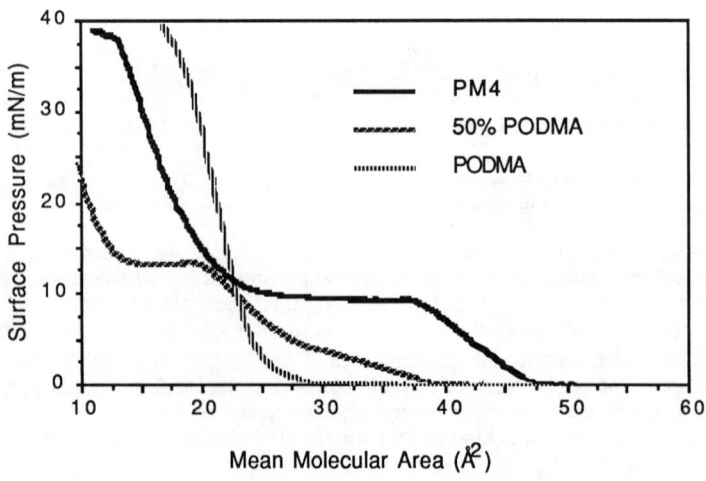

Figure 4b. Isotherms of PM4, PODMA, and a 1:1 blend.

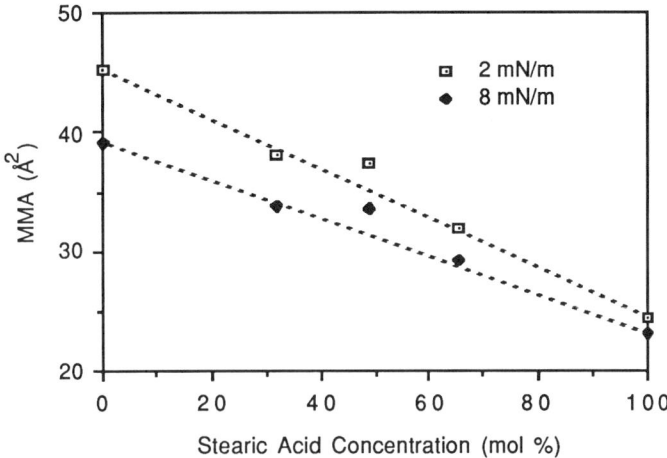

Figure 5a. Mean molecular area vs. component concentration for blends of PM4 with stearic acid. The dotted lines represent a theoretical linear mixing relationship.

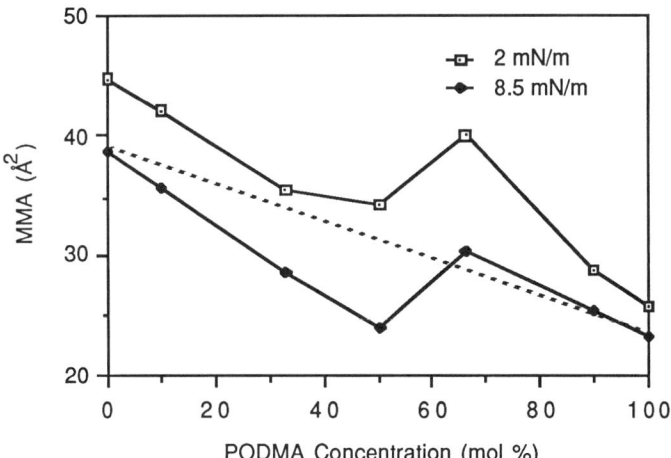

Figure 5b. Mean molecular area vs. concentration of PODMA for blends of PODMA with PM4. The dotted line represents a theoretical linear mixing relationship.

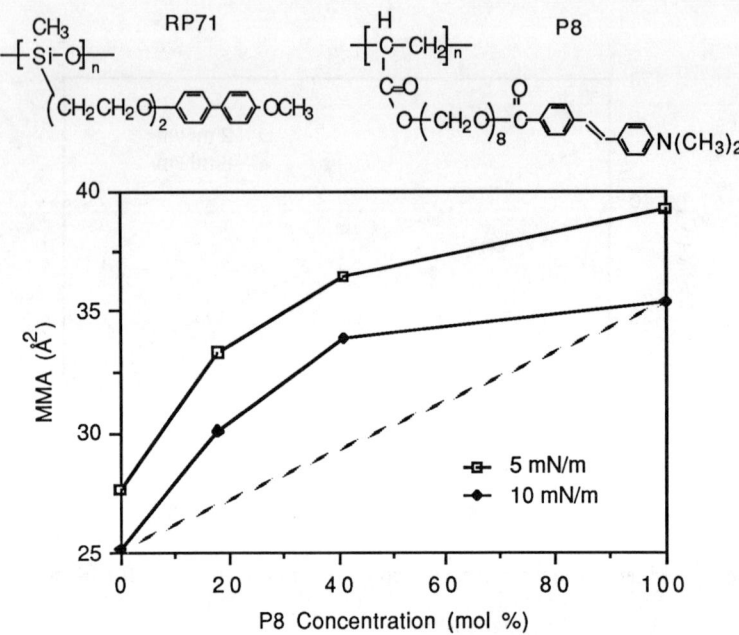

Figure 6. Immiscible polymer blend. Mean molecular area vs. concentration data of the polymer mixture whose components are shown above the graph.

Stability. In order to deposit monolayer films on a solid substrate by the conventional vertical dipping method, films must exhibit reasonable stability at the desired transfer pressure. To monitor this, films were compressed to a particular surface pressure and held at this pressure while monitoring the change in the mean molecular area versus time. Figure 7 shows the isobaric stability of a PM4/PODMA blend and the pure components at an applied surface pressure of 10 mN/m. At this pressure, only the blend appears to be reasonably stable within a short period of time. Although PODMA stability increases with time, an hour or more is required to achieve a film stable enough to acquire accurate transfer data. The most notable point of this analysis is the fact that the blend is more stable than either component. If the film were immiscible, an intermediate stability would be expected. This data infers there is interaction between the two polymers that produces this increased stability.

Figure 8a shows the isotherms of PODMA at various temperatures. Between temperatures of approximately 20 and 30°C, small deviations in temperature cause significant changes in the film behavior. The onset areas shift to higher values with increasing temperatures in this range. The PODMA films also appear to be more stable in the liquidous regions that are observed at elevated temperatures. As shown in Figure 8b, the PM4 films show little change with temperature. The most notable difference is that the surface pressure in the apparent "phase transition" appears to decrease with increasing temperature rather than increase as would be expected. Temperature dependent isotherms of the 50% blend are shown in Figure 8c. The behavior of the composite film with respect to temperature appears to be dominated by the PM4 component and shows little change. Were the mixture phase separated, the temperature dependence would be expected to be a weighted average of that of the two components.

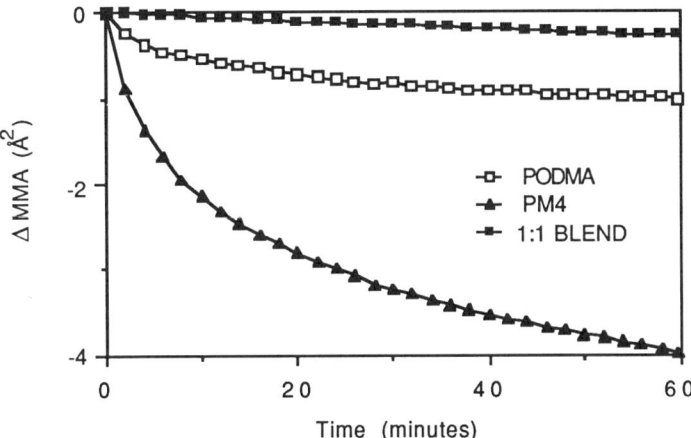

Figure 7. Stability data. The above graph shows the change in mean molecular area (ΔMMA) with time at a constant applied film pressure of 10 mN/m.

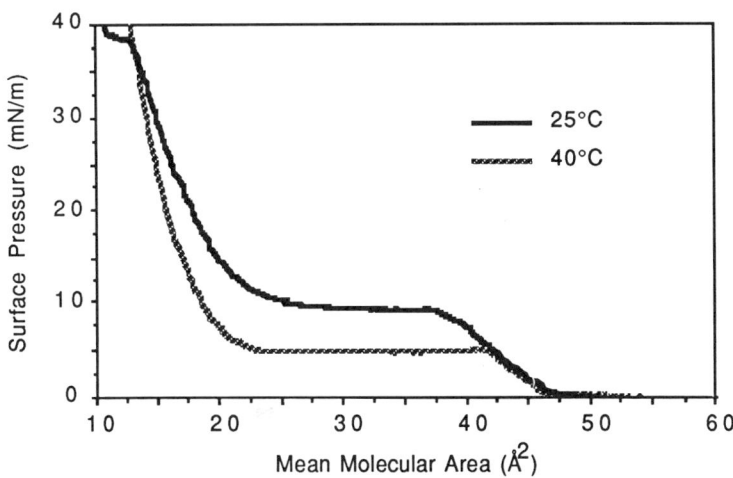

Figure 8a. Isotherms of PM4 homopolymer at different temperatures.

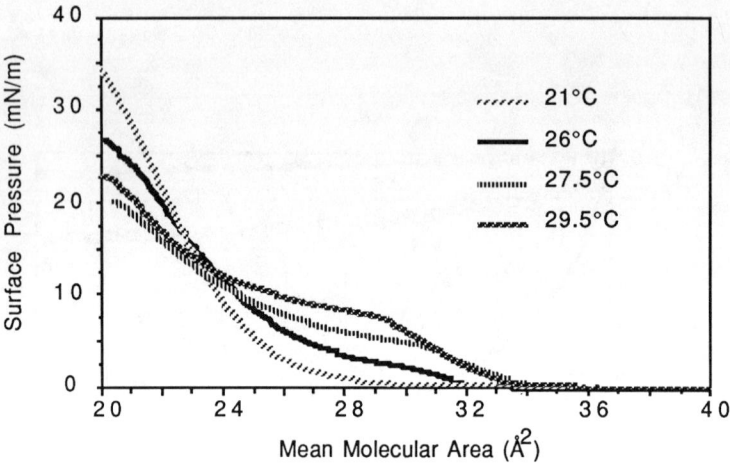

Figure 8b. Isotherms of PODMA at different temperatures.

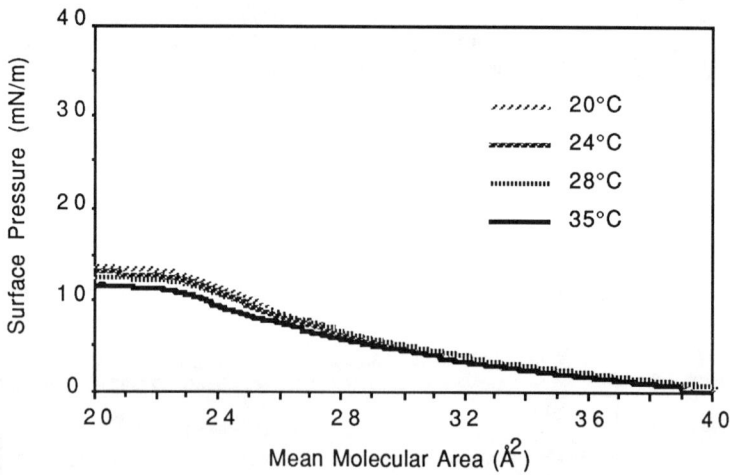

Figure 8c. Isotherms of the 1:1 PM4:PODMA blend at different temperatures.

Deposition. The PM4 polymer transferred to a hydrophilic substrate to form Z-type multilayer films of fifteen layers with transfer ratios exceeding 0.9 (Figure 9 shows five layers). UV data shows an approximately linear relationship of absorbance versus number of layers. In the UV spectra taken of the same film the next day, no noticeable change in absorbance is observed, illustrating the transferred multilayer film's stability for short periods of time. However, transfer conditions are at 5 mN/m and approximately 40 angstroms squared per monomer repeat unit. In the bulk, the closest packing area of the biphenyl group is 22 Å^2. Therefore, the side chains are far from being closely packed, and large gaps must be present in the film. These transfer conditions will most likely lead to films of lowered stability and mechanical strength.

Although other work cited has shown that poly(octadecyl methacrylate) transfers to hydrophilic substrates, we were unable to build up more than one monolayer, even in the liquid region at elevated temperatures. Deposition was attempted at 32°C at a dipping speed of 5 mm/min and an upper delay of at least 20 minutes. Transfer occurred on the upstrokes and appeared to quantitatively redeposit upon the water surface during subsequent downstrokes.

The blend shows greater stability at higher pressures than either pure polymer. This gave us reason to believe that the blends would transfer better than both the PM4 and PODMA. However, after two Z-type layers, the transfer ratio fell off dramatically. This may be improved by adjusting the deposition conditions and will be investigated in future experiments.

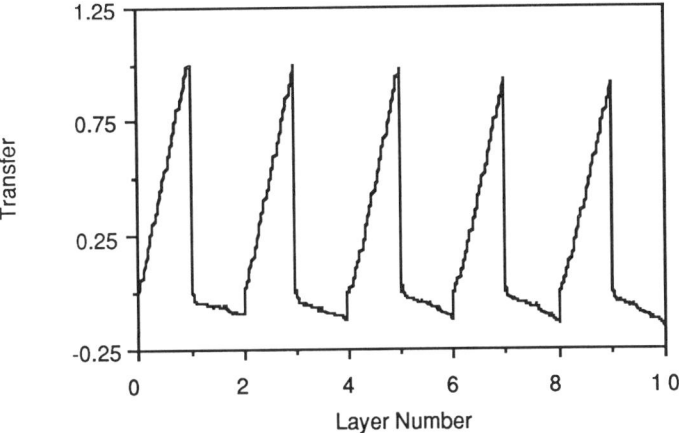

Figure 9. Transfer versus layer number for PM4 at 5 mN/m and 25°C. Odd number layers are the transfer down from air to below the water surface. Even number layers represent the upward transfer out of the water. There was only significant transfer on the upstroke representing Z-type deposition onto hydrophilic quartz.

Concluding Remarks

The stability of polymer monolayer films can be enhanced by mixing with low molecular weight and polymeric compounds of suitable structure. In the examples shown, stearic acid and PODMA mixtures with PM4 raised the collapse pressure and lowered the collapse area to a value closer to the optimum packing area in the bulk.

Comparison of isobaric mean molecular areas versus concentration (below collapse) gives an indication of the type of interaction between the side chains of the polymers analyzed. As opposed to other miscibility tests for bulk polymer blends (such as DSC) this method also visualizes the magnitude of these interactions, whether they are favorable or unfavorable.

Future work will analyze the miscibility of these polymers at the air-water interface. Transfer of the film to an appropriate substrate and analysis by small angle

X-ray diffraction should give an indication to the extent or existence of domain formation. If this correlates with the isotherm data, the analysis presented may be an alternative method of determining polymer blend compatibility with the added benefit of quantifying the interactions. Also, if the conformation of the polymers in the bulk state is different from that expected at the air-water interface, it may be a more appropriate method for determining miscibility of monolayer blends.

Indications from the isobaric analysis of the blend isotherms and stability data of the blend versus the pure polymers, are that the blends are miscible at concentrations of up to 50% PODMA and form monolayer films that are more mechanically and thermally stable than either of the polymers alone.

Acknowledgments

Acknowledgement is made to the Donors of the Petroleum Research Fund, administered by the American Chemical Society, for partial support of this research.

We would like to acknowledge financial support from AKZO America. We would also like to thank KSV Instruments Ltd., Helsinki, Finland for their cooperation and support with the LB troughs and software. One of us (AT) would like to acknowledge Christian Erdelen, a PhD. student in the laboratory of Dr. Ringsdorf, for his endless patience and assistance with their troughs and for his insight and discussions regarding this work. Also, thanks to everyone in the Ringsdorf research group for their help and hospitality. Dr. G. Friedmann of the ICS Strasbourg, France is acknowledged for the gift of the RP71 sample.

Literature Cited

1. Duran, R.; Guillon, D.; Gramain, Ph.; Skoulios, A., Makromol. Chem., Rapid Commun., (1987), 8, 181-186.
2. Duran, R.; and Strazielle, Macromolecules, (1987), 20, 2853.
3. Duran, R.; Guillon, D.; Gramain, Ph.; and Skoulios, A., J. Physique, (1987), 48, 2043.
4. Duran, R.; Guillon, D.; Gramain, Ph.; and Skoulios, A., J. Phys. France, (1988), 49, 121.
5. Duran, R. Guillon, D.; Gramain, Ph; and Skoulios, A., J. Phys. France, (1988), 49, 1455.
6. Gaines, "Insoluble Monolayers at the Gas-Water Interface"
7. Discussion with Andreas Schuster in reference to unpublished data at the time of this writing.
8. Mumby, S. J.; Swalen, J. D.; and Rabolt, J. F., Macromolecules, (1986), 19, 1054.
9. Duda, G; Schouten, A. J.: Arndt, T.; Lieser, G.; Schmidt, G. F.; Bubeck, G. and Wegner, G., Thin Solid Films, (1988), 159, 221-230.
10. Plate, N. A.; Shibaev, V. P.; Petrtukhin, B. S.; Zubov, Yu, A.; and Kargin, V. A., J. Polymer Science: Part A-1, (1971), 9, 2291-2298.
11. Hsieh, H. W. S.; Post, B. and Morawetz, H., J. Polymer Sci.: Poly. `Phys. Ed., (1976), 14, 1241-1255.
12. Gabrielli, G; Puggelli, M.; Ferroni, E., J. Colloid Interface Sci., (1974), 47, 145.
13. Gabrielli, G; Baglioni, P., J. Colloid Interface Sci. , (1980), 73, 582.
14. Gabrielli, G; Puggelli, M.; Baglioni, P., J. Colloid Interface Sci., (1982), 86, 485.
15. Kawauchi, M.; Nishida, R., Langmuir, (1990), 6, 492.

RECEIVED September 24, 1991

Chapter 4

Effect of Surface Pressure on Langmuir–Blodgett Polymerization of 2-Pentadecylaniline

Huanchun Zhou and R. S. Duran

Department of Chemistry, University of Florida, Gainesville, FL 32611

> The application of the Langmuir Blodgett (LB) technique to the polymerization of 2-pentadecyl aniline on the LB trough is discussed in this paper. The changes in the mean molecular area (Mma) during the polymerization is mainly due to the change of the alkyl sidechain conformation. The polymerization rate (PR) is deduced from the change of the Mma. The effect of the surface pressure on the polymerization is investigated. It is found that the effect of the surface pressure is mainly on the reaction rate constant, not on the surface concentration.

The interest in LB techniques has increased in recent years due to the special nonlinear optical and electronic characteristics of LB films (1,2). Because polyaniline is conductive and air stable, polymerizations of substituted anilines on LB troughs were tried and successfully realized (3-6).

Meanwhile, it was found that the mean molecular area was decreasing during the polymerization of 2-pentadecyl aniline. Also no polymer was found when the reaction was run at low applied surface pressure, such as 1 mN/m (4). However, it was not clear why Mma decreased during the reaction and why no polymer was formed at lower surface pressure. In this paper, the cause of Mma decrease is revealed and the effect of surface pressure is discussed. Moreover, it is shown that LB techniques may be a good probe to observe and to monitor polymerizations confined to surfaces on LB troughs, like dilatometry which is used to observe bulk polymerizations (7).

In the first section we present results from the monomer and selected model compounds at the air/water interface which are useful in interpreting the subsequent polymerization reaction. The second section describes the polymerization reaction and how the LB techniques are used to monitor the polymerization rate. The third section describes the effect of applied surface pressure on the polymerization reaction.

The experimental part is described in detail in the previous paper (5).

Results and Discussions

Isotherms of 2-Pentadecyl Aniline, Stearic Acid and 4-Hexadecyl Aniline. It was shown in a previous paper (4) that the onset Mma of the monomer (2-pentadecyl aniline) is about 80 Å^2 on a 0.5 M sulfuric acid subphase. However, the area of poly (2-pentadecyl aniline) per unit was only about 40 Å^2 (unpublished data). The area difference per unit between monomer and polymer, we suppose, is the cause of Mma decrease during the polymerization of 2-pentadecyl aniline. Compared with the area of a long alkyl sidechain or a benzene ring, the onset Mma of the monomer was exceptionally large. To understand this behavior, stearic acid and 4-pentadecyl aniline are used as model compounds for comparison.

The structures of 2-pentadecyl aniline, stearic acid and 4-hexadecyl aniline are shown in Figure 1.

Figures 2 and 3 show surface pressure vs surface area isotherms of 2-pentadecyl aniline on water (Fig. 2a) and on 0.5 M sulfuric acid (Fig. 3a). These are compared with stearic acid (Figs. 2b and 3b) and 4-hexadecyl aniline (Figs. 2c and 3c). All three compounds spread on both subphases to give repeatable isotherms which show little hysteresis upon subsequent decompression. The monolayers of these compounds are also stable as indicated by very small changes in mean molecular area with time under various constant applied surface pressures. From Figure 2 it can be seen that the 2-pentadecyl aniline isotherm is shifted to substantially higher surface areas than the other two isotherms on water, while its collapse pressure is considerably less than those of both stearic acid and 4-hexadecyl

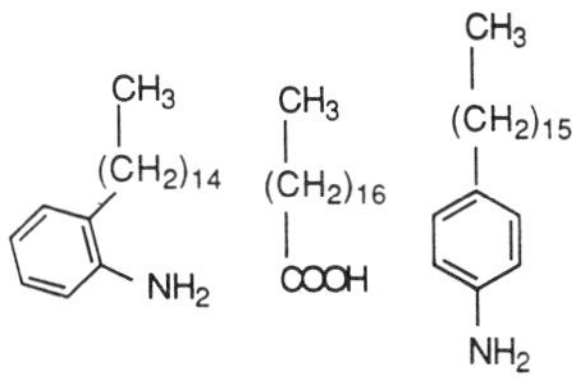

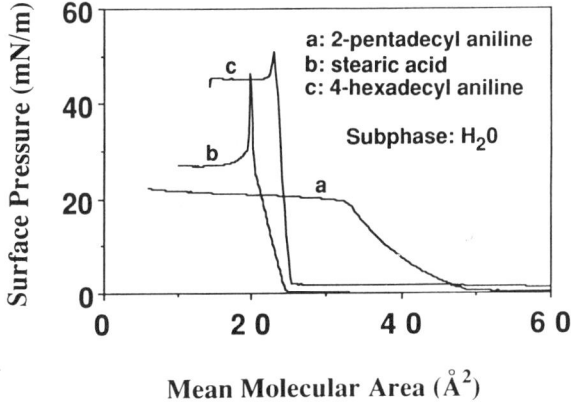

Figure 1. The structures of 2-pentadecyl aniline (a), stearic acid (b) and 4-hexadecyl aniline (c)

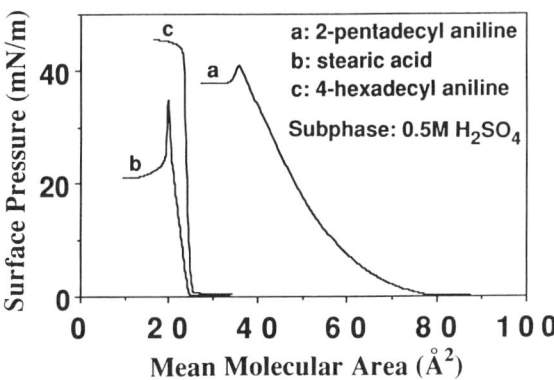

Figure 2. Surface pressure vs Mma isotherms of the three compounds on a pure water subphase, T = 23 °C, Barrier speed = 50 mm/min..

Figure 3. Surface pressure vs Mma isotherms of the three compounds on a 0.5 M H_2SO_4 subphase, T = 23 °C, Barrier speed = 50 mm/min.

aniline. Additionally, on the acidic subphase the isotherm of 2-pentadecyl aniline is shifted to still higher surface areas. Both the surface pressure onset areas and collapse points of the other compounds are considerably less effected by the acidity of the subphase.

It is well known that at high applied surface pressures, stearic acid side chains are largely in a *trans* conformation, close packed, and rather well ordered (*2*). Similarly, one can expect that the 4-hexadecyl aniline side chains are ordered at high pressures, but not as well packed as stearic acid due to the bulky aromatic head group. The surface pressure onsets and collapse points of stearic acid and 4-hexadecyl aniline on 0.5M sulfuric acid subphase are 24.6 $Å^2$ / 25.2 $Å^2$ and 19.8 $Å^2$ / 23.3 $Å^2$ respectively, which are nearly the same as those on the water subphase, 24.5 $Å^2$ / 25.1 $Å^2$ and 19.8 $Å^2$ / 23.0 $Å^2$. We suppose, this indicates that the conformation of the stearic acid or 4-hexadecyl aniline molecules is basically not effected by the acidity of the subphase. The proximity of the onsets also indicates that the head group of the 4-hexadecyl aniline is not strongly hydrated. For steric reasons, the aromatic head group is expected to be rather perpendicular to the air/water interface instead of laying flat at the surface. Figure 4a shows a pictorial representation of the conformation expected to be assumed by the 4-hexadecyl aniline.

The 2-pentadecyl aniline behaved entirely different from the other two compounds. The onset point of 2-pentadecyl aniline on 0.5M sulfuric acid, 76.5 $Å^2$, is much greater (27.6 $Å^2$ larger) than that on the water subphase, 48.9 $Å^2$. The collapse point of 2-pentadecyl aniline on 0.5M sulfuric acid, 35.8 $Å^2$, however, is only slightly larger (2.9 $Å^2$) than that on water, 32.9 $Å^2$. As 4-hexadecyl aniline and 2-pentadecyl aniline are chemically similar, the difference in the surface pressure onset is expected to be due largely to the substitution position rather than hydration. Additional insight into the behavior of 2-pentadecyl aniline was obtained by a study of the isotherms as a function of the subphase acid concentration shown in Figure 5.

From Figure 5, it is seen that there is little difference in the isotherms at pH 4 or higher. Below pH 4 the surface pressure onset points obviously increase to higher areas with decreasing pH. We suppose that the increase in the onset point may result from the dissociation of the amine group and its protonation

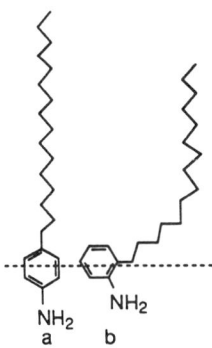

a: 2-pentadecyl aniline
b: 4-hexadecyl aniline

Figure 4. Pictorial view of the conformation of the substituted anilines at the air/aqueous interface.

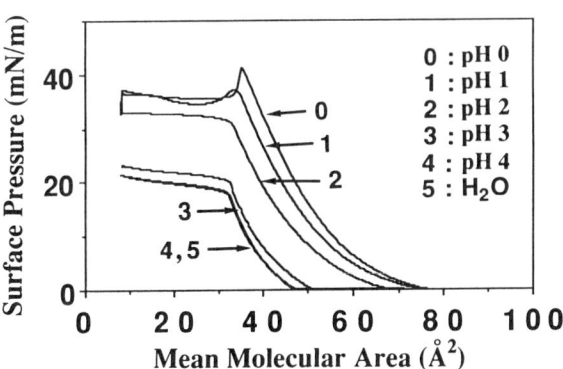

Figure 5. Surface pressure vs Mma isotherms of 2-pentadecyl aniline on different pH subphases, T = 23°C, Barrier speed = 50 mm/min.

reaction. Assuming that the onset point becomes larger as the concentration of protonated species increases, one can deduce that the maximum of $\partial Mma/\partial pH$, the maximum of Mma change/acidity change, occurs at $pK_a = pH$. From Figure 5 it can be qualitatively estimated that the maximum of $\partial Mma/\partial pH$ is in the range of pH 2 ~ 3. This is to say that the K_a of the 2-pentadecyl aniline monolayer is about 10^{-2} ~ 10^{-3} which is about 10^2 higher than that of aniline in solution(8) After the protonated species form, the conformation of 2-pentadecyl aniline molecules may change due to the static effect which pulls the amine cation downward as shown pictorially in Figure 4b. As shown in Figure 4b, as the amine group is tilted down towards the water, *gauche* conformations must be introduced in the alkyl side chain and a portion of the side chain may occupy the interface with the aromatic groups. This accounts for the large increase in the mean molecular areas between the isotherms of 2- and 4-substituted anilines. The large onset area of 76.5 Å^2 for 2-pentadecyl aniline on the 0.5M sulfuric acid surface, 51.3 Å^2 more than that of 4-hexadecyl aniline, means that in addition to the aromatic ring, at least several CH_2 units of the pentadecyl side chain contribute to the steric interactions on the surface.

In both Figures 2 and 3, the isotherms of 2-pentadecyl aniline have a considerably different shape than those of the other compounds. The long gentle curved shapes of these isotherms are in contrast to the sharply sloped, nearly linear isotherms of the other compounds. It is expected that as the monolayer of 2-pentadecyl aniline is compressed on 0.5M sulfuric acid, the benzene ring is turned and the side chain is somewhat straightened. In other words, as the molecule is compressed, considerable energy is put into changing its conformation at the surface.

In the case of 2-pentadecyl aniline, the work done upon compressing the monolayer, we suppose, is basically contributed to the "twist energy" due to straigtening of the pentadecyl chain. From Figure 2a, the twist energy is estimated in the range of one to several kilocalories per mole. However, in the cases of stearic acid and 4-hexadecyl aniline, the long side chain is supposed to be basically perpendicular during compressing.

From the above discussion one can see that the isotherms of all the compounds varied with different environmental conditions, especially those of 2-pentadecyl aniline. The changes in the isotherms may be related with the structure of monolayer.

The LB Polymerization of 2-Pentadecyl Aniline. When 2-pentadecyl aniline polymerizes, two hydrogen atoms per monomer are lost and covalent bonds link the monomers together. Analogous to vinyl polymerizations, the replacement of the two Van der Waals distances between unreacted monomer molecules by two shorter covalent bond distances results in a net densification of the compound upon polymerization. This allows for the polymerization reaction to be studied by the technique of dilatometry (7). Similar densification occurs in a monolayer at the Langmuir trough, allowing the polymerization to be studied by monitoring the mean molecular area or barrier speed as a function of reaction time. Typical curves obtained in the LB polymerization of 2-pentadecyl aniline are shown in Figure 6.

This polymerization reaction was performed at 27°C and under a constant applied surface pressure of 30 mN/m. The isotherm of the polymerized material is reproducible with little hysteresis upon subsequent decompression. During the reaction, the mean molecular area is observed to decrease monotonically from approximately 42 Å^2 to a value of 22 Å^2 after 34 min at the end of the polymerization as shown in Figure 6. This change in surface area is much larger than the bulk density change normally associated with a polymerization. We suppose that the surface area change consists of two components. One is the densification associated with forming covalent bonds and the other is due to conformation changes of the monomer upon polymerization. Aniline polymerizes primarily in the 1 and 4 positions. To polymerize, 2-pentadecyl aniline monomers must change conformation. Upon polymerization, the side chain tilt associated with the monomer conformation can easily straighten to form a more tightly packed conformation with more *trans* content.

The initial mean molecular area (Mma_o) on a rectangular trough can be expressed as follows:

$$Mma_o = 1.66 \times 10^{-10} (L \times W)/ M \quad (Å^2/molecule) \quad (1)$$

where L is the length of the subphase on the LB trough covered by the monomer in mm, W is the width of the subphase on the LB trough covered by the monolayer in mm, and M is the number of moles of monomer spread.

Assuming the Mma change is only due to polymerization and to simplify calculations assuming this change is a constant for all reaction steps under same polymerization conditions, the Mma (in Å^2/molecule) at a given reaction time may be calculated by:

$$Mma(t) = 1.66 \times 10^{-10}(L \times W - 10^{-14} n_t \times \Delta Mma)/ M \tag{2}$$

Where n_t is the number of the monomer molecules having polymerized at time, t, and ΔMma is the change in Mma (Å^2/molecule) due to polymerization.

The average barrier speed (BS), the time derivative of the barrier displacement needed to maintain constant pressure, is also shown in Figure 6. BS increases at the beginning of the reaction, then subsequently decreases to zero as the polymerization terminates.

BS can be written in the following form:

$$BS = \Delta l/\Delta t \quad (mm/min) \tag{3}$$

Where Δl represents the distance the barrier has moved in the time interval Δt. From equations 2 and 3, one may deduce the relationship between BS and Mma as follows The relationship between BS and Mma is differential:

$$BS = (L/Mma_o) \, \partial Mma/\partial t \tag{4}$$

BS is proportional to $\partial Mma/\partial t$ and inversely proportional to the Mma_o on a given trough. Figure 7 shows experimental data from the polymerization of three different spread volumes of 2-pentadecyl aniline. From Figure 7, it is seen that $\partial Mma/\partial t$ is independent of the amount of spread monomer, which is in accordance with equation 2.

The PR can then be written in the following form:

$$PR = \Delta n/\Delta t \quad (molecule/min) \tag{5}$$

where Δn represents the number of the monomer molecules having polymerized in Δt. From equations 2, 3 and 5, one may have

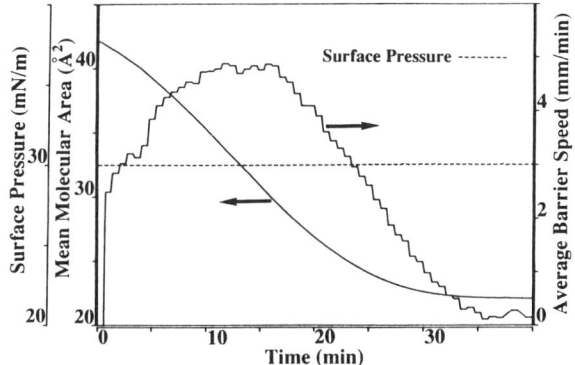

Figure 6. Surface pressure, mean molecular area and average barrier speed vs reaction time during the polymerization of 2-pentadecyl aniline, T = 27 °C, 0.5 M H2SO4, 0.05 M ammonium peroxydisulfate.

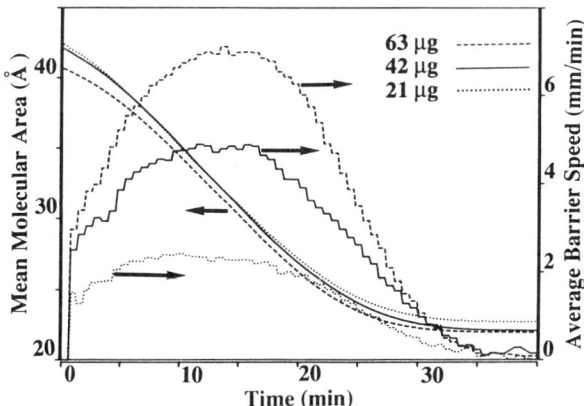

Figure 7. Mean molecular area and average barrier speed vs time for different spread amounts of 2-pentadecyl aniline, T = 27 °C, 0.5 M H2SO4, 0.05 M ammonium peroxydisulfate.

$$PR = -(10^{14} \, W) \, BS/\Delta Mma = -(6.025 \times 10^{23} \, M/\Delta Mma) \, \partial Mma/\partial t \quad (6)$$

Equation 6 expresses the relationship between the polymerization rate, barrier speed and mean molecular area. As a check on the relationship between BS and the Mma_o in equation 4, the barrier speed values from Figure 7 can be compared; if equation 4 is obeyed, a constant should be observed. The peak barrier speed values of curves a, b, and c in Figure 7 are 2.42, 4.86 and 7.13 mm/min respectively. The Mma_o values for these curves are 168.63, 84.31 and 56.21 $Å^2$ respectively. The products of the peak barrier speeds and Mma_o's for the curves in Figure 7 are 408, 410 and 401 $Å^2$ mm/min, which are in very close agreement with the constant predicted by equation 4.

Furthermore, from equation 6, one may estimate the polymerization rate. In the case of Curve b in Figure 7, the ΔMma is approximately 20 $Å^2$, as determined from the difference between Mma_o and the Mma at time 40 min. (W is 150mm for this trough). Then the polymerization rate at the peak barrier speed is 5.3×10^{15} (molecule/min).

The information such as that shown in Figure 6 may be useful in studying many other polymerization reactions. Assuming the ΔMma is constant under a given set of conditions and due only to polymerization, the polymerization rate is directly proportional to BS and can be conveniently estimated. Therefore, information about when a polymerization starts and finishes, when and how much the highest polymerization rate is, and how the polymerization rate changes during the polymerization may be gained from the barrier speed curves. The Mma change may be also used as a convenient measure of the conversion of monomer to polymer as a function of reaction time. Furthermore, LB polymerizations of monomers such as 2-pentadecyl aniline use only tens of micrograms of monomer per experiment and are fast, convenient, and highly reproducible.

The LB polymerization of 2-pentadecyl aniline also has some fundamental differences compared to typical bulk polymerization reactions. The first of these is that the reaction is confined to a surface. The reaction is not expected to be strictly two-dimensional in nature, however, as the ammonium peroxydisulfate oxidizing agent diffuses to the monolayer from the bulk subphase. The monomers are also likely to undergo significant vertical displacement at the surface due to thermal motion and other surface perturbations. Nonetheless, the initial conformation of the compressed monomer monolayer before polymerization starts is both considerably more anisotropic and ordered than that attained in a classical polymerization.

The different monomer conformation on the LB trough may affect the polymerization process. For example, in the solution polymerization of 2-alkyl anilines under the same polymerization conditions, the yields of poly(2-methylaniline), poly(2-ethylaniline) and poly(2-propylaniline) were observed to be 80%, 16% and 2% respectively (9). The cause of this big difference in yields was steric hindrance from the side chain. On the other hand, the yield of the LB polymerization of 2-pentadecyl aniline is larger than 90%. This may be due to the fact that the monomer is "pre-oriented" in sterically favorable conformations before the polymerization.

Finally, once the LB polymerization is in progress, the topological constraints on the growing chains are considerably different from what would be seen in a bulk polymerization reaction. This may affect the polymerization process.

Further investigations are in progress to investigate the above effects.

The Effect of Applied Surface Pressure. Previous studies (4) have found that the polymerization of 2-pentadecyl aniline occurred at high surface pressure, 30mN/m, but no polymer was found at lower surface pressure, 1 mN/m, in 34 min. The LB polymerization of 2-pentadecyl aniline can be done under conditions of certain constant applied surface pressure. It is therefore interesting to study the effect of surface pressure upon the polymerization reaction. Figure 8 shows the effect of different applied surface pressures on the Mma vs reaction time curves. The Mma change upon polymerization is seen to depend strongly on the applied surface pressure.

Figure 9 shows plots of BS vs reaction time for the polymerization at different applied surface pressure, π. It is seen that the peak values of the barrier speed, BS(P), increase, go through a maximum, and then decrease as a function of increasing surface pressure. However, as shown in equation 6, PR depends not only on the barrier speed, but also on ΔMma. Here, the change in ΔMma under different applied surface pressures is not

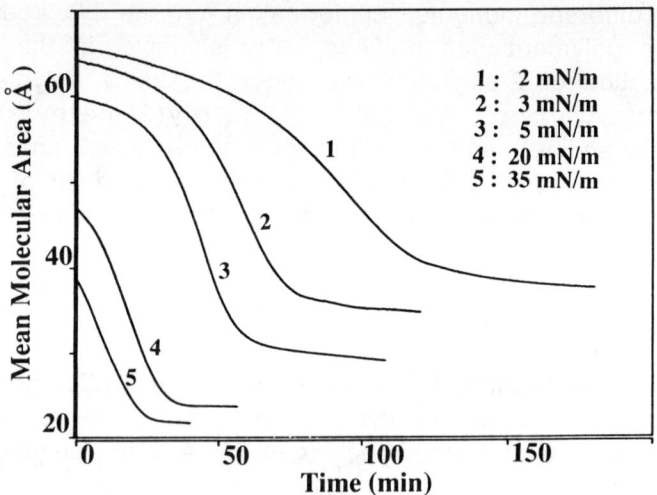

Figure 8. Mean molecular area vs time at different applied surface pressures during the polymerization, T = 27 °C, 0.5 M H2SO4, 0.05 M ammonium peroxydisulfate.

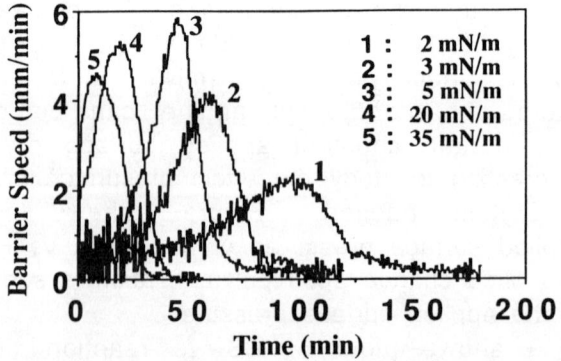

Figure 9. Barrier speed vs time at different applied surface pressures during the polymerization, T = 27 °C, 0.5 M H2SO4, 0.05 M ammonium peroxydisulfate.

negligible. The different ΔMma values were taken from Figure 8 and the peak polymerization rates, PR(P), are collected in Table I. This table shows that PR(P) do increase with the increase of surface pressure over the entire experimental range.

Table I Polymerization data at different surface pressures

π (mN/m)	0.5	1	2	3	5	10	20	30	35
ΔMma (-Å^2)	34.8	33.1	31.3	32.8	33.3	28.1	24.6	20.1	17.5
BS(P)(mm/min)	0.74	0.99	2.36	4.20	5.85	5.79	5.34	4.89	4.62
PR(P)*	3.20	4.50	11.3	19.2	26.3	30.9	32.5	36.6	39.5
t_c (min)	403	312	129	81.5	62.6	48.4	37.8	33.6	28.6
BS(S)(mm/min)	0.30	0.30	0.30	0.32	0.44	0.69	1.41	2.15	2.17
PR(S)*	1.29	1.36	1.44	1.45	1.98	3.68	8.60	16.1	18.6

* The units of PR(P) and PR(S) are 10^{14} molecules/min.

The time to complete the polymerization, t_c, was obtained from Figure 9 as the intercept of the tangent to the inflection point and a line through the baseline after complete reaction. These values are shown in Table I. The t_c increases as the surface pressure decreases, especially at low values of π. The initial barrier speed, BS(S), is also shown in Table I and increases with increasing π. All of the above trends may be expecte

The π is somewhat analogous to the pressure, P, in the bulk. The P has the unit of "force/area" in three dimensions while π has the unit of "force/length" in two dimensions. In common gas state reactions, the effect of pressure changes is mainly one of changing the reactant concentrations. However, in the LB polymerization of 2-pentadecyl aniline, the observed effect of applied surface pressure on the polymerization rate might not be primarily due to changing the average distance between reacting monomers and thus their collision frequency. Here on a trough, the surface area changes under the constant applied surface pressure during the polymerization. This indicates that the apparent activation energy for the polymerization reaction could be affected.

Polymerization experiments were performed on mixed monolayers to investigate these effects. From isotherm studies (8), stearic acid appeared to form a compatible mixture with 2-pentadecyl aniline. Furthermore, when stearic acid was spread on the Langmuir trough under conditions that would result in the polymerization of 2-pentadecyl aniline, no measurable surface area change was observed. Stearic acid was thus considered to be an "inert" blending agent.

A first mixture experiment involved polymerizing pure monomer and a mixture at the same applied surface pressure. If the polymerization rate were dominated by the average distance between aniline monomers, the polymerization rates would be expected to differ substantially. Figure 10 shows that the effect of polymerizing a 3:1 mol ratio (2-pentadecyl aniline : stearic acid) mixture compared to the pure monomer at a constant surface pressure of 30 mN/m. Both the mean molecular area curves and the barrier speed curves indicate that the difference in the polymerization rate is very small. Similar results were also seen for a 3:2 mol ratio mixture. The shift seen between the mean molecular areas of the mixture and the pure 2-pentadecyl aniline curves is due simply to the different areas occupied by the two components at the surface.

Another mixture experiment performed is shown in Figure 11. In this reaction, the applied surface pressure of the mixture was adjusted to be higher than that of the pure monomer so that the average distance between 2-pentadecyl aniline molecules in the pure monomer and mixture was nearly the same at the beginning of the reaction. If the polymerization reaction were dominated by concentration, the initial rates of both curves, as indicated by $\partial Mma/\partial t$, would be nearly the same. It is clearly observed, however that the initial rates are significantly different from each other.

The above mixture experiments indicate that the effect of applied surface pressure on the LB polymerization of 2-pentadecyl aniline is not dominated by simple concentration differences.

As discussed in the first section, the conformations adopted by 2-pentadecyl aniline may vary substantially with the applied surface pressure. As the surface pressure increases, the alkyl side chain is likely to be more perpendicular. This may help to

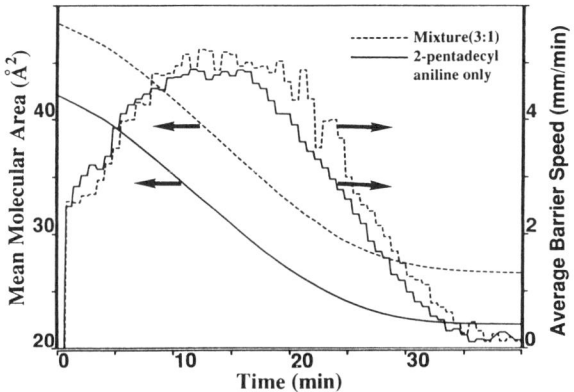

Figure 10. Mean molecular area and average barrier speed vs time during the polymerizations of 2-pentadecyl aniline and a 3 : 1 mole ratio blend with stearic acid, π = 30 mN/m, T = 27 °C, 0.5 M H2SO4, 0.05 M ammonium peroxydisulfate.

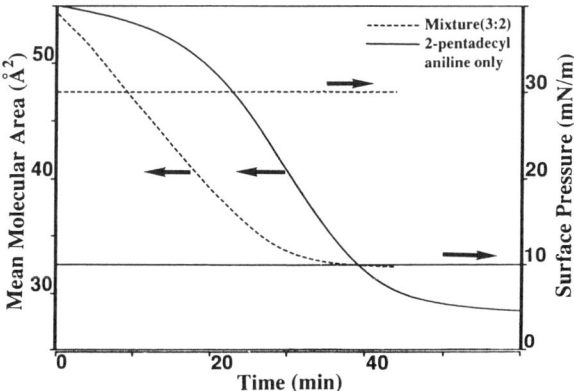

Figure 11. Mean molecular area vs time during the polymerizations of 2-pentadecyl aniline and a 3 : 2 mole ratio blend with stearic acid, T = 27 °C, 0.5 M H2SO4, 0.05 M ammonium peroxydisulfate.

overcome the steric hindrance of the long side chain and help the para position polymerization.

However, the relationships between the barrier speed, applied surface pressure, and t_c are not quantitatively simple. As discussed above and shown in Figure 10, the polymerization rate at constant π and different Mma does not change appreciably. It is useful to discuss these relationships from a kinetics point of view. For example, the relationship between BS(S), the average barrier speed at the start of the polymerization, and applied surface pressure will be discussed below.

In general, under constant surface pressure and constant temperature one may have:

$$d[M]/dt = -K[M]^X \tag{7}$$

where [M] is the surface concentration of the monomer in molecules/Å^2, K is a reaction rate constant and X is the number of the reactant monomer molecules. As is known, K is dependent on the activation energy. Here, the polymerization is carried out under given applied surface pressure and the surface area is changing as the polymerization proceeds. Thus the work, defined by the applied surface pressure times the change in surface area during the polymerization, will contribute to the reaction rate constant like the activation energy term. Then, one may have:

$$K = A\exp(W/kT) \tag{8}$$

where A is a constant and W is the work done during the polymerization. At the starting step of the polymerization, the change in [M] may be negligible and from equations 7 and 8 one may obtain:

$$d[M]/dt = -A[M]_o^X \exp(W/kT). \tag{9}$$

Then from equations 4 and 9:

$$BS(S) = LA[M]_o^{X-1} \exp(W/kT)$$

and

$$\ln(BS(S)[M]_o^{1-X}) = \ln(LA) + W/kT \tag{10}$$

Figure 12 shows the plot of $\ln(BS(S)[M]_o^{1-X})$ vs W at X = 1, 2 and 3. Higher values of X are not necessary because usually few

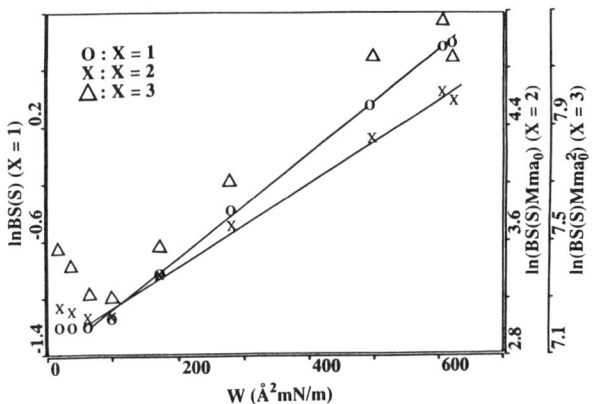

Figure 12. Kinetics plot of 2-pentadecyl aniline polymerizations.

reactions are of three or more molecules. From Figure 12, one may see that the relationship between $\ln(BS(S)[M]_o^{-2})$ and W is not linear at X = 3. The curves for both X = 1 and 2 are much more linear. Because of the experimental error one can not be sure whether the polymerization initiation reaction involves one monomer molecule or two. However, by comparing the curves in Figure 12, X=1 is preferable.

Acknomledgements

We gratefully acknowledge financial support from KSV Instruments Ltd., Helsinki, Finland, and the Office of Naval Research. We would also like to acknowledge Dr. R. Stern and Prof. C. Batich for the gift of the 2-pentadecyl aniline used in the initial part of this study and their encouragement.

Literature Cited

1. G. Robert, Ed., Chap. 7 in "*Langmuir-Blodgett Films*", Plenum Press, New York, 1990
2. G. L. Gaines, Jr., "*Insoluble Monolayers at Liquid-Gas interfaces*", Wiley-Interscience, New York, London and Sydney, 1966
3. H. Zhou, C. Batich, R. Stern, and R. S. Duran, *Polymer Preprints*, **1990**, *31*(2), 560
4. H. Zhou, C. Batich, R. Stern, and R. S. Duran, *Makromol. Chem., Rapid Commun.*, **1990, *11*,** 409
5. H. Zhou and R. S. Duran, *Polymer Preprints*, **1990**, *32*(1), 202
6. R. R. Bodalia and R. S. Duran, *Polymer Preprints*, **1991**, *32*(1), 248
7. G. Odian, "*Principals of Polymerization*", Second Ed., John Wiley Sons, Inc., New York, Chichester, Brishane, Toronto and Singapore, 1981
8. R. C. Weast, "*Handbook of Chemistry and Physics*", 51st Edtion, D-117, The Chemical Rubber Co., 1970
9. L. H. Dao, M. Leclerc, J. Guay and J. W. Chevalier, *Synth. Met.*, **1989,***29*, E377

RECEIVED September 24, 1991

Chapter 5

Monolayer and Thin-Film Crystallization of Isotactic Poly(methyl methacrylate)

R. H. G. Brinkhuis and A. J. Schouten

Laboratory of Polymer Chemistry, University of Groningen, Nijenborgh 16, 9747 AG Groningen, Netherlands

Monolayers of isotactic poly(methylmethacrylate) exhibit a pressure induced crystallization process, in which double helical structures are formed. The mechanism of this crystallization process is shown to exhibit many similarities with a normal melt crystallization process, and is suggested to be characterized by a twodimensional lamellar growth mechanism. The transferred monolayers can be used to build highly oriented multilayers of these materials, in which the lateral orientation of the helical structures can be regulated and changed from parallel to, to perpendicular to the transfer direction. Finally, an approach is suggested, using these crystallized LB monolayers as crystallization nuclei to prepare highly oriented thin films up to thicknesses of several microns.

Langmuir Blodgett monolayers of poly(methylmethacrylate) have received a lot of attention from researchers over the last decades, both in fundamental monolayer studies, PMMA being a standard polymeric material with a good spreading behaviour (1), as well as in the context of applications of PMMA thin films in lithography (2), or as matrix materials for dyes (3). Most of these studies were limited to conventional PMMA materials, with little attention devoted to the matter of the tacticity of these polymers. In this article, we will elaborate on some novel aspects pertaining to the monolayer behaviour of isotactic PMMA.

Monolayer Behaviour of Isotactic PMMA.

As was reported already by Beredjick and Ries (4), the monolayer behaviour of isotactic PMMA deviates strongly from that of the predominantly syndiotactic material. Whereas the latter material forms a highly condensed monolayer on the water surface, the isotactic polymer exhibits an isotherm characteristic for an expanded monolayer, building up a surface pressure even at large areas (Figure 1). The differences in monolayer behaviour can be attributed to strong differences in the lateral cohesive interactions in the monolayer (5, 6).

A puzzling aspect of the pressure area isotherm of i-PMMA, is the transition observed at approximately 8 mN/m, that is accompanied by a temporary fall in surface pressure during compression. This 'dip' was not explicitly reported by other authors publishing isotherms of this material (4, 7). As we will show in the following pages, this intriguing transition is associated with a very interesting monolayer process.

The Structural Nature of the Transition. The most important clues for the elucidation of the structural nature of this monolayer transition follow from infrared experiments on transferred LB layers on solid substrates (8). When the i-PMMA monolayers are transferred at surface pressures corresponding to the expanded regime in the isotherm (before the transition has taken place), the infrared characteristics of the transferred layers do not deviate significantly from those expected for a completely amorphous, isotropic thin film (8). If, on the other hand, the monolayers are transferred at a surface pressure of 12 mN/m, corresponding to the post-transition condition of the monolayer, the grazing incidence reflection infrared spectrum does exhibit some characteristic anomalous features, as illustrated by Figure 2. The differences between the two spectra in this figure perfectly correspond to differences in the spectra of amorphous and crystalline isotactic PMMA (8), strongly suggesting that the transition observed in the monolayer isotherm reflects a monolayer crystallization process analogous to that in the melt. When these multilayers are heated to a temperature above the Tg of i-PMMA, the crystalline features are retained, and when annealed at 120°C, these features are even strongly enhanced (8).

The crystal structure of isotactic PMMA, which it supposedly also acquires upon compression in the monolayer, has been suggested to be a 10_1 *double helix* (9, 10). An important confirmation for this proposed double helical nature of the post transition monolayer stems from the limiting area: this area (14.5 Å^2/monomeric unit upon extrapolation to zero pressure) is in very good agreement with a high conversion to the double helix structure as proposed for bulk crystalline i-PMMA, with these double helices lying flat on the water surface (area calculated to be 13.2 Å^2/monomeric unit). Surface potential measurements on these monolayers show a distinct discontinuity in the rise of the surface potential upon compression, exactly at the onset of the suggested crystallization process, leading to the start of a sharp fall of the perpendicular dipole moment per segment, $\mu_\perp$: this observation can be understood in terms of a change of the amphiphilic orientation of the segments in the expanded condition of the monolayer, to the double helical conformation, restricting the freedom of the segments to orient their polar parts with respect to the interface (5).

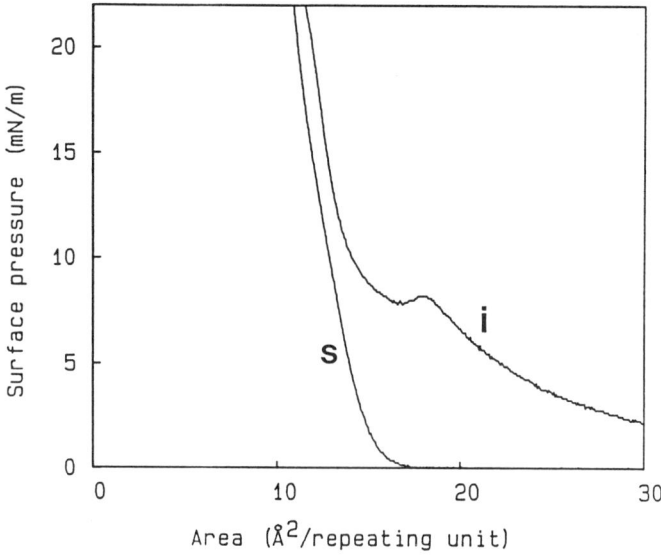

Figure 1. Pressure area isotherms of syndiotactic (s) and isotactic (i) PMMA (21°C).

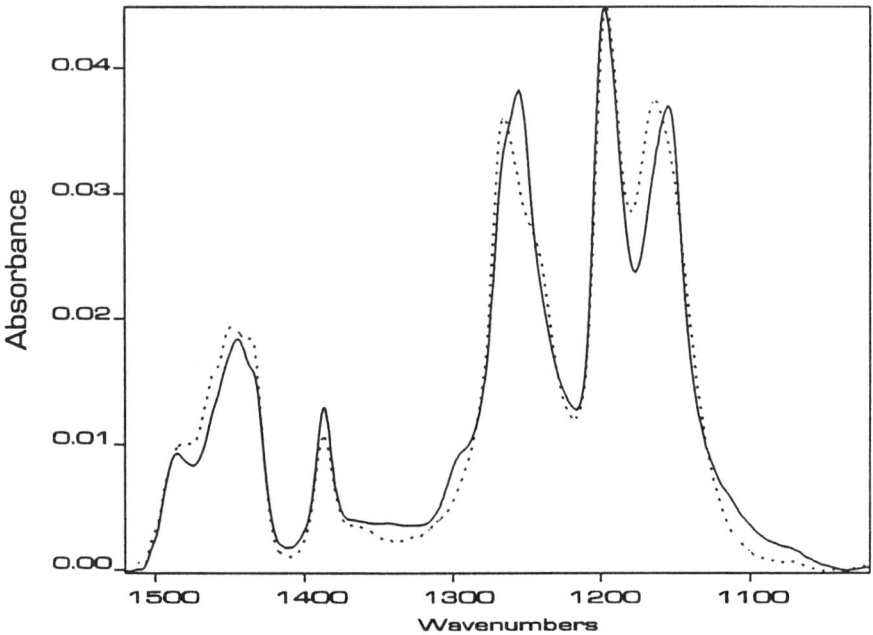

Figure 2. Grazing incidence reflection infrared spectra of transferred LB layers of i-PMMA on gold substrates; transfer pressure 5 mN/m (dashed line) and 12 mN/m (solid line).

All experimental results thus support the interpretation of the monolayer transition in terms of a crystallization process, induced by the surface pressure build up upon compression, through a strongly favorable $\Pi \Delta A$ contribution to the free energy of the transition, ΔA being negative. Isotactic PMMA, to our knowledge, presents the first example of a double helical structure based on a synthetic polymer at the water surface, and of a polymer exhibiting a random coil to helix transition in monolayers at the air water interface.

The monolayer crystallization (associated with a negative entropy) is strongly dependent on temperature, the transition shifting to higher surface pressures at elevated temperatures; it is also strongly dependent on the compression speed, the crystallization rate being directly determined by the surface pressure.

Monolayer Crystallization Kinetics. The 'dip' observed in the compression isotherms already illustrated an important effect pertaining to the kinetics of the proposed monolayer crystallization process, i.e. its *auto-accelerating* character. When performing isobaric stabilization experiments at surface pressures close to the transition pressure, this auto-accelerating effect leads to a slow start of the conversion process, with the rate gradually increasing until it levels off at high conversions, as illustrated in Figure 3. This observation is reminiscent of analogous effects that are standardly observed in melt crystallization processes of polymers, and strongly suggest a nucleation stage to be associated with the monolayer crystallization process. The isobaric stabilization experiments can be analyzed in terms of an Avrami analysis, which, for a series of surface pressures studied, nicely yields straight lines, with a characteristic Avrami exponent of close to 2 (6).

These results lead us to propose a mechanism for the formation of the crystalline monolayer, very much analogous to 'normal' melt polymer crystallization processes. In this model, the formation of a stable crystalline nucleus is required, which can then exhibit a one-dimensional growth (leading to an Avrami exponent of 2). This formation of monolayer crystallites may be best represented by a lamellar growth type mechanism, in which new (double) helical sequences are formed parallel to the already existing helices of the crystallites, so as to optimize the lateral helix-helix contacts (Figure 4); the growth direction of the crystallites is then oriented perpendicular to the helix axes, and growth may continue until it is blocked by another crystallite. This mechanism can explain the high (although not complete) conversion attained in the monolayer, and is indirectly supported by several observations, as discussed later on.

The variable stability of the crystalline monolayer upon decompression also appears to lend support to the model suggested above: upon decompression from the post transition condition, the crystalline monolayer eventually 'melts', and completely re-attains the initial expanded conformation. In this respect, the surface pressure associated with this melting process provides a clue for the stability of the crystallites formed, more stable crystallites melting at lower surface pressures. Crystallites that were formed at low surface pres-

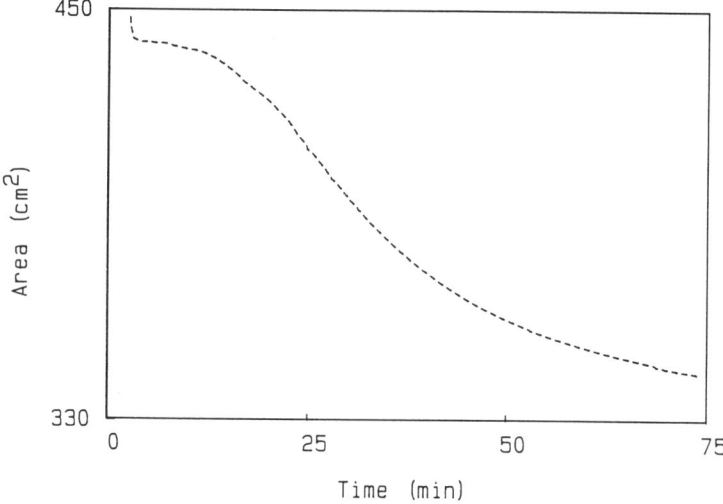

Figure 3. Typical isobaric stabilization experiment of a monolayer of isotactic PMMA at 6.5 mN/m and 22°C.

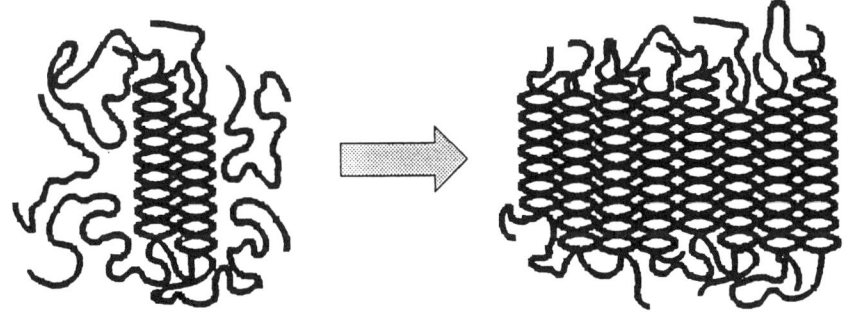

Figure 4. Schematic representation of the suggested monolayer crystallization process.

sures can be observed to be significantly more stable than crystallites formed at high surface pressures (6), an observation that can be directly compared to the fact that, in melt crystallizations, the use of a higher thermodynamic driving force for crystallization (a larger degree of undercooling) leads to less stable structures relative to conditions representing a lower thermodynamic driving force (higher crystallization temperatures). At lower surface pressures, more extensive lamellar structures need to be formed in order to be stable, which can withstand lower surface pressures before 'melting' (6).

Finally, it should be mentioned that the ease of the monolayer crystallization process stands in sharp contrast with the melt crystallization process of i-PMMA, which is notoriously slow.

Molecular Weight Effects. Some interesting observations can be made pertaining to the monolayer crystallization process, when varying the molecular weight of the i-PMMA samples. In Figure 5, the monolayer compression isotherms are given for three narrow molecular weight fractions of i-PMMA, corresponding to low (2800), intermediate (36,000) and high (1.3 million) molecular weights. All samples exhibit an expanded monolayer behaviour at low surface pressures, but, in contrast to the higher molecular weights, the lowest molecular weight sample does not show any sign of the monolayer crystallization process discussed above. In fact, a suppression of this crystallization process can be seen to gradually develop when using samples of molecular weights under 25,000, with, at 21°C, a complete inhibition being observed for molecular weights under 5,000. This observation points to the existence of a *critical chain length*, required for the formation of stable crystalline structures. This concept is well known from e.g. polymer complexation studies, and reflects the relatively high contribution of an initial entropy loss when combining short chains into double helices, or more extensive lamellar crystallites, compared to the gain in free energy per chain, which is proportional to the length of the helices that can be formed. In melt crystallization studies of isotactic PMMA, a similar critical chain length effect can be observed (5).

For molecular weights above this critical regime, up to molecular weights of over one million, all pressure area isotherms are practically identical. This observation is more remarkable than it may seem at first sight, especially when it is compared to molecular weight effects in the melt crystallization of i-PMMA. When raising the molecular weight of the i-PMMA, the melt crystallization rate can be observed to decrease dramatically, which implies an almost complete inhibition of crystallization for the highest molecular weight fractions (5). In the monolayer process, the effects of using extremely high molecular weights are minimal. The absence of a clear molecular weight effect in this regime can be attributed to the high mobility of the polymer chains in the expanded monolayer, in combination with the absence of restrictive entanglements, due to the twodimensional character of the monolayer (11), which proves to be a very favorable feature for the monolayer crystallization process, allowing very high molecular weight samples to be effectively crystallized.

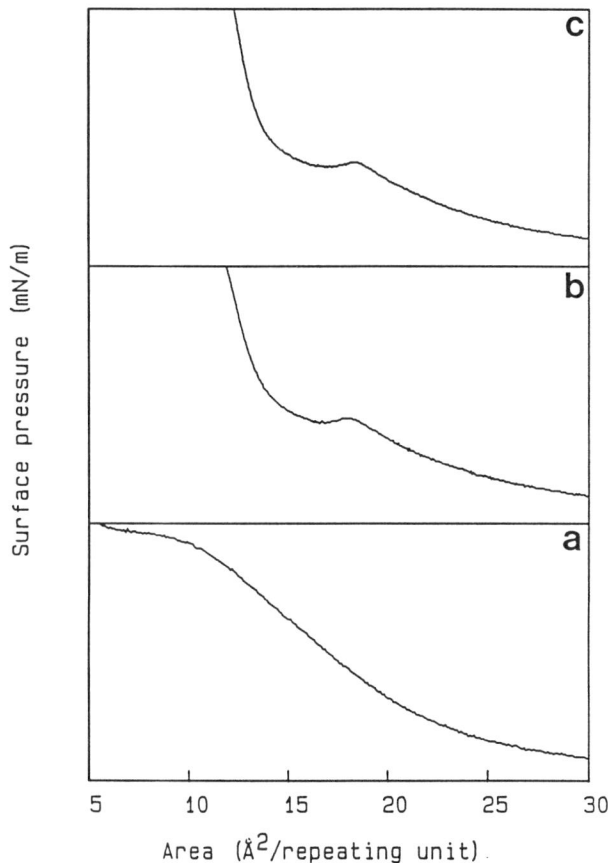

Figure 5. Pressure area isotherms of i-PMMA (21°C): $\overline{M}n = 2800$ (a), 36,000 (b) and 1.3 million (c). Isotherms are shown from 0-20 mN/m.

Thin Film Behaviour.

The crystallized monolayers of i-PMMA can be easily transferred to solid substrates to build multilayers, a possibility that was already hinted at before, when discussing the IR characteristics of these thin films. This method allows for a way to prepare thin films of i-PMMA, relevant for e.g. lithographic or optical applications. As will be shown, the crystallized monolayers can be used to prepare highly crystalline, uniaxially oriented thin films, using two approaches, either building multilayers or through a surface nucleation technique.

Multilayers. Multilayers can easily be built by repetitive transfer of the LB monolayers to solid substrates, using a conventional vertical dipping method. The LB layers, transferred in the crystalline condition, and characterized by a loose packing of the as deposited helical units (8), rapidly attain a highly crystalline structure upon annealing at 120°C, a process that is completed in several hours time. This way, it is possible to efficiently prepare highly crystalline thin films of isotactic PMMA even for the highest molecular weight samples, something that is almost impossible to do starting from amorphous thin films because of the highly restrictive effect of the geometric constraints of the thin film (12). The films prepared this way (those being thicker than approximately 60Å), have melting points not far from those of melt crystallized i-PMMA samples (± 155°C).

The crystalline films prepared with the LB deposition technique prove to be strongly oriented; this orientation can be inferred from the dichroic effects observed in the infrared spectra of these films (8). The dichroic effects observed are in complete agreement with the presence of helical structures in the thin film, which are oriented parallel to the substrate, but also, in many cases, with a very strong preferential orientation within the XY plane, parallel to this substrate surface. The first observation appears to be trivial, considering the fact that the helices are transferred layer by layer; the latter observation corresponds to reports of other helical structures, exhibiting an orientation in the dipping direction upon transfer (13, 14), caused by the flow in the monolayer that is associated with the deposition process: in this respect, the results agree with the proposed helical nature of the crystalline monolayer structures. Still, there is a very important and intriguing difference with the behaviour of other, 'regular' helical substances: the monolayers of isotactic PMMA can exhibit an orientation of the helical structures parallel to the dipping direction, but may also, under the proper conditions, exhibit a preferential orientation *perpendicular* to the transfer direction, as illustrated by the negative values for $L_{C=O}$ (8) in Figure 6. It may be clear that, if we consider the rheology of the monolayer as that of a collection of rigid helical structures, with maybe some kind of local nematic ordering present, this last observation cannot easily be explained.

Bearing in mind the model proposed for the crystallization process, we can explain a possible orientation of the helices perpendicular to the transfer direction, if we assume that not the individual helices, but rather the *crystal-*

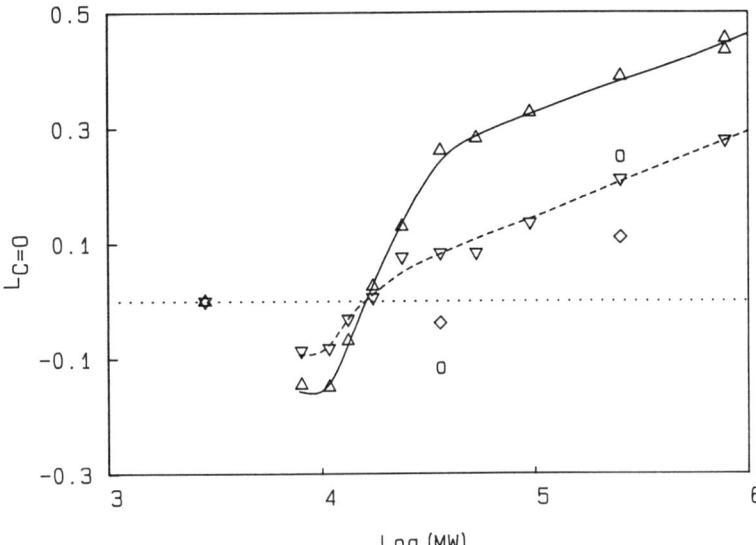

Figure 6. Lateral orientation parameter (8) as a function of molecular weight and crystallization conditions. Rapid compression to 12 mN/m, as deposited (▽), and post-crystallized at 120°C (△); crystallization pressure 6.5 mN/m, as deposited (◇) and post-crystallized (o).

lites that those helices are part of, behave as the rheological entities being oriented in the flow direction. This assumption requires that the aspect ratio of the crystallites can be such, that the longest axis of the crystallite is oriented perpendicular to the helix axes (as illustrated by Figure 7), and that the lateral coherence within the crystallites is very strong, prohibing the mobility of the individual helices in the crystallites with respect to each other. This latter requirement can be met e.g. by the suggested lamellar nature of the crystallites, with possibly fairly tight folds present. The system under study here thus probably deviates from other systems containing helical molecules by this strong lateral coherence; in this respect, there is a clear analogy with the difference in rheological behaviour of nematically ordered substances (with the mesogens being oriented in the flow direction) and of molecules in layer like smectic phases, with a strong lateral coherence, that tend to be oriented with the individual mesogens perpendicular to the flow direction (*15*).

Accepting the idea that we are dealing with an orientation of monolayer crystallites instead of individual helices, we can expect that the aspect ratio of the crystallites will be an important parameter, a perpendicular orientation requiring broad crystallites to be present. Assuming that crystallite growth is onedimensional as suggested above, such an effect may be obtained when conditions are used, leading to a low nucleation density, so that the nulei formed can grow extensively before running into another growing crystallite. The experimental conditions under which a helical orientation perpendicular to the transfer direction was observed, can indeed be correlated to a low nucleation density. For i-PMMA samples of up to intermediate molecular weights, such experimental conditions include (*8*):
- allowing the crystallization process to proceed at a relatively low surface pressure; this directly implies a low thermodynamic driving force, leading to a low nucleation density;
- using low molecular weight materials in the 'critical' molecular weight regime; in this case, the relatively high entropy loss also lowers the thermodynamic driving force, with the same result;
- mixing the i-PMMA with a low molecular weight non-crystallizable diluent (e.g. its own oligomer); in this case, for crystallization to occur, a demixing process must take place, introducing an unfavorable entropy term to the free energy of the transition, and a suppressed crystallization is observed to result. Also in this case, this effect can be argued to be responsible for a preferential orientation of the helices perpendicular to the transfer direction.

The correlation with the anticipated aspect ratio of the crystallites works quite well for this molecular weight regime, although it should be emphasized that considering the crystallites as isolated units being oriented in the monolayer, may also be a clear oversimplification, neglecting e.g. possible effects due to tie molecules connecting the crystallites. When using very high molecular weight samples, such tie molecules can be expected to become important in determining the rheological behaviour of the monolayer, enhancing the propensity for an orientation of the helices parallel to the transfer direction. Upon raising the molecular weight up to one million, very strong lateral orientations can be achieved, leading to thin films in which the preference for a lateral orientation in the dipping direction is as strong as the preference for an orientation parallel to the substrate. Also, for these high molecular weight samples, monolayer crystallization at a low surface pressure no longer leads to a complete inversal of the preferred lateral orientation, presumably due to the presence of tie molecules.

It may be clear that the crystallization conditions and the sample characteristics provide enough handles to be able to regulate the orientation characteristics of the multilayers built this way.

Surface Nucleated Thin Films. A very important limitation of the Langmuir Blodgett technique is the fact that the preparation of films thicker than several hundreds of Ångstroms is extremely time consuming. Thicknesses in the micron range are not realistic, although this thickness regime is very relevant in terms of applications in lithography and as optical waveguides. Still, it is possible to prepare highly oriented thin films of isotactic PMMA in the micron range, using the crystallized monolayers discussed above as *surface nucleation agents*. The idea of Langmuir Blodgett films functioning as nucleation agents is not new; it is e.g. known that monolayers of long chain aliphatic alcohols can nucleate ice formation in the subphase due to a similarity in the crystal lattices of the monolayer and ice (*16*). In this case, we want to use the crystallized monolayers of i-PMMA to induce crystallization in amorphous i-PMMA.

If we start from a thin film of amorphous isotactic PMMA, deposit several crystallized LB monolayers of isotactic PMMA on top of it, and anneal the film for a period of time at 120°C (as presented schematically in Figure 8), we observe that crystallization is nucleated by the overlayer, and that the entire film can be crystallized fairly rapidly, with a linear crystallization rate identical to the normal spherulitical growth rate, leading to crystallinity levels similar to spontaneously melt crystallized samples (*5*). Moreover, when the overlayers, applied to the surface of the film, are strongly oriented (either parallel to, or perpendicular to the transfer direction, as discussed above), the crystallization process in the thin film can be observed to *copy* these orientational characteristics (Figure 9). By this process of 'epitaxial' growth, it is possible to very effectively prepare highly oriented thin films by a simple deposition of a few monolayers on top of an amorphous film. In some cases, the lateral orientation parameter determined for these films can be higher than the highest achieved for a thin film completely built from deposited LB layers. Similar orienting effects induced by surface layers, have been reported for thin films of liquid crystalline materials (*17*).

When following the growth of the crystalline phase in the nucleated films, it can be observed that the orientational characteristics imposed by the overlayer are slowly lost upon progress of the crystallization front (*5*). Still, the approach can efficiently induce oriented crystallization up to distances of several microns from the surface. The orienting effect of the LB overlayers is maximal for overlayer thicknesses of more than 40Å, but even for monolayers (12Å thick) deposited on the film surfaces, clear orientation effects can be

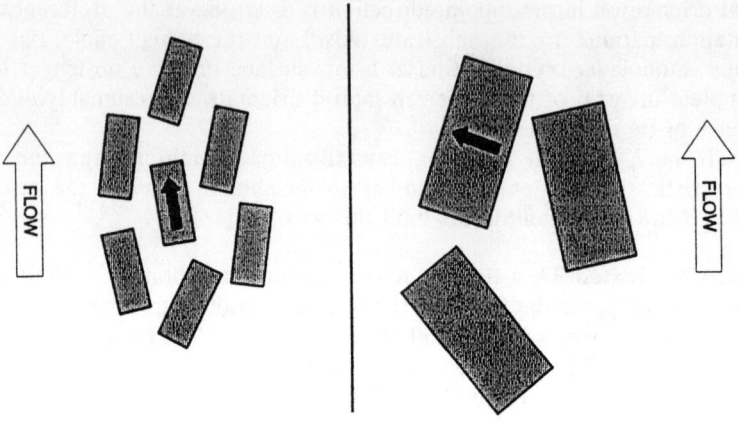

Figure 7. Schematic representation of the flow induced orientation process as a function of the aspect ratio of the crystallites. Arrow indicates direction of helix axes within crystallite.

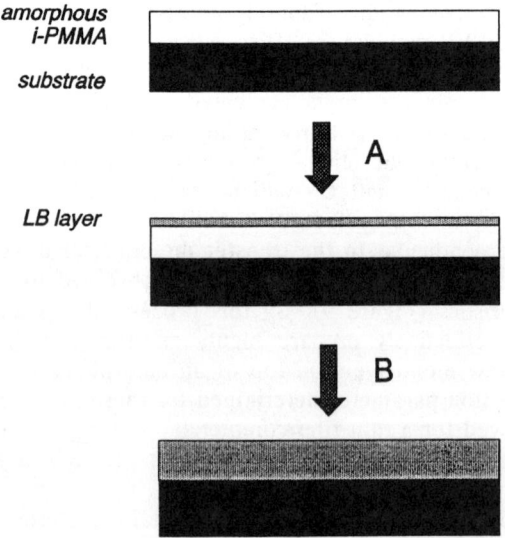

Figure 8. Schematic representation of the use of LB layers as surface crystallization nuclei: crystallized LB monolayers are deposited on top of an amorphous film (A), which is subsequently crystallized at 120°C (B).

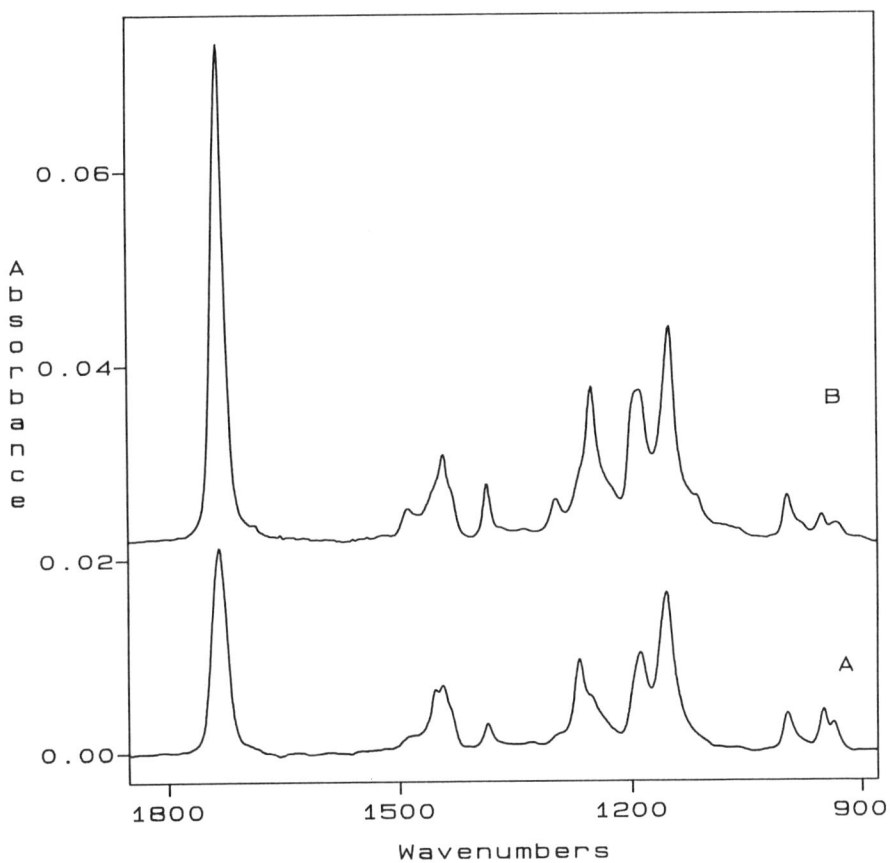

Figure 9. Polarized transmission spectra of an isotropic amorphous film (300Å) of i-PMMA covered with a 60Å nucleating overlayer of crystalline i-PMMA LB-layers ($\overline{M}n=770K$), after 2 hours at 120°C. Polarization along (A) and perpendicular to (B) transfer direction.

observed to be induced in the film. The fact that the orientation of the monolayers is copied by the crystallizing thin film provides a handle to effectively visualize the orientation of the monolayer as it was deposited (5). The orientation of the crystalline structures in the film can easily be assessed with a polarization microscope, on the basis of the birefringence characteristics. In doing so, it was observed that the orientation of a transferred monolayer was not homogeneous; for one complete dipping cycle (with negligible transfer during the first downstroke) a maximal orientation effect was only observed several millimeters from the top meniscus line on the substrate (corresponding to the beginning of the monolayer transfer). Especially when a monolayer was crystallized with the film supporting substrate already present in the interface (thus locally forming an extra barrier for the monolayer), it can be seen that at the start of the transfer process, extending up to 10 mm from the top meniscus line, a preferential orientation of the helices perpendicular to the dipping direction was observed following transfer, for a monolayer that normally exhibits a helical flow orientation parallel to the transfer direction. Evidently, this observation must be the result of an initial preferential orientation in the monolayer on the water surface close to the substrate barrier, the helices being aligned parallel to it upon compression. It also implies that the flow induced orientation is not an instantaneous meniscus process.

Disadvantages of the overlayer approach (compared to the multilayer approach) include the fact that the surface nucleated crystallization of the thin film leads to severe deformations, with a rugged surface relief developing due to the stresses in the film as a result of the anisotropic crystallization shrinkage, and the fact that, since the linear crystallization speed is equal to the normal melt crystallization speed, for the process to be efficient, it is necessary to use fairly favorable molecular weights for the matrix material.

Conclusions.

Monolayers of isotactic PMMA were shown to exhibit a surface pressure induced crystallization process, in which, starting from an amorphous expanded conformation, crystallites appear to be formed, consisting of double helical structures of PMMA, oriented flat on the water surface. The crystallization proces was tentatively interpreted in terms of a lamellar growth mechanism, following an activated nucleation stage. In this respect, the behaviour of i-PMMA appears to be unique in that it presents an actual monolayer crystallization process involving a polymer backbone, instead of a low molecular weight material, or merely polymer side groups.

The crystallized monolayers can be transferred to solid substrates, yielding crystalline thin films which can be very strongly oriented in the XY plane. In contrast to the normal behaviour of helical substances reported sofar, the helices of i-PMMA can be transferred with a preferential orientation parallel to, as well as perpendicular to the dipping direction, indicating that the rheological units being oriented are in fact the crystallites, rather than the individual helices.

Highly oriented films can also be effectively prepared according to a procedure, in which the crystallized monolayers are used as surface nucleation agents for amorphous thin films, in which the orientation of the nucleating layers is copied in the resulting crystalline structure. This method extends the usefulness of the LB technique up to thin films in the micron range.

Literature Cited

1. Poupinet, D., Vilanove, R., Rondelez, F., *Macromolecules* **1989**, *22*, 2491.
2. Kuan, S.W.J., Frank, C.W., Fu, C.C., Allee, D.R., Maccagno, P., Pease, R.F.W., *J. Vac. Sci. Technol. B.* **1988**, *6(6)*, 2274.
3. Stroeve, P., Srinivasan, M.P., Higgins, B.G., Kowel, S.T., *Thin Solid Films* **1987**, *146*, 209.
4. Beredjick, N., Ahlbeck, R.A., Kwei, T.K., Ries Jr., H.E., *J. Polym. Sci.* **1960**, *46*, 268.
5. Brinkhuis, R.H.G., Schouten, A.J., manuscript in preparation.
6. Brinkhuis, R.H.G., Schouten, A.J., *Macromolecules* **1991**, *24*, 1487.
7. Henderson, J.A., Richards, R.W., *Polym. Prepr.* **1990**, *31(2)*, 83.
8. Brinkhuis, R.H.G., Schouten, A.J., *Macromolecules* **1991**, *24*, 1496.
9. Kusanaga, H., Tadokoro, H., Chatani, Y., *Macromolecules* **1976**, *9*, 531.
10. Bosscher, F., Ten Brinke, G., Challa, G., *Macromolecules* **1982**, *15*, 1364.
11. Carmesin, I., Kremer, K., *J. Phys. France* **1990**, *51*, 915.
12. Billon, N., Haudin, J.M., *Colloid Polym. Sci.* **1989**, *267*, 1064.
13. Duda, G., Thesis, Max Planck Institut für Polymerforschung, Mainz, BRD, **1988**.
14. Orthmann, E., Wegner, G., *Angew. Chem. Int. Ed. Engl.* **1986**, *25*, 1105.
15. Zentel, R., Wu, J., *Makromol. Chem.* **1986**, *187*, 1727.
16. Gavish, M., Popovitz-Biro, R., Lahav, M., Leiserowitz, L., *Science* **1990**, *250*, 973.
17. Ito, S., Kanno, K., Ohmori, S., Onogi, Y., Yamamoto, M., *Macromolecules* **1991**, *24*, 659.

RECEIVED September 24, 1991

Chapter 6

Preparation of Multicomponent Langmuir–Blodgett Thin Films Composed of Poly(3-hexylthiophene) and 3-Octadecanoylpyrrole

M. Rikukawa and M. F. Rubner

Department of Materials Science and Engineering, Massachusetts Institute of Technology, Cambridge, MA 02139

> Langmuir-Blodgett multilayer thin films were fabricated from mixed monolayers containing poly(3-hexyl thiophene) and 3-octadecanoyl pyrrole. The mixed monolayers were found to be stable at the air-water interface and could be readily deposited onto solid substrates as Y-type films by the vertical lifting method. The structure of the mixed multilayer thin films was probed by visible absorption, thickness, X-ray diffraction, and FTIR measurements. Evidence for preferential orientation of both the polymer chains and the surface active molecules was found. The conductivities of the mixed LB films doped with strong oxidizing agents such as $NOPF_6$, $FeCl_3$, H_2PtCl_6 and $SbCl_5$ were in the range of 10^{-3}-1.0 S/cm. The temperature dependence of the conductivity was found to fit a logs varies as $T^{-1/2}$ dependence.

The desire to manipulate electroactive polymers into thin film structures with well defined and controllable molecular organizations has lead to the use of the Langmuir-Blodgett technique and the development of mixed monolayer systems comprised of a surface active component and a nonsurface active electroactive polymer (1). Through the use of mixed monolayer systems, it is possible to fabricate multilayer thin films and superlattices from a wide variety of conjugated polymers that would not be otherwise amenable to LB manipulation. The final molecular organization of these mixed multilayer thin films and hence their electrical properties ultimately depend on the structures of the molecules being manipulated and the nature of their intermolecular interactions. By suitable choice of the surface active component and conjugated polymer, it should be possible to create films with morphologies ranging from microphase separated multidomain systems to completely compatible single phase systems. Depending on the particular combination of electrical, thermal, and mechanical properties required from these new thin films, either of these two extremes may be desirable. Thus, multicomponent systems provide an additional avenue for controlling the properties of electroactive multilayer thin films.

In this paper, we describe the preparation of Langmuir-Blodgett thin films comprised of poly(3-hexyl thiophene) (PHT) and 3-octadecanoyl pyrrole (3ODOP). 3ODOP in this case constitutes the surface active component of the mixed LB film whereas PHT is representative of a nonsurface active conjugated polymer. We have previously shown

(2) that LB films of the poly(3-alkyl thiophenes) can be fabricated by mixing the polymer with cadmium stearate. In this case, it was found that the two components form phase separated microdomains when the alkyl sidechain of the polymer is relatively short (8 or less carbons) but became significantly more phase mixed when the polymer sidechain is replaced with an octadecyl unit (3). For films formed with PHT, the phase separated polymer domains were dispersed in a matrix of highly ordered domains of cadmium stearate in which the cadmium stearate molecules assumed their normal bilayer molecular organization. The 3ODOP multilayer films were fabricated to study the effect that a different surface active agent has on the molecular organization of PHT based LB films and to create electroactive LB films that were free of ionizable head groups. For the fabrication of thin film microelectronic devices and sensors it is clearly desirable to use a surface active component that does not contain inorganic ions or labile protons that can migrate under the influence of an electric field. This paper describes the structures of these new thin films and some of their electrical properties.

EXPERIMENTAL

Poly(3-hexyl thiophene) (PHT) was prepared according to the method of Tamao et al (4). The synthesis of the 3-octadecanoyl (3ODOP) used in this study has been described in detail elsewhere (5). Figure 1 shows the structures of these materials. Monolayers were spread from chloroform solutions (concentration: 1 mg of mixture of PHT and 3ODOP per ml of solvent) onto a water subphase that was purified with a Milli-Q purification system (Millipore Corp.). The surface pressure-area isotherms were measured on a Lauda film balance at 20°C with a compression speed of 5 Å^2 molecule^{-1}min^{-1}. Monolayers are defined as stable when they hold an essentially constant area at a constant pressure of 20 mN/m for at least 10 hrs.

Y-Type multilayer films were built up by the vertical dipping method at 20 °C and 20 mN/m. A dipping speed of 1 mm/min was used for the first and second layer whereas the upstroke dipping speed was increased to 5 mm/min for subsequent dips. Drying times of 1h were used after depositing the second layer and this time was reduced to 15 min for all subsequent dips. The mixed LB films were built onto hydrophobic glass slides (prepared by treatment with 1,1,1,3,3,3,-hexamethyldisilazane) for optical absorption and X-ray measurements. Zinc selenide plates and platinum-coated slides were used for FTIR measurements. Visible absorption spectra were recorded using an Oriel Instaspec System 250 multichannel spectrophotometer. Small angle X-ray diffraction measurements were made with a Rigaku Routalex RU 300 unit (copper Ka target). FTIR spectroscopy was done with a Digilab FTS-40 FTIR spectrometer with a Spectra Tech Model 500 spectra reflectance accessory. The grazing incident reflection spectra were generated on LB films consisted of 10 layers deposited onto platinum-coated slides, whereas zinc selenide plates deposited with 10 layers on one side were used for the transmission measurements. The LB films were chemically doped with $NOPF_6$, $FeCl_3$, H_2PtCl_6 and $SbCl_5$. In the case of $NOPF_6$, the LB films were immersed in acetonitrile solutions containing 1 mg/ml of $NOPF_6$ for 5 min. (for $FeCl_3$ and H_2PtCl_6, the LB films were immersed in nitromethane solutions in the same way) For $SbCl_5$ doping, the LB films were exposed to the dopant vapor for 10 min under partial vacuum. In-plane conductivity measurements were made using a four point Van der Pauw configuration with a Keithley model 602 electrometer. Contacts were made to the corners of the LB film using conductive electrodag paste.

RESULTS AND DISCUSSIONS

Surface Pressure-Area Isotherms

The surface pressure-area isotherms of PHT, 3ODOP, and various mixtures of PHT and 3ODOP with different molar ratios (based on the molecular weight of the PHT repeat unit) are presented in Figure 2. As indicated by the essentially nonexistent surface activity of pure PHT, stable condensed monolayers can not be formed directly from this polymer due to its lack of a strong hydrophilic component. The isotherm of pure 3ODOP, on the other hand, shows that this material readily forms stable, condensed monolayers with a limiting area per molecule of about 28 $Å^2$. CPK models indicate that the 3ODOP molecule in its fully extended form has an end-to-end distance of about 28 Å and a head group area of about 7 Å ± 4 Å. These values in conjunction with the isotherm indicate that at high surface pressures, the pyrrole head groups are anchored in the water subphase and the hydrophobic tail groups are tightly packed and oriented essentially normal to the plane of the subphase. The low surface pressure shoulder observed in the isotherm of 3ODOP suggests that a change in the molecular packing of the head groups occurs during film compression. We surmise that at low surface pressures the head groups are better capable of maximizing their hydrogen bonding interactions thereby creating a more expanded head group area.

We have previously demonstrated (2) that high quality LB films of the alkyl substituted polythiophenes can be realized by forming mixed monolayers with suitable portions of stearic acid. As indicated by Figure 2, stable mixed monolayers can also be obtained when 3ODOP is used as the surface active molecule. This figure shows that condensed monolayers are formed from mixtures containing as much as 90 mole % of the conjugated polymer (in the case of the mixed monolayers, the area per molecule axis is based solely on the number of 3ODOP molecules spread onto the subphase in order to determine the effect that the added polymer has on the isotherms of this material). Also note that the isotherms of the mixed monolayers of PHT and 3ODOP exhibit what appear to be phase transitions at elevated pressures (between 25 and 35 mN/m). In the case of the PHT/cadmium stearate mixed monolayers, this type of behavior was attributed to a rejection of the phase separated polymer molecules from the cadmium stearate monolayer (2). The transitions observed in this latter case, however, were significantly better defined than those observed in the mixed monolayers of PHT and 3ODOP. The more diffuse nature of these transitions suggests that a greater level of phase mixing occurs in the mixed monolayers of PHT and 3ODOP. The structural consequences of this higher level of molecular interaction will be become apparent shortly.

Structure of the Multilayers

Monolayers of 3ODOP (6) and mixtures of 3ODOP and PHT in different molar ratios can be readily transfered onto solid substrates at 20 mN/m and 20°C as Y-type LB films. Figure 3 shows the hydrocarbon stretching region of the FTIR transmission and reflection spectra of the 3ODOP LB film and a mixed LB film with a 5/1 mole ratio of PHT/3ODOP. In the case of the multilayer thin film of 3ODOP, the NH stretching vibration at 3200 cm^{-1} and the C=O stretching vibration at 1630 cm^{-1} (not shown) of the head group are stronger in the transmission spectrum than in the reflection spectrum. In addition, the intensity of the asymmetric CH_2 (2920 cm^{-1}) stretching vibration is significantly stronger in the reflection spectrum than in the transmission spectrum, whereas the intensity of the symmetric CH_2 (2852 cm^{-1}) stretching vibration is weakest in the reflection spectrum. These results indicate that the pyrrole rings of the 3ODOP molecules lie nearly parallel to the substrate surface and that their aliphatic tail

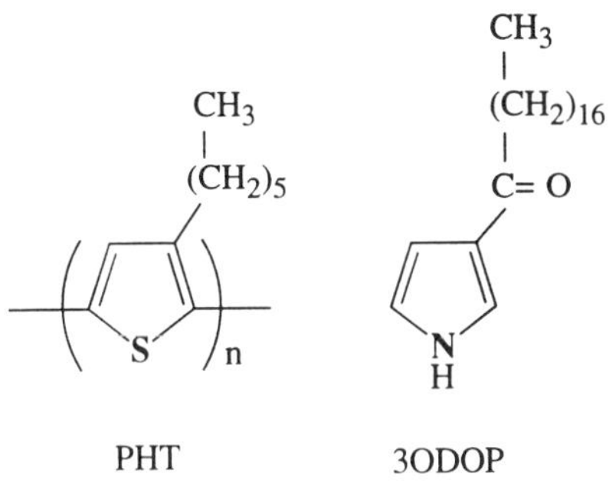

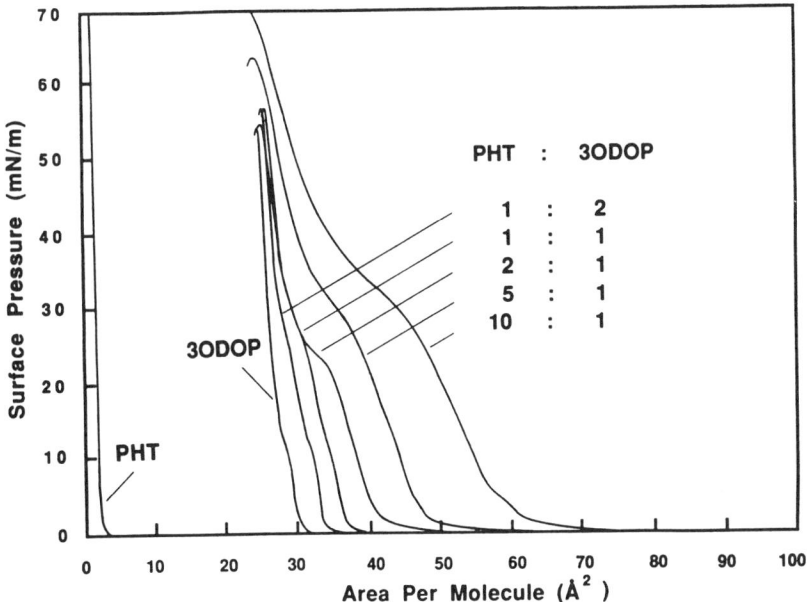

Figure 1. Structure of poly(3-hexyl thiophene) and 3-octadecanoyl pyrrole.

Figure 2. Pressure-area isotherms of PHT, 3ODOP, and mixed monolayers containing PHT and 3ODOP in different molar ratios.

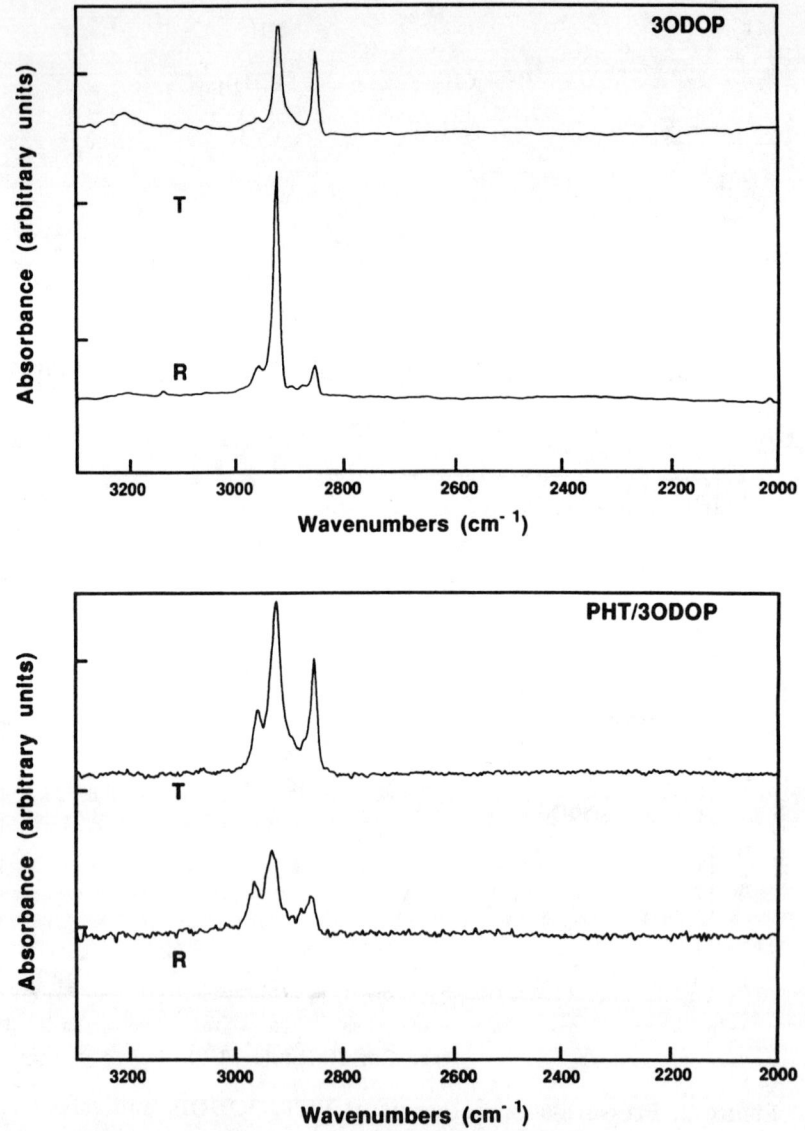

Figure 3. Transmission (T) and reflection (R) FTIR spectra of 3ODOP and a mixed LB film with a 5/1 (PHT/3ODOP) mole ratio.

groups tilt away from the surface normal with a fairly large tilt angle. Assuming that the hydrocarbon chains are in their fully extended all-trans form, an analysis of the polarization dependence of this system (7) reveals that the tail groups are oriented about 30° from the surface normal.

In sharp contrast, quite different results were observed with the 5/1 mixed LB film of PHT and 3ODOP. In this case, the asymmetric (2920 cm^{-1}) and symmetric (2852 cm^{-1}) CH_2 stretching vibrations both exhibit significantly stronger intensities in the transmission mode than in the reflection mode. This again demonstrates that the multilayer film is comprised of highly oriented molecules, however, tilt angle calculations carried out using Umemura's techniques (8) indicate that the alkyl groups of the 3ODOP molecules and the polythiophene chains on average are oriented about 10° from the surface normal. This value should be viewed as only a rough approximation as it is not possible to independently probe the orientation of the alkyl groups of the two components present in the film. In any event, these results show that the alkyl groups of the polymer and the surface active component are preferentially oriented within the film and exhibit a strong tendency to align, on average, close to the surface normal. Similar results were obtained from films containing a 2/1 mole ratio of PHT/3ODOP thereby suggesting that the strong polarization dependence observed in these films is not a simple superposition of disordered PHT alkyl groups and 3ODOP molecules oriented with their usual 30° tilt angle. In addition, the fingerprint region of the spectrum (not shown) clearly shows that the 3ODOP pyrrole head groups do not adopt the level and type of orientation found in the pristine multilayer films of 3ODOP. Thus, the organization of the 3ODOP molecules is radically altered in the presence of PHT. This is quite different from the mixed films fabricated from PHT and cadmium stearate where it was found that the cadmium stearate molecules assemble with their normal bilayer organization even in films containing a 5/1 mole ratio of polymer to surfactant.

The small angle X-ray diffraction patterns of a multilayer thin film of 3ODOP (60 layers) and a mixed film containing a 5/1 mole ratio of PHT/3ODOP are shown in Figure 4 (100 layers). The diffraction pattern of the 3ODOP multilayer exhibits a number of (00l) Bragg reflections that give a bilayer repeat distance of about 28 Å for a Y-type LB film (the first order reflection is very intense and occurs at about 2θ=3°). A thickness per layer of about 14 Å was also obtained from ellipsometry and profilometer measurements. In sharp contrast, the mixed LB film exhibits only a single diffraction peak at a 2θ value of about 4.9° (d =18 Å). Both diffraction patterns also exhibit a broad amorphous halo centered at about 24° arising from the supporting glass substrate.

For the 3ODOP multilayer film, the bilayer d-spacing calculated from the diffraction pattern and the implied thickness per monolayer of only 14 Å suggests that the aliphatic tail groups of the 3ODOP molecules tilt at an angle of approximately 55° from the surface normal. This is significantly larger than the value calculated from the FTIR spectra thereby suggesting that the hydrogen bonded pyrrole head groups from each adjoining monolayer are interdigitated (7). Nevertheless, the absence of an equivalent diffraction pattern in the mixed LB film shows that it is not possible for the 3ODOP molecules to assemble into their normal bilayer organization. Indeed, the single peak observed in the diffraction pattern of the mixed LB film can be attributed to the interchain packing distance of the polythiophene molecules as opposed to the molecular stacking of 3ODOP molecules. As has been reported by Winoker et al. (9) and Gustafsson and Inganas (10), poly(3-hexyl thiophene) forms a layered structure in which the main chains are stacked on top of each other forming parallel planes. In addition, the alkyl side chains adopt their fully extended planar zig-zag form and are confined to the plane of the polythiophene backbone. The layer spacing of the PHT

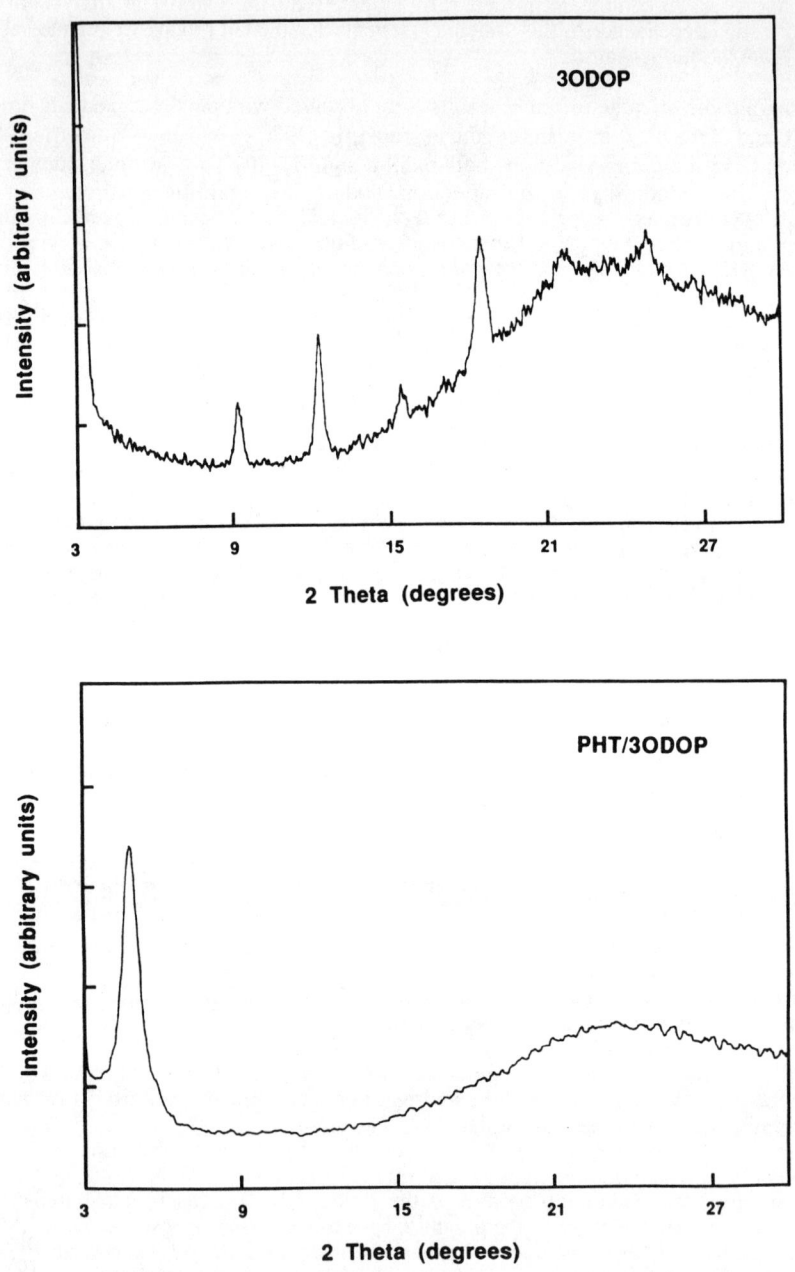

Figure 4. X-ray diffraction patterns of the 3ODOP LB film and a mixed LB film with a 5/1 (PHT/3ODOP) mole ratio.

chains was found to be about 16.8 Å which is very close to the 18 Å value that is obtained from the diffraction pattern of the mixed PHT/3ODOP film. The proposed origin of this peak is also consistent with the fact that the peak intensity decreases as the amount of PHT in the mixed LB film decreases. It is also interesting to note that the intensity of this peak in the 5/1 mole ratio film is significantly stronger than that observed in solution cast films with either the same mole ratio of components or pure PHT. Thus, the layer-by-layer deposition of the mixed monolayers has a significant impact on the orientation and organization of the polymer molecules.

These results suggest that the polymer chains are aligned with their long axes in the plane of the substrate and with their hydrocarbon sidegroups oriented (on average) normal to this direction as depicted schematically in Figure 5. Polarization experiments carried out in the visible region of the spectrum would seem to further support this hypothesis. For example, Figure 6 shows the visible spectra of a 5/1 mixed LB film recorded with s- and p-polarized light with the substrate of the film oriented at a 45° incidence angle. This figure shows that the primary absorption band of the $\pi-\pi^*$ transition of the polythiophene conjugated backbone at about 480 nm is more intense when observed with s-polarized light (electric vector parallel to the substrate surface) than with p-polarized light (electric vector oriented 45° to the substrate surface). Since this particular excitation is polarized along the polymer backbone, the polythiophene chains must exhibit a tendency to orient parallel to the plane of the substrate.

The picture that emerges is summarized by the molecular organizations illustrated in Figure 5. This figure is presented to provide a simplified comparison of the important differences between the pure and mixed system and is not meant to be a detailed depiction of their structures. In the 3ODOP multilayer thin film, the molecules adopt a Y-type structure in which the hydrocarbon tail groups assume a tilt angle of at least 30° from the surface normal (60° from the substrate surface). It is also highly probable that the head groups from adjoining monolayers are interdigitated due to interlayer hydrogen bonding. When mixed with PHT molecules, the organization of the 3ODOP molecules is completely disrupted. In this case, unlike what was found for cadmium stearate based mixed LB films, the 3ODOP molecules do not organize into well ordered molecular stacks but rather form a more random supporting template for the polymer chains in which the average orientation of their hydrocarbons tails is close to the surface normal. The polymer chains, on the other hand, tend to orient parallel to the surface substrate but with their hydrocarbon tail groups orienting in a direction similar to the tail groups of the 3ODOP surface active molecules. This type of molecular organization is most likely the consequence of a higher level of molecular interaction between the the polymer and the host 3ODOP molecules and perhaps even more importantly, the weaker driving force (as compared to cadmium stearate molecules) of the 3ODOP molecules to assemble into well ordered molecular domains. In other words, the cadmium stearate molecules exhibit a greater tendency to crystallize in the presence of PHT than do the 3ODOP molecules.

Electrical Properties of the Mixed LB Films

The mixed PHT-based LB films can be doped with strong oxidizing agents such as $NOPF_6$, $FeCl_3$, $SbCl_5$, and H_2PtCl_6. The highest conductivity observed to date is about 1 S/cm (see Table I). The in-plane conductivities of these doped samples are relatively stable in vacuum or nitrogen, however, the conductivity decreases rapidly in air (particularly moist air). Similar results were obtained with the mixed LB films of PHT and cadmium stearate (11). Figure 7 shows the absorption spectra of the mixed

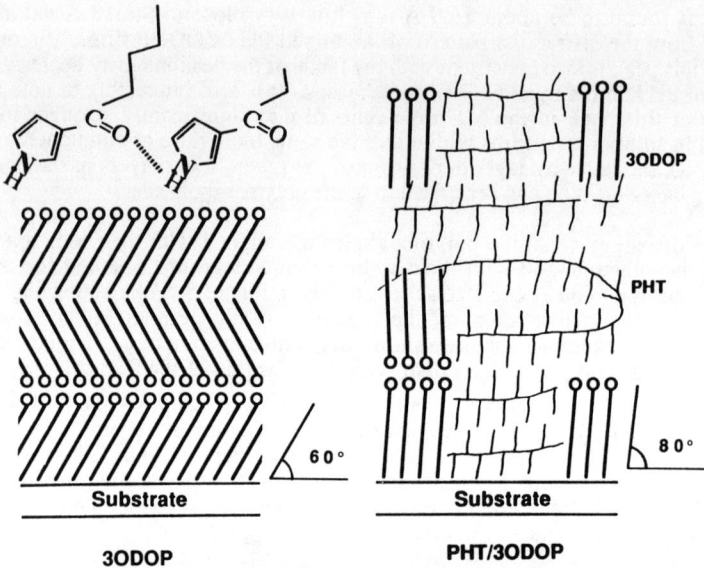

Figure 5. Simplified models illustrating the important structural differences between a neat 3ODOP LB film and a mixed PHT/3ODOP LB film.

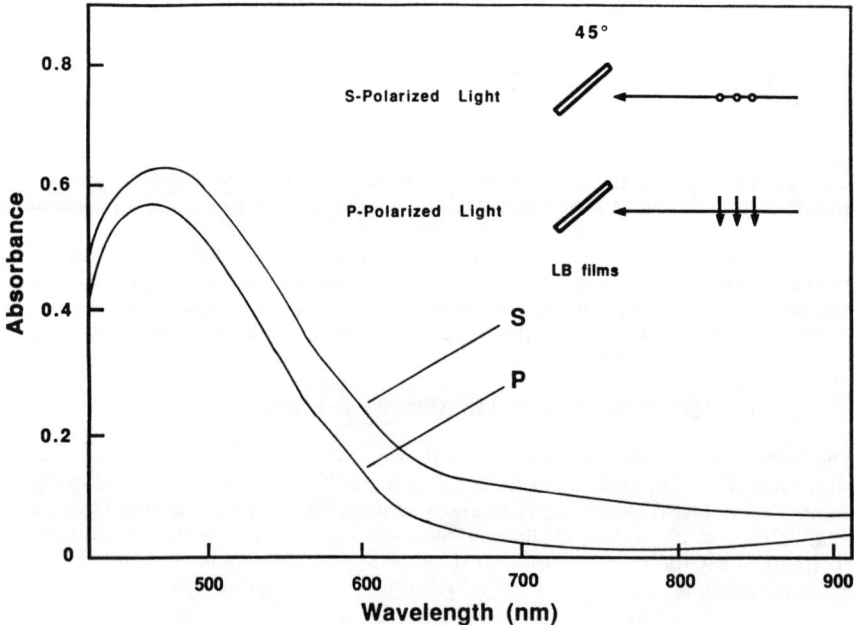

Figure 6. Polarized absorption spectra of a mixed LB film with a 5/1 (PHT/3ODOP) mole ratio.

LB film before and after doping with FeCl3. After doping, the absorption band of the π–π* electronic transition of the polythiophene conjugated backbone at about 480 nm is eliminated and a new lower energy excitation at longer wavelengths appears. These changes are generally attributed to the creation of localized charged defects in the form of bipolarons (12).

Table I. In-plane conductivities of the doped mixed LB films with a 5/1 (PHT/3ODOP) mole ratio

Dopant	Method	Conductivity(S/cm)
$NOPF_6$	Solution doping	6.8×10^{-1}
H_2PtCl_6	Solution doping	9.4×10^{-3}
$FeCl_3$	Solution doping	1.0
$SbCl_5$	Gas doping	2.4×10^{-1}

Previously, we have shown (13) that the conduction mechanism active in the PHT/3ODOP mixed LB films doped with $NOPF_6$ is best described by the theory of charging energy limited tunnelling between conductive islands (14). Similarly, the temperature dependence of the conductivity for the PHT/3ODOP mixed LB films doped with a variety of different oxidizing agents was also found to fit a log s varies as $T^{-1/2}$ dependence. Figure 8 displays the temperature dependence of the conductivity of mixed LB films doped with $NOPF_6$, $FeCl_3$, and $SbCl_5$. The dramatic decrease in the conductivity at temperatures above 50 °C is due to the thermal dedoping process that is well documented in the literature (15). This latter phenomenon is irreversible (i.e, the conductivity change is irreversible with temperature).

CONCLUSION

We have demonstrated that mixed LB films of the poly(3-alkyl thiophenes) can be fabricated without the use of traditional fatty acids or their metal ion salts. By using 3-octadecanoyl pyrrole as the surface active agent, it is possible to form ordered Y-type multilayer thin films containing as much as 90% of the nonsurface active conjugated polymer. The multilayer structures of these new mixed LB films vary considerably from their fatty acid counterparts. Most notably, the use of 3ODOP promotes a higher level of molecular level mixing and the establishment of oriented polymer chains. The mixed LB films can be easily rendered electrically conductive by doping with strong oxidizing agents. The conductivities of the doped samples are stable in vacuum or nitrogen. The mechanism of conduction active in the doped films was found to best fit the theory of charging energy limited tunnelling between conductive islands. This system may prove useful in the fabrication of thin film devices such as field effect transistors as it is now possible to fabricate poly(3-alkyl thiophene) based LB films that are not contaminated with mobile ions and protons. The use of 3ODOP also opens up the possibility for additional polymerization chemistry at the air-water interface or in the multilayer thin film (5).

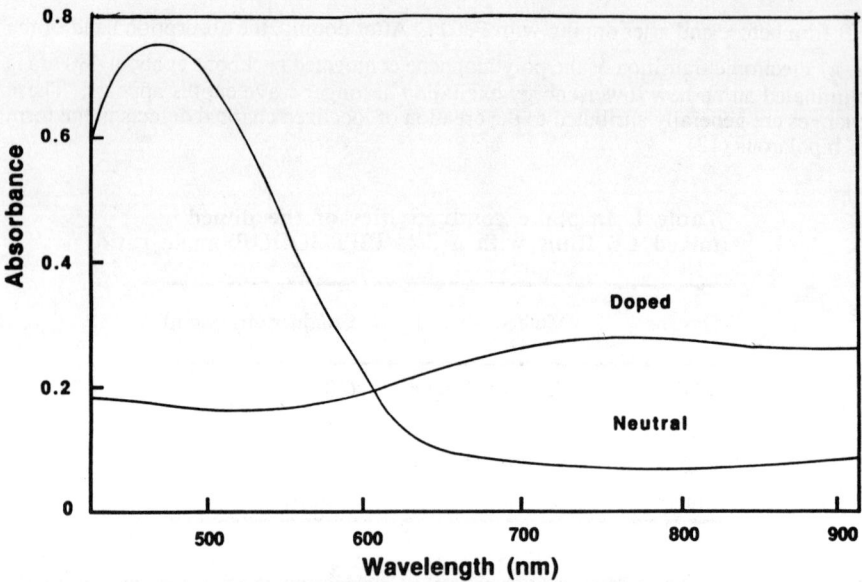

Figure 7. Absorption spectra of a 5/1 (PHT/3ODOP) mixed LB film recorded before and after doping with FeCl$_3$.

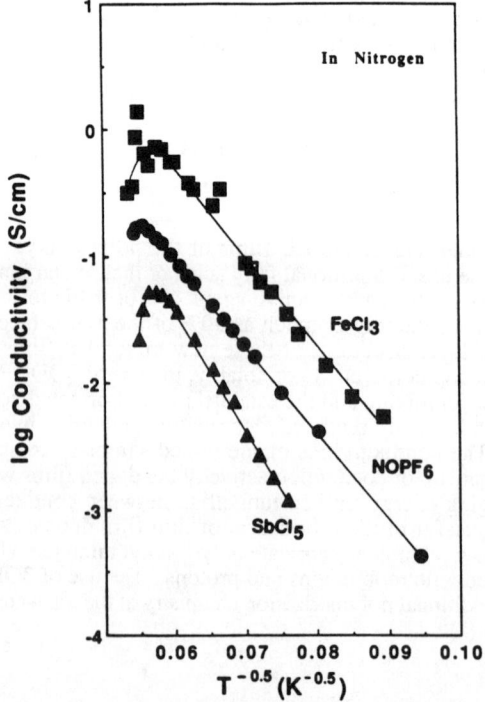

Figure 8. The temperature dependence of the in-plane conductivity of 5/1 (PHT/3ODOP) mixed LB films oxidized with different dopants.

ACKNOWLEDGEMENTS

The authors would like to acknowledge Ms Josephine H. Cheung, Mr Robert B. Rosner, Mr A. Tim Royappa and Dr Yading Wang of MIT for their contributions to this research. Partial support of this research was provided by the National Science Foundation and the MIT Center for Materials Science and Engineering. In addition, the authors thank Hitachi Chemical Co. Ltd. for support of M. R.

REFERENCES

1. (a) Nakahara, H.; Nakayama, J.; Hoshino, M.; Fukuda, K. *Thin Solid Films* **1988**, *160*, 87. (b) Watanabe, I.; Hong, K.; Rubner, M. F. *Chem.Comm.* No. 2, 123. (c) Yli-Lahti, P.; Punkka, E.; Stub, H.; Kuivalainen, P.; Laakso, J. *Thin Solid Films* **1989**, *179*, 221.
2. Watanabe, I.; Hong, K.; Rubner, M. F. *Langmuir* **1990**, *6*, 1164.
3. Watanabe, I.; Rubner, M. F. *J. Phys. Chem.* **1990**, *94*, 8715.
4. Tamao, K.; Kodama, S.; Nakajima, I.; Kumada, M.; Minato, M.; Suzuki, K *Tetrahedron* **1982**, *38*, 3347.
5. Hong, K.; Rubner, M. F. *Thin Solid Films* **1988**, *160*, 187. Hong, K.; Rubner,
   M. F. *Chemistry of Materials* **1990**, *2*, 82.
6. Cheung, J.; Rubner, M. F. *Polymer Preprints* **1990**, *31*, 404.
7. Cheung, J.; Rubner, M. F. to be published.
8. Umemura, J.; Kamata, T.; Kawai, T.; Takenaka, T. *J. Phys. Chem.* **1990**, *94*, 62.
9. Winokur, M. J.; Spiegel, D.; Kim, Y.; Hotta, S.; Heeger, A. J. *Synth. Met.* **1989**, *28*, C419.
10. Gustafsson, G.; Inganäs, O. to be published.
11. Watanabe, I.; Rubner, M. F. *British Polymer J.* **1990**, *23*, 165.
12. *Handbook of Conducting Polymers;* Skotheim, T. J., Ed.; Marcel Dekker: New York, **1986**, Vol. 1 and 2.
13. Punkka, E.; Rubner, M. F.; Hettinger, J. D.; Brooks, J. S.; Hannahs, S. T. *Phys. Rev. B* **1991**, *43*, 9076.
14. (a) Sheng, P.; Abeles, B. *Phys. Rev. Lett.* **1973**, *31*, 44. (b) Abeles, B.; Sheng, P.; Coutts, M. D.; Arie, Y. *Adv. Phys.* **1975**, *24*, 407.
15. Wang, Y.; Rubner, M. F. *Synth. Met.* **1990**, *39*, 153.

RECEIVED September 24, 1991

Chapter 7

Langmuir–Blodgett Films of Novel Polyion Complexes of Conducting Polymers

A. T. Royappa and M. F. Rubner

Department of Materials Science and Engineering, Massachusetts Institute of Technology, Cambridge, MA 02139

A novel method of incorporating polyion complexes of conducting polymers into Langmuir-Blodgett films has been studied. The method involves forming polyion complexes of preformed conducting polymers with stearylamine. The polyion complexes form stable, easily transferable, expanded monolayers on the water surface. The resulting Langmuir-Blodgett films are highly ordered and Y-type in structure, with interdigitated stearylamine alkyl tails. The polymer chains are sandwiched in planes between stearylamine layers. The conductivity exhibited by these films is on the order of 10^{-2} S/cm.

Since the discovery of electrically conducting polymers more than a decade ago (1), a significant portion of the research on conducting polymers has been focused on those that are readily processable and environmentally stable, eg, the polythiophenes, polypyrroles and polyanilines. This paper deals with some novel methods of fabricating conducting Langmuir-Blodgett (LB) films of polythiophenes and polyanilines functionalized with ionizable side groups. The LB technique is aptly suited for the study and manipulation of conducting polymers because it allows for fine control of the molecular architecture of very thin films, yields a very high degree of anisotropic ordering, and holds out numerous possibilities for the construction of electronic devices and sensors (2). The most serious hurdle to processing any material by the LB method is a lack of surface activity on the part of the material. In this work, we present a novel approach to surmounting this hurdle for non-surface active conducting polymers, namely by forming surface active polyion complexes between the polymers and stearylamine, a commonly available surfactant. Although other workers have synthesized LB films containing polyion complexes of conducting polymer precursors (3), this research marks the first successful fabrication of LB films of a polyion complex of a preformed conjugated polymer. It should be noted at this point that this method is quite general, in that it is not restricted to conjugated polymers (3). The success of this method with the two disparate polymers mentioned above would seem to imply that any non-surface active polymer that can be functionalized with ionic groups can be reacted with a suitable surface-active agent to form surface active polyion complexes.

EXPERIMENTAL

Poly(thiophene-3-acetic acid) (PTAA) was prepared as follows: commercially available ethyl thiophene-3-acetate (Lancaster Synthesis) was polymerized by the $FeCl_3$ suspension method (4), and the resulting polymer was converted to PTAA by acid hydrolysis using aqueous HCl. The weight average molecular weight of the resulting polymer as determined by GPC (polystyrene equivalent calibration) was $\approx$ 1700 g/mol, which corresponds to a degree of polymerization of $\approx$ 12. This material should therefore be viewed more as an oligomer than a true high polymer. The polyion complex of PTAA and stearylamine was synthesized as follows: PTAA was dissolved in dimethylacetamide (DMAc), to which a solution of stearylamine in benzene was added dropwise, with stirring, to form the amine salt of PTAA. The structure of the complex is shown below:

$$\text{structure: thiophene ring with } CH_2COO^-\ {}^+H_3N\text{-}C_{18}H_{37} \text{ substituent, repeat unit } x$$

Sulfonated polyaniline (PAn) of number average molecular weight approximately 50,000 g/mol was obtained from Prof. A.J. Epstein of Ohio State University. The synthesis and properties of sulfonated polyaniline are detailed elsewhere (5). The polymer (PAn) provided to us was in the emeraldine salt form. The polyion complex of PAn was synthesized in the exact manner described above, resulting in the formation of the following amine salt of PAn:

$$\text{structure: sulfonated polyaniline with } {}^+H_3N\text{-}C_{18}H_{37} \text{ counterions on } SO_3^- \text{ groups}$$

The sulfonated polyaniline produces a dark green solution in DMAc, which turns to indigo-violet as the reaction with stearylamine takes place. This behavior is exactly identical to the reaction of the sulfonated polyaniline with NaOH in water.

All LB results were obtained using a modified Lauda film balance. To study the behavior of these PTAA-stearylamine (PTAA-StNH$_2$) salts on the LB trough, a set of solutions were made up, containing various mole ratios (1:1, 2:1 and 5:1) of thiophene structural units to stearylamine molecules. The only PAn-stearylamine (PAn-StNH$_2$) complex examined had one stearylamine molecule per sulfonate group, ie, one stearylamine molecule for every other aniline subunit. All solutions contained 1.0 mg/ml of stearylamine, and the solvent was a 3:1 (v/v) mixture of benzene and DMAc, respectively. This mixture has been found to be a good solvent for a wide range of materials and a good spreading agent for monolayer formation (6). In order to make LB films these solutions were spread onto an ultra-high purity Milli-Q water (>18 Mohm-cm) subphase and compressed to form condensed monolayers, which in turn were used to build the multilayer thin films. The vertical dipping method was used (dipping speed: 10 mm/min.), and the films were held at a constant surface pressure of about 20 mN/m and a temperature of 20.0°C.

For X-ray diffraction studies, UV-visible spectra, profilometry and conductivity measurements, the film substrates used were transparent hydrophobicized glass slides. X-ray patterns were measured on a Rigaku Rotaflex RU 300 automated X-ray diffractometer. Four-point van der Pauw conductivity measurements were made in a sealed glass flask containing the sample in dopant vapor, using a Keithley DC voltmeter and programmable current source. UV-visible spectra of film samples were recorded with an Oriel 250 mm multichannel spectrometer. A Sloan Dektak II was used for profilometry.

For infrared spectra of the LB films, IR-transparent zinc selenide plates were used as substrates in the transmission mode. In the reflectance mode, platinum coated glass slides were used in conjunction with a Spectra Tech model 500 specular reflectance attachment, capable of tilting specimens down to grazing angles of 5°. All reflectance spectra were measured at an incident angle of 8°. FT-IR spectra were recorded using a Nicolet 510P spectrometer with a liq. N_2 cooled MCT detector. Platinum coated glass slides were also used for ellipsometry, done on a Gaertner model L117.

RESULTS

Pressure-area isotherms on pure water at 20.0°C were recorded for the various polythiophene and polyaniline salts mentioned above, as well as for stearylamine. In every case, the limiting area per stearylamine molecule in the salt form was greater than that of pure stearylamine, as expected (see Table I, below). The abcissa was calibrated to the concentration of stearylamine in the spreading solution. All the complexes are best described as expanded monolayers on the water surface. It is of some interest to note that the limiting area per molecule is the same for the 2:1 polythiophene complex and the polyaniline complex. This is consistent with the fact that half of all the aniline rings have a stearylamine molecule attached at a sulfonic acid site, analogous to the 2:1 polythiophene complex, in which half the thiophene rings have a stearylamine molecule attached at the carboxylic acid side group.

All the polyion complexes were found to be exceptionally stable (up to 12 hrs. in most cases) to film collapse on the water surface when held at a surface pressure of about 20 mN/m. This is in contrast to pure stearylamine which is unstable, possibly due to its partial solubility in water. The 2:1 and 5:1 polythiophene salts actually expanded in area by up to 5% over time when held at constant pressure, which we surmise is due to the redistribution of the stearylamine molecules along the polymer backbone by ion hopping. All of these salts transferred easily onto the substrates listed above, the LB films formed being Y-type in structure (equal transfer on both up and down strokes).

Table I. Limiting area per molecule for various surface active materials

Surface Active Material	Limiting Area per Molecule ($Å^2$)
Stearylamine	18
1:1 PTAA-StNH$_2$	46
2:1 PTAA-StNH$_2$	28
5:1 PTAA-StNH$_2$	24
PAn-StNH$_2$	28

The transfer ratios did not vary significantly from layer to layer, indicating even, reproducible transfer throughout the deposition process. Figure 1, below, shows the UV-visible absorbance of LB films of these complexes as a function of the number of layers transferred.

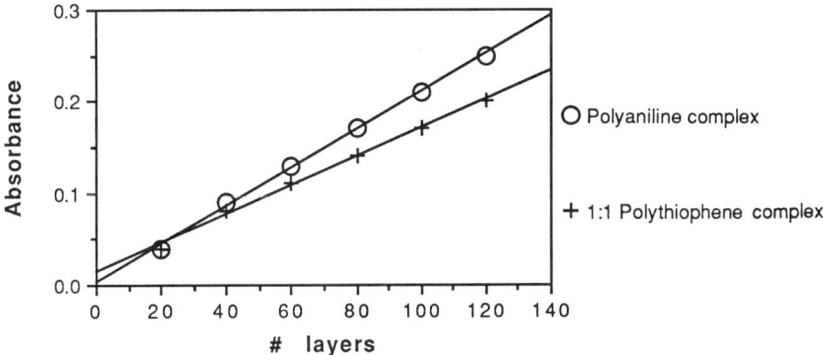

Figure 1. Absorbance as a function of #layers in LB films of PAn-StNH$_2$ (at 600 nm) and 1:1 PTAA-StNH$_2$ (at 425 nm).

The LB films used in these experiments have an equal number of layers on each side of the substrate. The high degree of linearity in the above plots further supports the evenness of deposition in these LB films, even up to 120 layers.

Polythiophene Multilayers

The LB films of the 1:1 PTAA-StNH$_2$ complex were clear yellow and free of gross structural defects such as streaks and spots, as viewed by the naked eye. However, the films of the 2:1 and 5:1 PTAA-StNH$_2$ complexes showed some streakiness and cloudiness, the 5:1 being more disordered than the 2:1. UV-visible spectroscopy of the PTAA-StNH$_2$ films showed the canonical π-π^* transition of conjugated systems at around 420 nm, which shows that the polythiophene molecules were incorporated into the LB film. Infrared spectra of LB films of the salts showed that the free acid C=O peak (1700 cm^{-1}) and the free acid O-H peak (3100 cm^{-1}) diminished as the mole ratio decreased from 5:1 to 1:1, concomitant with an increase in the amine salt C=O peak (1575 cm^{-1}), as would be expected. The other significant feature of the infrared spectra was the presence of very strong C-H peaks (2850 and 2920 cm^{-1}) corresponding to the long alkyl chains of the stearylamine.

A comparison of the FT-IR spectra of 1:1 PTAA-StNH$_2$ in reflection and transmission modes showed significant dichroism, indicating a high level of molecular orientation in the LB films. Measurements of the tilt angle (relative to the substrate normal) of the hydrocarbon tails of stearylamine in the LB films were accomplished by analyzing the differences between the transmission and reflectance FT-IR spectra, according to the methods described in (7) and (8). The tilt angle of these tails in the 1:1 PTAA-StNH$_2$ complex was found to be 10°±5°.

X-ray diffraction scans revealed signs of layer ordering in the LB films. As judged by the number of peaks appearing in the X-ray scans, the degree of order varied as follows: 1:1 > 2:1 > 5:1, confirming the visual observation mentioned above. For the 1:1 salt, the 2q peaks observed were at 2.6° and 5.2°. Assuming that these two peaks correspond to n=1 and n=2 Bragg reflections, the d-spacing for the 1:1 PTAA-StNH$_2$ complex was found to be ca. 28Å. This spacing corresponds to the bilayer repeat distance, and is consistent with measurements made on the same system by ellipsometry and profilometry. This means that each layer contributes approximately 14Å to the film thickness.

Preliminary doping studies of cast films of the PTAA-StNH$_2$ salts showed that they can be successfully doped either in solution (using NOPF$_6$ in acetonitrile or FeCl$_3$ in nitromethane) or in vapor (using gaseous SbCl$_5$). The films turned greenish upon doping, and the UV-visible spectra of the doped films showed the growth of the bipolaron band at around 750 nm accompanied by a marked decrease in the intensity of the π-π^* peak. The maximum conductivity achieved for these LB films is about 10^{-2} S/cm for the 2:1 salt doped with SbCl$_5$. The conductivity of these doped samples decayed rapidly when exposed to air, dropping to less than 10^{-5} S/cm in about 10 minutes.

Polyaniline Multilayers

The LB films made from the PAn-StNH$_2$ complex were clear violet and also of very high optical quality. The UV-visible spectra were identical to the spectrum of the sodium salt of sulfonated polyaniline, which confirms that we do indeed form a polyion complex, and that the polyaniline is transferred onto the LB substrate. The infrared spectra of LB films of this salt showed N-H peaks (3300 cm^{-1}) and the strong stearylamine C-H peaks (2850 and 2920 cm^{-1}) observed before. The tilt angle of the hydrocarbon tails in the PAn-StNH$_2$ complex was found by the same method as above to be 4°±5°.

The X-ray diffraction patterns were remarkably similar to those of the PTAA-StNH$_2$ salts; the 2q peak positions are 2.6° and 5.4°. Assuming that these peaks are the n=1 and n=2 Bragg reflections as before, the d-spacing for this system comes out to be about 32Å. Again, this value of the bilayer thickness was confirmed using ellipsometry and profilometry, which gave consistent results with X-ray diffraction. This similarity between the two polyion complexes leads us to believe that the structure in their respective LB films is dictated by the stearylamine alkyl chains attached to the polymer backbone.

The LB films of PAn-StNH$_2$ could be successfully doped in HCl solution. The violet films turned green upon immersion for 15 min., and the UV-visible spectrum of the doped film showed an increase in the polaron peak intensity at 800 nm and above, and a concomitant decrease in the HOMO-LUMO excitonic peak at 600 nm (5). The conductivity could then be measured after placing the samples in a stream of dry nitrogen and/or dynamic vacuum to remove the excess HCl solution adhering to the samples. The conductivity of these LB films was found to be 5x10^{-2} S/cm (measured in a moist environment), which is equal to the conductivity reported for the parent material (5). The conductivity is very sensitive to moisture, dropping to less than 10^{-5} S/cm when the sample is dried, and rising back to the above mentioned levels when it is remoistened, eg, by suspending the sample in a flask filled with water vapor.

Dichroic behavior observed while recording UV-visible spectra of LB films oriented at 45° (see Figure 2a below) to the polarized incident beam indicated that the polymer chains lie in planes parallel to the substrate.

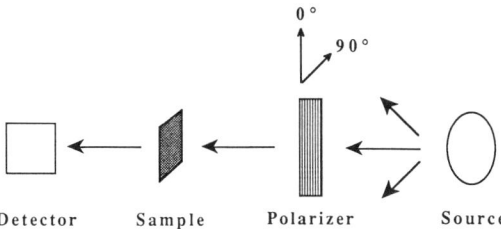

Figure 2a. Experimental set-up for UV-visible dichroic studies, showing the sample oriented at 45° to the plane of the polarized incident beam.

Figure 2b shows the UV-visible spectra of PAn-StNH$_2$ for 0° and 90° polarizations (electric field vector in the plane and partially out of the plane, respectively). The dichroic effect was more pronounced for the polyaniline complex than for the 1:1 polythiophene complex, probably due to the much greater chain length of the former. However, there is no indication of any preferred orientation of the chains within these planes (eg, in the dipping direction).

DISCUSSION

To account for a) the anomalously small monolayer thicknesses observed for both systems studied and b) the dichroic behavior indicating polymer chains sandwiched in planes between stearylamine layers, we propose that the hydrocarbon tails of the stearylamine in our LB films are interdigitated, with the polymer chains ionically bound to the stearylamine molecules. Since the length of a fully extended stearylamine molecule is about 25Å, mere tail tilting of 4°-10° is not sufficient to account for this low thickness value. Since we employ a surface pressure during film deposition that is very much lower than the film collapse pressure, interdigitation is not an unreasonable proposition, because the stearylamine tails would be loosely packed at the pressure used for dipping. Figure 3 is a schematic depiction of the proposed structure for the LB films of these complexes.

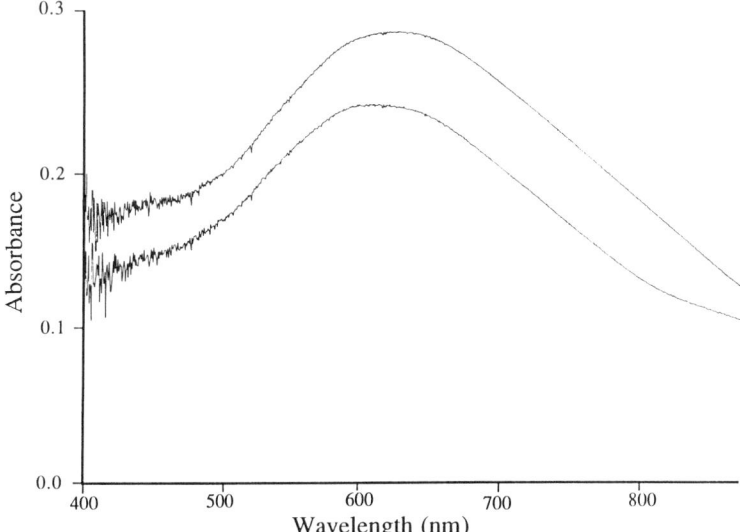

Figure 2b. Dichroism in the UV-visible spectra of PAn-StNH$_2$ for 0° (top) and 90° (bottom) polarizations.

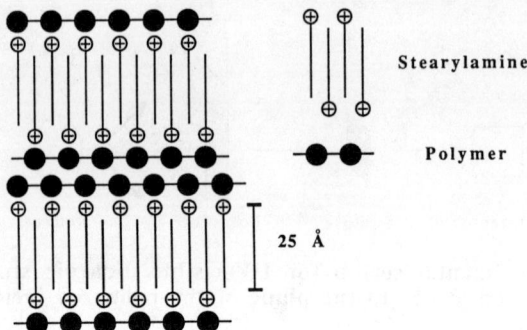

Figure 3. Proposed structure for LB films of polyion complexes showing interdigitation of stearylamine tails.

We have built CPK atomic scale models of the proposed structure, and we find that the thicknesses per monolayer reported above are consistent with the structure described in Figure 3.

CONCLUSIONS

A novel method of incorporating non-surface active conducting polymers into LB films is described. The conducting polymers used in this study (poly(thiophene acetic acid) and sulfonated polyaniline) have been functionalized with ionizable side groups and subsequently treated with a commonly available surfactant (stearylamine) to render them surface active. They can then be easily manipulated into very good quality LB films. The resulting LB films are Y-type in structure with interdigitated alkyl tails and polymer chains that are sandwiched in planes parallel to the LB substrate. These films can be doped using conventional techniques to conductivities on the order of 10^{-2} S/cm.

LITERATURE CITED

1. Shirakawa, H.; Louis, E.J.; MacDiarmid, A.G.; Chiang, C.R.; Heeger, A.J. *J. Chem. Soc., Chem. Commun.* **1977**, p.578.
2. See, for example: Punkka, E.; Rubner, M. F.; Hettinger, J. D.; Brooks, J. S.; Hannahs, S. T.; *Phys. Rev. B* **1991**, *43*, p.9076. Rosner, R. B.; Rubner, M. F.; *Mat. Res. Soc. Symp. Proc.* **1990**, p.363. Hong, K.; Rosner, R. B.; Rubner, M. F.; *Chemistry of Materials* **1990**, *2*, p.82. Watanabe, I.; Hong, K.; Rubner, M. F.; *Langmuir* **1990**, *6*, p.1164.
3. See, for example: Nishikata, Y.; Kakimoto, M-A.; Imai, Y. *Thin Solid Films* **1989**, *179*, p.191. Era, M.; Kamiyama, K.; Yoshiura, K.; Momii, T.; Murata, H.; Tokito, S.; Tsutsui, T.; Saito, S. *Thin Solid Films,* **1989**, *179*, p.1.
4. Sugimoto, R.; Takeda, S.; Gu, H.B.; Yoshino, K.; *Chemistry Express* **1986**, *1*, p.635.
5. Yue, J.; Epstein, A. J.; Macdiarmid, A. G.; *Mol. Cryst. Liq. Cryst.* **1990**, *189*, p.255. Yue, J.; Epstein, A.J.; *J. Am. Chem. Soc.* **1990**, *112*, p.2800.
6. Nishikata, Y.; Kakimoto, M-A.; Morikawa, A.; Imai, Y.; *Thin Solid Films* **1988**, *160*, p.15.
7. Umemura, J.;Kamata, T.;Kawai, T.;Takenaka, T.;*J. Phys. Chem.* **1990**, *94*, p.62.
8. Watanabe, I.; Cheung, J. H.; Rubner, M. F.; *J. Phys. Chem.* **1990**, *94*, p.8715.

RECEIVED September 24, 1991

Chapter 8

Fluorinated, Main-Chain Chromophoric Polymer
Langmuir Layer and Fourier Transform Infrared Spectroscopic Studies

Pieter Stroeve[1,2], Roni Koren[1,2], L. B. Coleman[2,3], and J. D. Stenger-Smith[4]

[1]Department of Chemical Engineering, [2]Organized Research Program on Polymeric Ultrathin Film Systems, and [3]Department of Physics, University of California, Davis, CA 95616
[4]Research Department, Naval Weapons Center, China Lake, CA 93555-6001

A nonlinear optical polymer, which belongs to a new class of fluorinated, main-chain chromophoric polymers has been investigated by equilibrium isotherm studies of the Langmuir layer and FTIR studies of Langmuir-Blodgett films. The polymer arranges in an accordion-like manner on the air/water interface, when compressed to moderate surface pressures (12 mN m^{-1}), and can be transferred in this configuration as Y-type LB multilayers. At higher surface pressures (> 12 mN m^{-1}) the Langmuir layer forms either bilayers or rearranges structurally. The monolayer collapses irreversibly at 20-25 mN m^{-1}. Analysis of FTIR studies confirm a accordion-like structure. A chemical change occurs in the polymer after aging for one day at the air/water interface.

For future optoelectronic applications, polymeric chromophores appear to be attractive materials because of the robustness of polymeric films. Much effort has been directed at the synthesis of polymeric chromophores composed of chromophoric side-chains from a polymer backbone (1-3). Deposition of non-centrosymmetric multilayer Langmuir-Blodgett (LB) films shows quadratic enhancement in second harmonic generation (SHG) with the number of chromophore layers (4). However, improved polymer materials with larger second order susceptibilities are necessary to make practical devices feasible (5). Recently Lindsay et al. (6) synthesized fluorinated, main-chain chromophoric, optically active polymers for LB deposition. Chromophores were incorporated into the polymer backbone in a head-to-head configuration separated by flexible fluorinated spacer groups. The sequence of spacers and chromophores should allow folding of the polymer main-chain in an accordion-like manner out of the interface to facilitate chromophore orientation, which can lead to effective SHG in LB multilayer structures (7). Hoover et al. (7) observed SHG generation in LB films of the accordion-type polymers.

0097-6156/92/0493-0083$06.00/0
© 1992 American Chemical Society

In this work we have investigated the equilibrium isotherm behavior of an accordion-like polymer and we have performed FTIR studies of LB films of the polymer. The purpose of the work is to assess the structure of the polymer at the air/water interface and in LB films.

Experimental Details

The molecular structure of the polymer is shown in Figure 1. Its synthesis has been described by Hoover et al. (7). The polymer was purified by preparative gel permeation chromatography in chloroform solution.

A Joyce-Loebl Langmuir trough (Model IV), housed in a class 100 laminar flow hood, was used for the equilibrium isotherm studies of the Langmuir layer and for the deposition of LB films. For the subphase, filtered and de-ionized water was passed through an activated carbon adsorber and then distilled in an all-glass still. The polymer was spread on the subphase from a chloroform solution at a concentration of approximately 1 mg ml^{-1}. The compression-expansion speed of the barrier in all experiments was 0.5 cm^2 sec^{-1}. Langmuir-Blodgett layers were deposited on aluminized glass slides and ZnSe wafers. The deposition speed was 5 mm min^{-1}.

Infrared spectra were obtained with a Nicolet 510 P spectrometer equipped with a room temperature DTGS detector. For infrared reflection-absorption spectroscopy (IRAS) studies a fixed angle (80°) Spectratech reflectance accessory was used with the LB films deposited on aluminized glass. Spectra were obtained with a resolution of 4 cm^{-1} by co-adding 256 interferograms. Transmission studies were conducted with LB films on ZnSe wafers.

Results and Discussion

The Langmuir Layer. The equilibrium isotherm for the Langmuir layer of fluorinated accordion-type polymer on a distilled water subphase is shown in Figure 2. The temperature in this experiment was 23± 0.5°C. Several compression and expansion cycles are shown. In the first cycle the Langmuir layer was compressed to

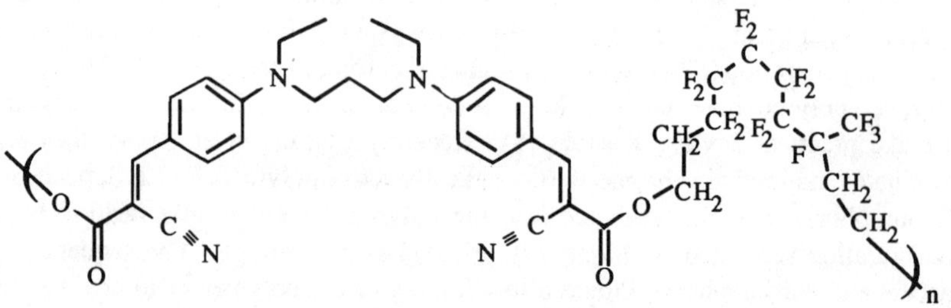

Figure 1. Structure of the repeat unit of the fluorinated, main-chain chromophoric, optically active polymer.

5.0 mN m^{-1}. At this point (a) the area was held constant for 5 minutes while the pressure was monitored. The surface pressure remained constant. Expansion of the monolayer retraced the compression curve exactly. The same procedure for a compression to 10.0 mN m^{-1} (b) gave a Langmuir layer with a stable surface pressure for 5 minutes. The expansion curve was identical to the compression curve. The second compression (b) accurately retraced the first compression (a) to 5.0 mN m^{-1}. The third compression was to 15.0 mN m^{-1} (c). From 12 to 15 mN m^{-1} a plateau-like region was observed with the specific area per repeat unit decreasing from 78 to about 48 o(°,A)2 per repeat unit. At point c the surface pressure at a constant specific area decreased from 15.0 to 12.1 mN m^{-1} in 5 minutes. The compression curve retraced compression b, but the expansion curve was different. The procedure was repeated for 20.0 (d), 25.0 (e), 30.0 (f), and 45.0 mN m^{-1}. The compression curves were retraceable if 20 mN m^{-1} was not exceeded. Up to point d compression e was identical. Compression f did not retrace compression e. The surface pressure of the Langmuir layer was increasingly unstable at constant surface area from point c to g. For example, after compression to point d and then the maintenance of constant surface area, the surface pressure dropped from 20.0 to 14.3 mN m^{-1} in 5 minutes. At point e the surface pressure dropped from 25.0 to 14.8 mN m^{-1} and at point f the surface pressure decreased from 30.0 to 15.3 mN m^{-1} in 5 minutes. The expansion curves were very different from the compression curves. From these data we conclude that the compression isotherm is reversible from 0 to 20 mN m^{-1}. After 20 mN m^{-1}, irreversible changes in the Langmuir layer take place. From an analysis of the data, the irreversible collapse pressure occurs between 20-25 mN m^{-1}.

Between 10 and 20 mN m^{-1} the surface pressure is not stable (at constant area) and monolayer material may be pushed into the subphase or out into the air phase. However, this is a reversible collapse since, after expansion of the Langmuir layer, full recovery of the compression isotherm is obtained upon recompression.

The stability of the surface pressure was investigated above 10.0 mN m^{-1} in compression-expansion experiments shown in Figure 3. A drop of surface pressure occurs over a 5 minutes period for initial surface pressure values of 13.0, 14.0, and 15.0 mN m^{-1} (b,c, and d). At 12.0 mN m^{-1} (a) the surface pressure remained constant for 5 minutes at constant surface area. Thus, reversible collapse takes place above 12.0 mN m^{-1}.

The specific surface area at 10 mN m^{-1} is about 78 Å^2 per repeat unit. When the isotherm is extrapolated to zero mN m^{-1}, the specific area is 95 Å^2. This is reasonably close to the sum of the specific cross-sectional areas of the backbone if each repeat unit is oriented in four segments sticking in the air-water interface in an accordion-like manner as depicted in Figure 1. The segments are: chromophore, chromophore (reversed), hydro-fluorocarbon, and fluoro-hydrocarbon. The cross-sectional area of the two chromophoric parts of the chain is about 25 Å^2 each and the fluoro-hydrocarbon portions of the backbone are about 20 Å^2 each. When the Langmuir layer is further compressed, a large reduction of the specific surface area takes place at a modest change of surface pressure. It is significant that the reduction in the specific area is approximately one half. This suggests that either a bilayer is

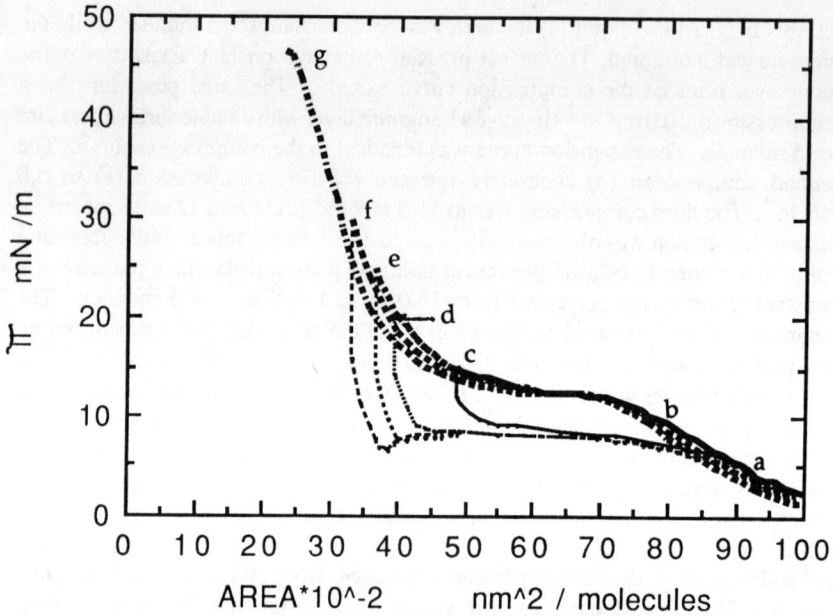

Figure 2. Successive compression-expansion curves for the Langmuir layer at 23 °C from 5 to 45 mN m^{-1}.

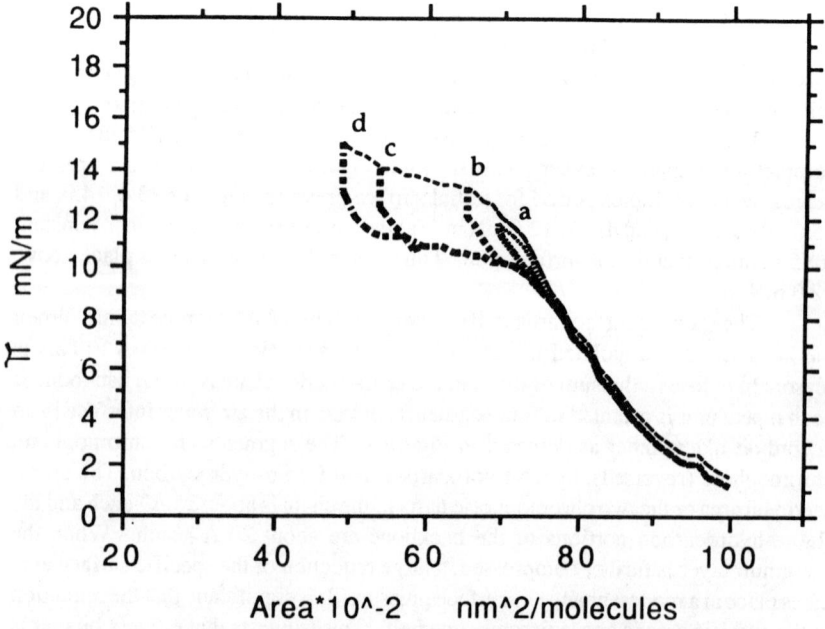

Figure 3. Successive compression-expansion curves for the Langmuir layer from 10 to 15 mN m^{-1} at 23 °C.

formed, with the monolayer buckling upon itself, or that the four segments are rearranged into two segments. In the latter case, both the chromophoric parts and the fluoro-hydrocarbon parts could form one segment each, again in an accordion-like structure. Alternately, each of the two segments could have a chromophore and part of the fluoro-hydrocarbon section of the backbone arranged in an accordion-like manner. Whatever the structure is, it is not stable as the surface pressure changes at constant area.

Equilibrium isotherms and the expansion curves are given in Figure 4 for 16.5, 23 and 29 ± 0.5 °C. At 29 °C the plateau-like region is very broad and occurs at a lower surface pressure than at 23 °C, suggesting that the higher the temperature the less stable the Langmuir layer is to buckling or rearrangement.

The Aged Langmuir Layer. We have also investigated the aging of the Langmuir layer. The polymer was spread on distilled water, compressed to 12 mN m^{-1} and then expanded completely. After 24 hours the equilibrium isotherm was measured. Figure 5 shows compression and expansion curves to 10.0 (a), 25.0 (b), 30.0 (c), 35.0 (d), and 42.0 (e) mN m^{-1}. Up to 35 mN m^{-1} the compression curves are retraceable and the expansion curves are identical to the compression curves. The shape of equilibrium isotherm is very different than that observed in Figure 2. From separate experiments, irreversible collapse was found to occur at 40 mN m^{-1} where the specific area per repeat unit is 20 Å^2. The monotonic increase of surface pressure with specific area, exhibited in Figure 4, suggests that no major phase transitions or monolayer rearrangement takes place. Further, up to 30 mN m^{-1} the surface pressure was constant at constant surface area for 5 minutes. These results suggest that a major change has occurred in the polymer molecule. Further discussion will be given in the analysis of the FTIR data on LB films.

FTIR of Langmuir-Blodgett Films. Deposition on a substrate was started with an upstroke. For a fresh Langmuir layer, deposition at 9.0 mN m^{-1} gave Z type LB layers with a transfer ratio of 1.07 on the upstroke. Deposition at 12.0 mN m^{-1} was Y-type with transfer ratios of 0.98 with the upstroke and 0.85 with the downstroke. Higher surface pressures were not used here because of the instability of surface pressure at constant area and consequently the instability of area at constant surface pressure. When the area is not constant, transfer ratios can not be measured accurately.

For the aged Langmuir layer Z-type deposition was also found at 9.0 mN m-1 with a transfer ratio of 0.90. Langmuir-Blodgett deposition at 17.0 mN m^{-1} gave Y-type LB layers with a transfer ration of 1.04 with the upstroke and 0.85 with the downstroke.

Figure 6 shows the FTIR reflection-adsorption (IRAS) spectrum of a cast film of the polymer on an aluminized microscope slide. The cast film was made by dripping a chloroform solution of the polymer on the slide followed by drying. The thickness of the cast film is approximately equivalent to the LB films studied here. Figure 7 gives the IRAS of five, Y-type, LB layers on aluminized glass. Comparisons of the figures show that the CH$_2$/CH$_3$ stretch bands are much more intense for the

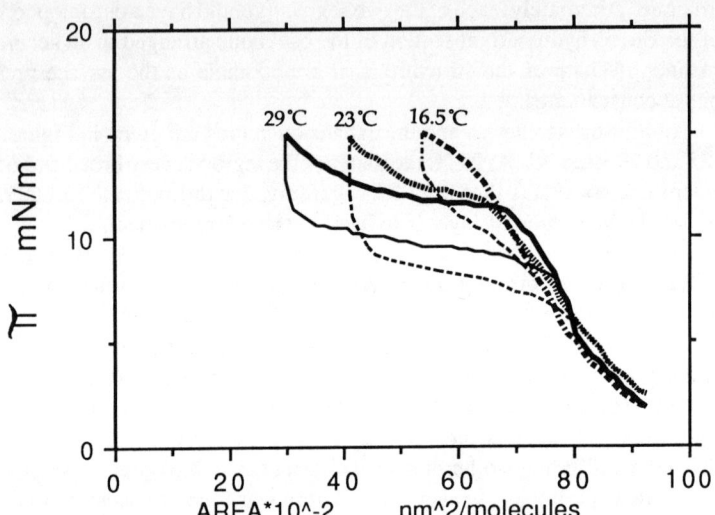

Figure 4. Compression-expansion curves at 16.5, 23, and 29 °C.

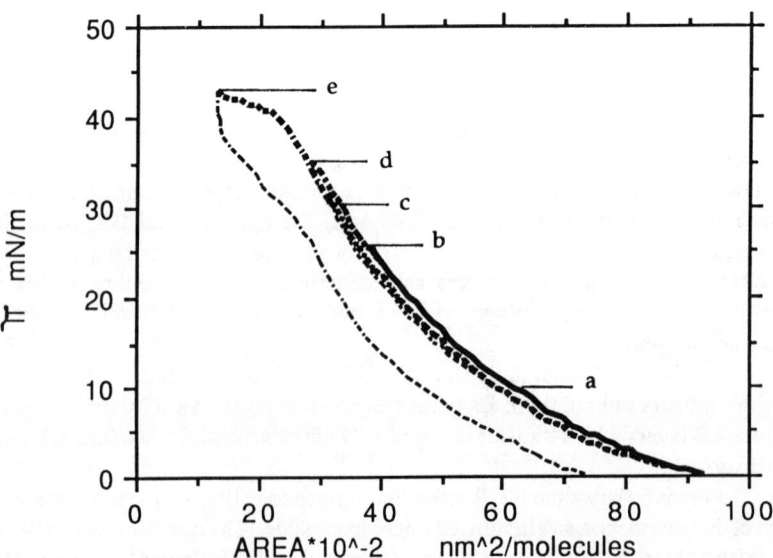

Figure 5. Compression-expansion curves for an "aged" Langmuir layer at 23 °C.

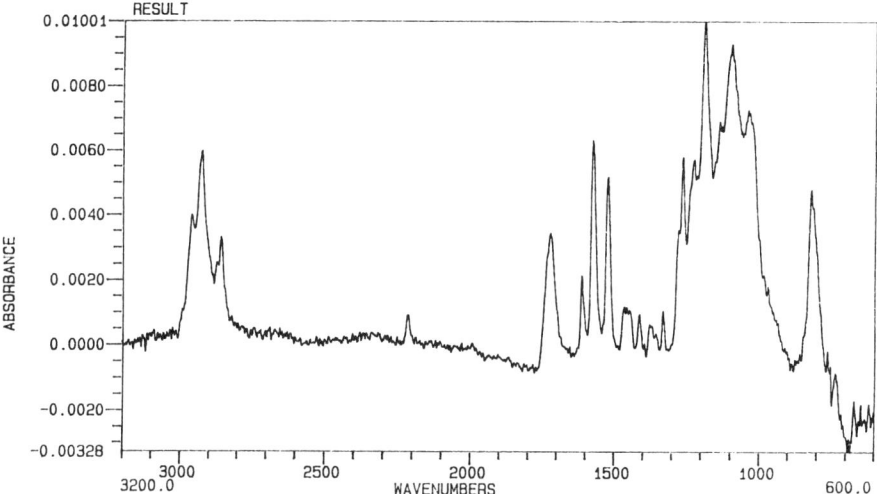

Figure 6. Infrared reflection-adsorption spectrum of a cast film of the polymer on an aluminized glass slide.

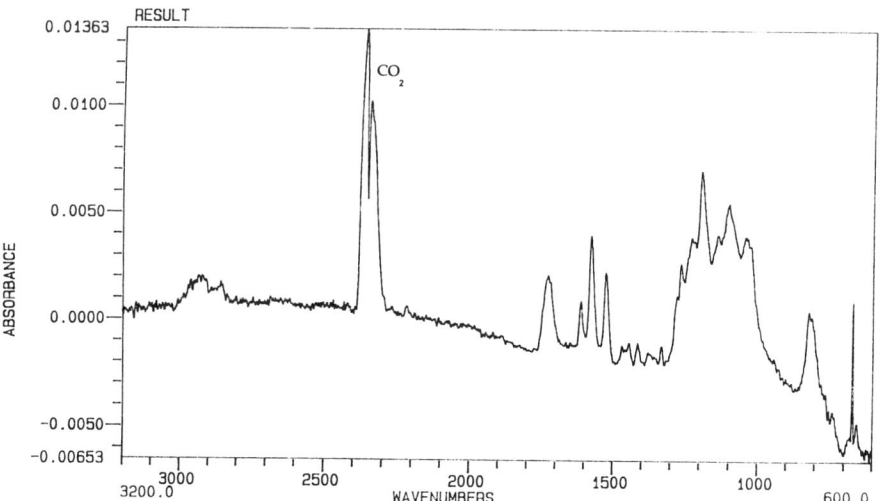

Figure 7. Infrared reflection-absorption spectrum of 5 Y-type LB layers of the polymer deposited at 12.0 mN m^{-1} on an aluminized glass slide.

cast film than the LB film. The results indicate that for the cast film the CH_2 and CH_3 dipoles are parallel to the substrate plane while for the LB films the dipoles are not parallel. The hydrocarbon portion of the chain lies flat on the surface in the case of the cast film.

Figure 8 shows the infrared transmission spectrum of eleven, Y-type, LB layers on a ZnSe wafer. The assignment of the infrared frequencies are given in Table I for reflection and transmission. These assignments are consistent with those found in the literature (8-11). The CH_2/CH_3 stretch bands are better defined in the transmission spectrum, although they are not large. Comparison of Figures 7 and 8 suggests that the CH_2 and CH_3 dipoles are neither parallel not perpendicular to the substrate plane, but

Table I: Infrared Band Assignment of LB Films in Reflection and Transmission ("fresh" layers deposited at $\pi = 12.0$ mN m^{-1}; resolution is 4 cm^{-1})

Reflection Frequency (cm^{-1})	Transmission Frequency (cm^{-1})	Assignment
2960 (w)	2962 (w)	ν_a (CH$_3$)
2926 (w)	2928 (m)	ν_a (CH$_2$)
2853 (w)	2857 (w)	ν_s (CH$_2$)
2216 (w)	2216 (w)	C≡N
1723 (w)	1716 (s)	ν (C=O) ester
1610 (w)	1613 (s)	ν (C=C) ring (ϕ)
1576 (w)	1570 (s)	ν (C=C) ring (ϕ)
1522 (w)	1520 (s)	ν (ϕ)
1462 (w)	1468 (m)	δ (CH$_2$)
1370 (w)	1371 (m)	ν (CF$_2$)
1334 (w)	1331 (m)	ν (CF$_2$)
1272 (s)	1277 (s)	ϕ - N
1227 (s)	1226 (s)	CF$_n$ stretch
1191 (s)	1184 (s)	ν (C-O-C) ester
1100 (s)	1094 (s)	CF$_n$ stretch
1138 (w)	1138 (s)	CF or C - ϕ ?
818 (w)	819 (m)	CH ($>$C - CH-)

rather at an angle, consistent with an accordion-like structure. A precise analysis of the CH_2/CH_3 orientation is very difficult because these groups do not only appear in the main chains but also as side-chains in the polymer. The transmission spectrum also shows that the C = C stretches of the phenyl groups and the C-O-C ester stretch are more intense than in reflection-absorption. These results indicate that the phenyl groups and the ester groups are oriented at an angle with the substrate, again consistent with an accordion-like structure for the polymer. In Figures 6-9, stretches assigned to CF_2 and CF_3 are relatively high. For Figures 7 and 8, the CF_n stretch is highest in reflection but the stretch in transmission is also very strong. If the hydrofluorocarbon portions of the chains are oriented out of the surface as indicated in Figure 1, for example, the chain would have to bend over bringing some CF_2 groups parallel to the surface. This orientation would give a strong absorbance in both reflection and in transmission as is seen in the spectra.

Figure 9 shows the IRAS spectrum of 10, Z-type, LB layers of the aged polymer. Deposition was carried out after the Langmuir layer was 24 hours old. Comparison of the spectrum to those in Figures 7 and 8 shows that the $C \equiv N$ stretch has disappeared which indicates that a chemical reaction has taken place at that location of the polymer molecule. It is possible that the organic nitrile group gets hydrolized into ammonia and organic acid. For LB films the hydrolysis reaction could be caused by water vapor in the air, or bound water in the LB film, or both. The monotonic increase in the equilibrium isotherm could be due to scission of the polymer backbone at the $C \equiv N$ attachment, because no phase changes are observed for the aged Langmuir layer. For the fresh polymer, phase changes were observed. For the aged polymer, a minimum specific area per repeat unit of about 20 Å^2 is obtained in a monotonic fashion. Since the crossectional area of the backbone of the polymer is from 20-25 Å^2, the data can be explained if the cleaved remnants of the polymer are stacking perpendicular to the interface. In addition, the relative weakness of the CH_2/CH_3 stretch bands means that the CH_2 and CH_3 portions of the molecule are more perpendicular than parallel to the substrate surface, which is consistent with the hypothesis of perpendicular stacking.

The change in the chemistry of the polymer should have negative implications for the use of this polymer in nonlinear optical devices. We have observed that IRAS spectra of LB films of the fresh polymer change within a few weeks and show similar spectra to the one shown in Figure 9. Since the nitrile group in the polymer is necessary for resonance in the chromophore, the chemical change would lead to a reduction of SHG with time in LB films.

Conclusion

Hoover et al. (7) have shown that accordion-type polymers give an SHG response when deposited in LB films. Obviously, for these accordion-type polymers to be used in optical devices they must be stable to chemical transformation. Our studies show that the polymer does appear to arrange in an accordion-type structure, but also that the polymer is chemically not stable. These results imply that the chemistry of the polymer needs to be modified to impart chemical stability. Nevertheless, our studies indicate that equilibrium isotherms of Langmuir layers and FTIR studies of LB films are very useful tools to quickly assess new materials and their applicability in non-linear applications of thin films.

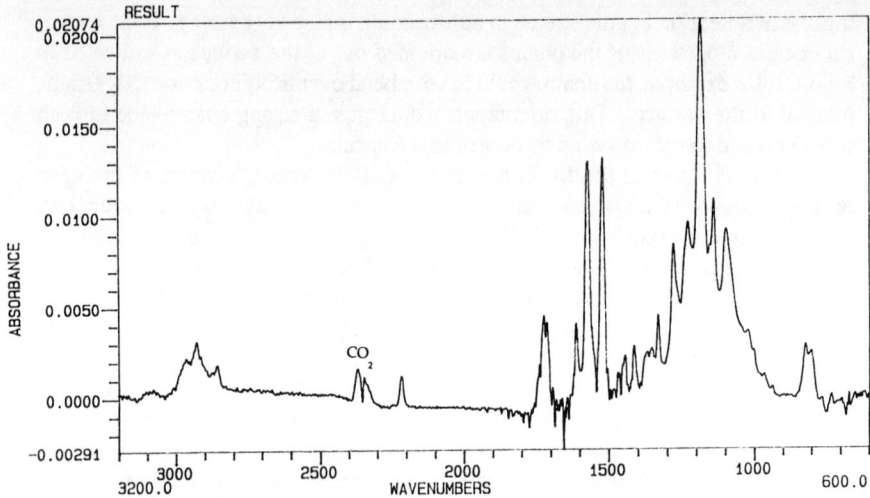

Figure 8. Infrared transmission spectrum of 11 Y-type LB layers of the polymer deposited at 12.0 mN m^{-1} on a ZnSe wafer.

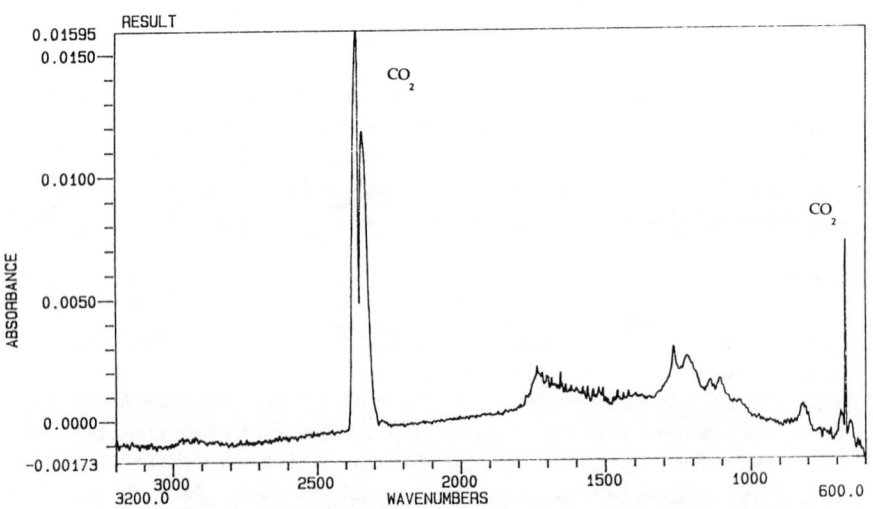

Figure 9. Infrared reflection-absorption spectrum of 10 Z-type LB layers of the aged polymer deposited at 9.0 mN m^{-1} on an aluminized glass slide.

An alternative, but more radical hypothesis, is that the aged polymer is not cleaved but it forms a more complex structure on the air/water interface, for example, helices (*12-13*). Additional experiments are necessary to deduce the structure.

Acknowledgment

Funding for this research was provided by the National Science Division (Grant CBT-8720282) and a UC Faculty Research Grant to Pieter Stroeve.

Literature Cited

1. Hall, R.C., Lindsay, G.A., Kowel, S.T., Hayden, L.M., Anderson, B.L., Higgins, B.G., Stroeve, P., and Srinivasan, M.P. Proceedings of the Society of Photo-Optical Instrumentation Engineers SPIE, 1988, 824:121.
2. Hall, R.C., Lindsay, G.A., Anderson, B.L., Kowel, S.T., Higgins, B.G., and Stroeve, P., Proceedings of the Materials Research Society Symposium, 1989, 109:351.
3. Anderson, B.L., Hall, R.C., Higgins, B.G., Lindsay, G.A., Stroeve, P., and Kowel, S.T., Synthetic Metals, 1989, 28:D683.
4. Anderson, B.L., Hoover, J.M., Lindsay, G.A., Higgins, B.G., Stroeve, P., and S.T. Kowel, Thin Solid Films, 1989, 179: 413.
5. Garito, A.F., Wu, J.W., Lipscomb, G.F., Lytel, R., Proceedings of Materials Research Society, 1990, 173: 467.
6. Lindsay, G.A., Fisher, J.W., Henry, R.A., Hoover, J.M., Kubin, R.F., Seltzer, M.D., Stenger-Smith, J.D., Journal of Polymer Science, Part A, 1991, in press.
7. Hoover, J.M., Henry, R.A., Lindsay, G.A., Nadler, M.P., Nee, S.F., Seltzer, M.D., Stenger-Smith, J.D., ACS Symposium Series, 1991, in press.
8. Schneider, J., Ringsdorf, H., Rabolt, J.F., Macromolecules, 1989, 22: 205.
9. Schneider, J., Erdelen, C., Ringsdorf, H., Rabolt, J.R., Macromolecules, 1989, 22:3475.
10. Stroeve, P., Saperstein, D.D., and Rabolt, J.F. Journal of Chemical Physics, 1990 92:6958.
11. Stroeve, P., Spooner, G.J., Bruinsma, P.J., Coleman, L.B, Erdelen, C.H. and Ringsdorf, H. , ACS Symposium Series, 1991, 447, 177-191.
12. Brinkhuis, R.H.G, and Schouten, A.J. Macromolecules, 1991, 24: 1487.
13. Brinkhuis, R.H.G, and Schouten, A.J. Macromolecules, 1991, 24: 1496.

RECEIVED October 1, 1991

Chapter 9

Langmuir—Blodgett Multilayers of Fluorinated, Main-Chain Chromophoric, Optically Nonlinear Polymers

J. M. Hoover, R. A. Henry, G. A. Lindsay, M. P. Nadler, S. F. Nee, M. D. Seltzer, and J. D. Stenger-Smith

Research Department, Naval Weapons Center, China Lake, CA 93555—6001

Alternating copolymers of syndioregic (head-to-head), bis(α-cyanocinnamate) monomers and fluorinated diols have been synthesized, characterized and fabricated into Langmuir-Blodgett multilayer films that exhibit second-order nonlinear optical behavior. The regiospecific configuration of these novel polymers is intended to permit the polymer chains to fold in an accordion-like architecture at the air-water interface, thereby promoting the organization of the chromophores into a noncentrosymmetric monolayer assembly. Pressure-area isotherms for one of the samples gave an area per chromophore of less than 36 Å^2 at a surface pressure of 10 mN/m. Multilayer films of up to 100 Y-type or 100 Z-type layers have been prepared, and both types of deposition yielded noncentrosymmetric films as indicated by second harmonic generation experiments at 1064 nm.

Advancements in the development of optical computing and communications technologies will require innovations in low-cost, processable, highly nonlinear optical materials and their fabrication into a variety of optical and opto-electronic devices (1). Nonlinear optical (NLO) polymers are especially attractive candidate materials that contain polarizable chromophoric groups which may be oriented by one or more of several techniques. Electric-field poling has been used at many laboratories to orient chromophoric groups that are (a) incorporated into polymers as side-chain, main-chain or cross-link components through covalent bonds or (b) simply dissolved in polymers to form (guest-host) solid solutions (2-5). Langmuir-Blodgett (LB) film technology (6,7) has also been applied at many laboratories to the fabrication of noncentrosymmetric films of chromophoric molecules (8-12). However, chromophoric polymer films have received much less attention with this technique (13,14).

Polymer films prepared as Langmuir layers and deposited by the LB technique can be especially useful in the study of molecular assembly at interfaces. Through the use of well designed amphiphilic polymer molecules, LB technology offers excellent control in the fabrication of highly ordered macromolecular assemblies. The air-water interface provides a mobility constraint to a two-dimensional system of molecules that can undergo kinetically controlled assembly by the adjustment of surface pressure. Characterization of the molecules can be accomplished on the interface and/or after the assembled molecules have been transferred to a solid substrate. Macromolecule assembly at interfaces can also provide the polymer scientist with a very convenient route to well defined, perhaps even model films having very significant second-order nonlinear optical properties. Additionally, these LB assemblies can incorporate ionic chromophores having exceptionally large molecular polarizabilities (*13,14*).

An essential characteristic of all second-order NLO materials is a noncentrosymmetric arrangement of polarizable species (*15*). In amorphous polymers, high levels of long-range molecular order are not entropically favorable. The relaxation of oriented dipoles (polarization or anisotropy) is especially evident in poled guest-host polymers. Consequently, much research has dealt with schemes to reduce chromophore (dye) mobility by physical aging or by chemically attaching the dye molecule to the polymer through one more covalent bonds. Many reports have appeared in the literature on synthesizing side-chain chromophoric polymers, chromophoric networks, and liquid-crystalline polymers. In these materials, chemical bonding of the chromophore to the polymer and enthalpic effects (heats of fusion and/or specific interactions such as hydrogen-bonding) improve the temporal polarization stability of the chromophoric polymers over simple solid solutions.

In attempts to reduce chromophore mobility through chemical bonding while retaining processability, we have recently synthesized a series of main-chain chromophore polymers having a unique configuration of dye species (*16,17*). Chromophores were incorporated into the polymer backbone in a head-to-head, tail-to-tail (syndioregic) configuration. In some samples, chromophores were separated by flexible hydrophobic spacer groups (Figure 1). The amphophilic characteristics of these molecules have been adjusted by varying the chromophoric and spacer groups.

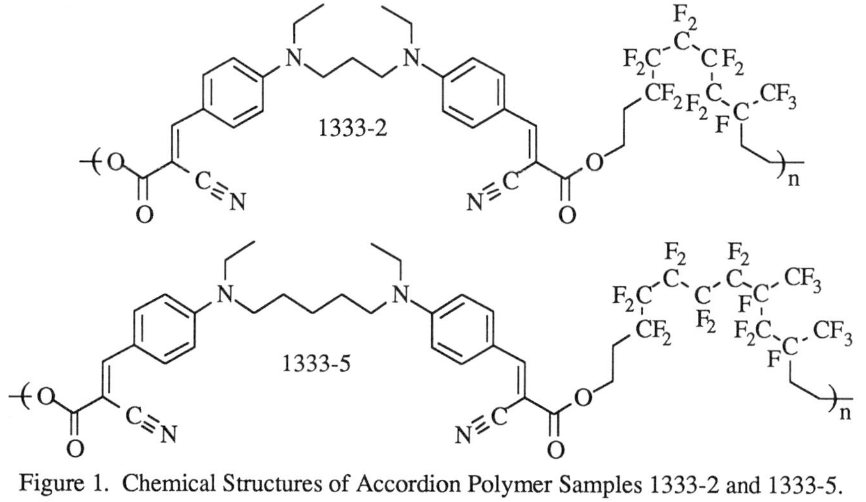

Figure 1. Chemical Structures of Accordion Polymer Samples 1333-2 and 1333-5.

This sequence of spacers and chromophores should allow folding of the polymer mainchain in an accordion-like manner to facilitate chromophore orientation while permitting a noncentrosymmetric arrangement of chromophores to be achieved.

Experimental

Main-chain α-cyanocinnamate-fluorocarbon copolyesters (Figure 1) were synthesized from fluorocarbon diols (provided by Fluorochem Inc.) and syndioregic, bis(α-cyanocinnamate) ethyl esters in melt (160°C) transesterification condensation-type polymerization using dibutyltin dilaurate as the catalyst (details to be published). The polymers were purified by preparative gel permeation chromatography (GPC, Waters Delta-Prep 3000) in chloroform solution using Styragel columns together with differential refractive index and UV/visible detection.

The polymers were characterized by differential scanning calorimetry (DSC, DuPont 1090), UV/visible spectroscopy (Beckman DU-7), Fourier transform infrared (FTIR) spectroscopy (Nicolet 60SX), analytical GPC, and nuclear magnetic resonance (NMR) spectroscopy (Bruker AMX-400 and IBM NR-80).

Polymeric monolayers were prepared by spreading 1.0 mg/mL chloroform solutions on 18 MΩ (Millipore Milli-Q system) water (pH 5.6) in a Nima Technology dual compartment LB trough (Model TKB 2410A). The monolayers were compressed at 0-30 cm^2/min (ca. 500 cm^2 initial area) to a surface pressure (Π) of 10 mN/m and deposited onto various substrates at 1.08 cm/min (5.4 cm^2/min) at Π = 9.8 mN/m. The substrates included fuzed quartz or glass microscope slides, p-type [100] silicon wafers and ZnSe attenuated total reflectance (ATR) crystals (45° cut). Polymeric multilayers were prepared by repeated monolayer depositions in X-type (downstroke), Z-type (upstroke) and Y-type (alternating upstroke and downstroke) architectures.

The resulting polymeric multilayer films were characterized by UV/visible spectroscopy, ellipsometry at 0.9 and 1.0 μm wavelengths, ATR-FTIR spectroscopy, and optical second harmonic generation (SHG) with a fundamental wavelength of 1064 nm.

Results and Discussion

Several characteristics of the two polymers described above are listed below in Table I. Polymer 1333-2 has a shorter hydrocarbon linkage between amino nitrogens and a shorter fluorocarbon ester linkage. The glass transition temperature (T_g) of each polymer is lower than desired for most electro-optic device applications and is strongly influenced by the large amount of fluorocarbon linkage between the relatively small chromophores. When the fluorocarbon and hydrocarbon spacers in 1333-5 were decreased in size to 14 fluorine atoms and 3 methylenes per repeat unit (1333-2), the T_g was observed to increase significantly (75°C).

Langmuir-Blodgett Films. Qualitatively, these α-cyanocinnamate accordion polymers formed monolayer films having good spreading characteristics (highly compliant with uniform color at the air-water interface), very low compression-expansion hysteresis and surface pressures with low sensitivity to barrier speed.

Typical compression isotherms for 1333-2 and 1333-5 are provided below in

Table I. Results of GPC, UV/Vis and DSC Characterization

Polymer Characteristic	1333-2	1333-5
$<M_n>$/1000 Daltons	15.4	22.9
$<M_w>$/1000 Daltons	28.6	55.0
polydispersity	1.9	2.4
Absorbance Max. (nm)	436.0	436.8
Absorbtivity (L/mol cm)	83,000	120,000
T_g (°C)	74	45

Figure 2. For sample 1333-2, these data indicate a collapse pressure of 11 mN/m, a compressed area per repeat unit (chromophore pair) of 72 Å^2 (at Π = 10 mN/m), and an extrapolated area per repeat unit of approximately 82 Å^2 (at Π = 0 mN/m). Since each repeat unit consists of two chromophores and a fluorocarbon ester linkage, each chromophore should occupy somewhat less than 36 Å^2. Each fluorocarbon ester group should occupy some small but significant area, leaving an area per chromophore on the order of 30 Å^2. Previous work by this author and others has indicated that a reasonable area for well-aligned aromatic amphiphiles should be closer to 25 Å^2. This qualitative but reasonable allocation of occupied space for compressed molecules of 1333-2 indicates that the chromophores are very likely aligned somewhat normal to the interface. [Future experiments to detect reflected SHG from a monolayer as a function of Π and determine the angle of chromophore alignment with respect to the surface normal have been planned.] The 1333-5 molecules have a compressed area per repeat unit of approximately 80 Å^2 at collapse and a significantly lower collapse pressure of approximately 8.5 mN/m. These molecules contain about 50% more fluorocarbon

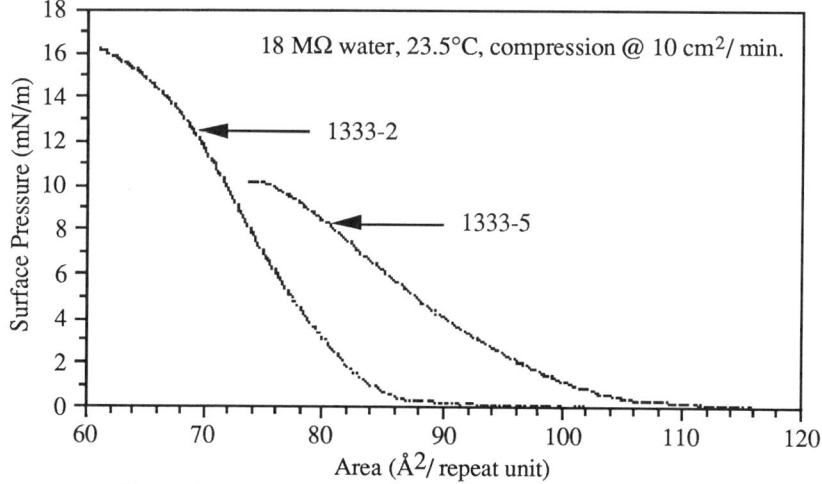

Figure 2. Pressure-Area Isotherms of Accordion Polymers

material in their ester linkage and occupy significantly more area (~15%) at the same surface pressure than 1333-2 molecules. In general, the 1333-5 molecules formed less desirable monolayers and their deposition characteristics are not reported here.

Monolayers of 1333-2 were readily transferred to hydrophilic substrates such as fuzed quartz slides for UV/visible spectroscopy, optical glass slides for SHG measurements, silicon wafers for ellipsometry, and ZnSe ATR crystals for FTIR spectroscopy. At a deposition surface pressure of 9.8 mN/m, transfer ratios for monolayers were typically 1.00 ± 0.05 for all Z-type (upstroke) layers and 0.85 ± 0.10 for all X-type (downstroke) layers. As many as 100 Y-type layers (alternating X and Z-type) or 100 Z-type layers have been deposited onto glass and silicon substrates to date.

UV/Visible Spectroscopy. A series of UV/visible spectra for increasing numbers of Z-type layers is shown below in Figure 3. Apparently, two distinct oscillators having charge-transfer band absorptions approximately 20-25 nm apart have affected the shape of the main absorbance peak as a function of the number of layers deposited. The absorbance due to the higher energy oscillator (415-420 nm) increases at a faster rate than that for the lower energy oscillator (440 nm) and is most likely due to the formation of relatively higher energy aggregates by forcing adjacent chromophores (within the same layer or adjacent layers) into close proximity. It can be seen from the UV/visible spectra (Figure 4) that the dilute solution absorbance peak has been affected much more by the lower energy (presumedly) unaggregated oscillator.

The UV/visible spectra of multilayer films stored in air were observed to change with time. The intensities of all spectral features were observed to slowly decrease with very little, if any, change in shape as a function of storage time (room temperature, ~0% relative humidity). This slow bleaching (over several days) was accelerated by exposure to UV light (low intensity, 365 nm). Chemical degradation of the chromophore was suspected and substantiated by a simple photo-imaging experiment.

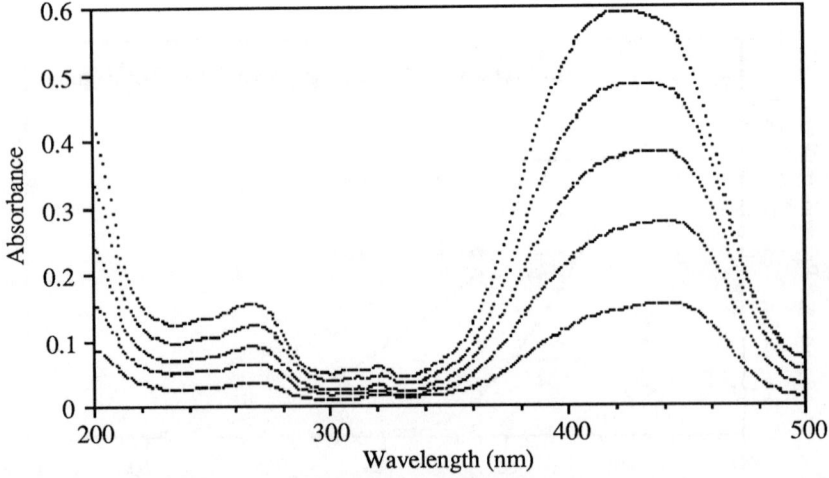

Figure 3. UV/Visible Spectra of 10, 20, 30, 40 and 50 Z-type Layers of 1333-2 on Quartz

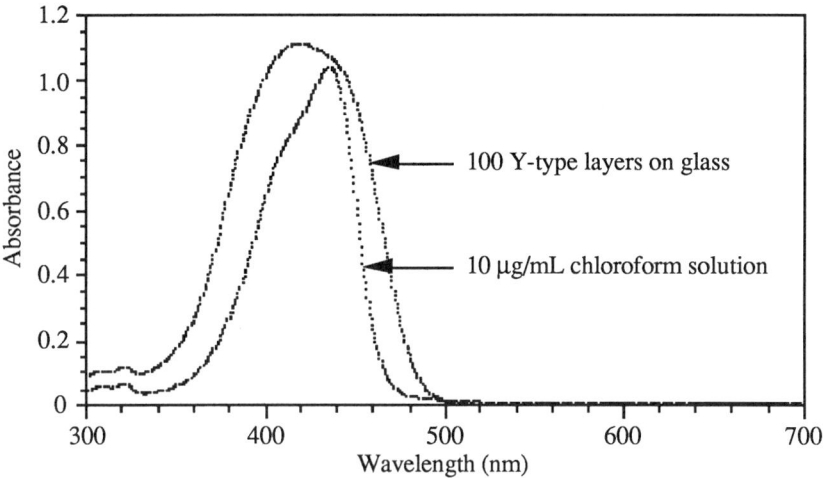

Figure 4. UV/Visible Spectra of Sample 1333-2

A 50 layer film was deposited on a silicon wafer and exposed through a foil mask to low-intensity, 365 nm light for 1 hour. The film was then dipped into a 50% solution of aqueous acetone and the exposed area readily dissolved away, leaving the unexposed area intact. Experiments on films stored in an inert atmosphere and on multilayer films containing layers of antioxidant molecules have been planned.

Ellipsometry. Multilayer films built from exclusively Z-type or Y-type layers contained noncentrosymmetric molecular dipole assemblies (anisotropy) as determined from near-infrared, null ellipsometry experiments at 0.9 µm and 1.0 µm wavelengths. The results of ellipsometry at $\lambda = 1.0$ µm on four multilayer films of 1333-2 are summarized in Table II where n_i is the refractive index, d is the film thickness, σ_i is the residual error in the measurement with an instrument precision of 0.01°; and n, q and f are the refractive index, depolarization factor (effective shape) and volume fraction of the chromophore.

Table II. Results of Ellipsometry Characterization at $\lambda = 1.0$ µm

Sample No.	1295-46	1295-52	1295-48	1295-47
Type	Y	Z	Z	Y
Layers	10	30	50	50
$n_{x,y}$	1.5781	1.5906	1.5861	1.5804
n_z	1.5974	1.6026	1.5970	1.5972
d (nm)	12.774	40.836	69.502	64.187
σ_Ψ (0.01°)	1.434	0.633	1.163	1.494
σ_Δ (0.01°)	1.035	2.165	2.662	2.257
n	1.8315	1.8346	1.8625	1.8668
q	0.1042	0.1931	0.2181	0.1557
f	0.5379	0.5196	0.5575	0.5687

Several pieces of very important information were obtained from ellipsometry measurements and model-fitting calculations for Y-type (10 and 50 layer) and Z-type (30 and 50 layer) films deposited onto single crystal silicon wafers. Model-fitting calculations (*18*) based on Maxwell Garnett effective medium theory were able to compute the anisotropic refractive index or birefringence ($n_{x,y}$ vs n_z) present in the multilayer films; and the depolarization factor (q), refractive index (n) and volume fraction (f) of the polarizable chromophores present within the layers (assuming two-dimensional symmetry). The average chromophore shape appeared to be elliptical with the long axis normal to the surface. The depolarization factor (at 1 μm) ranged from about 0.1 (for a 10 layer film) to 0.22 (for a 50 layer film) indicating that the anisotropy or polarization was somewhat lower for films having a greater number of layers (q < 1/3 for a prolate ellipsoid, q = 1/3 for a sphere and q > 1/3 for an oblate ellipsoid). The depolarization factor was also significantly lower for the Y-type films indicating that the chromophore alignment was closer to the film normal in these samples. The chromophore refractive index ranged from 1.83 to 1.87 assuming that n for pure fluorocarbon material at 1 μm may be approximated by 1.376 (*19*), and the chromophore volume fraction ranged from 0.52 to 0.57. [For comparison, the chromophores account for ~53% of the mass of each repeat unit.] The normal refractive index (n_z) for multilayer films was calculated to be ~1.60 (at 1 μm) with a birefringence of only 0.01.

Near-infrared ellipsometry also enabled the measurement of film thickness (Figure 5) and showed that the monolayer thickness (~13 to 14 Å) would be consistent with an oriented α-cyanocinnamate chromophore structure if the chains were folded in an accordion-like fashion similar to that depicted above in Figure 1. The Y-type multilayer films were slightly thinner than Z-type films presumedly due to the lower transfer ratios obtained for downstroke depositions.

Ellipsometry measurements were also sensitive to film age and indicated a temporal decrease in anisotropy (perhaps in part due to degradation) which was consistent with other types of analysis (UV/visible spectroscopy and SHG).

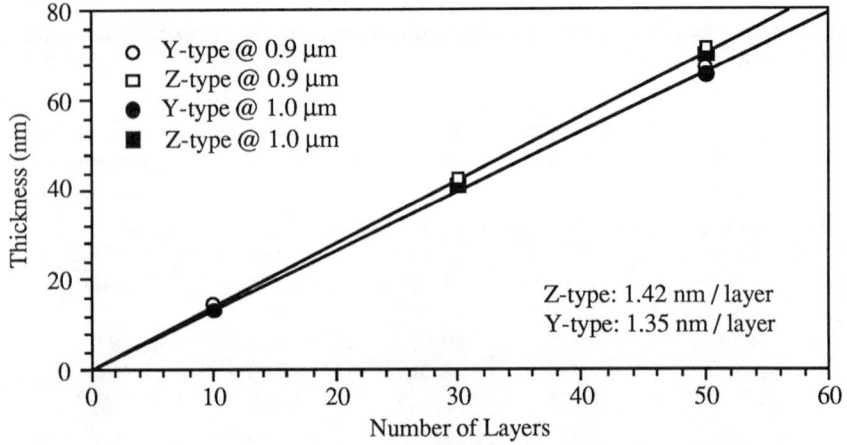

Figure 5. Measurement of Multilayer Film Thickness by Null Ellipsometry

FTIR Spectroscopy. Multilayer films were deposited onto ZnSe ATR crystals and examined in transmission and in polarized ATR immediately after fabrication of the films and again after several days of aging the films in air and/or after exposure of the films to UV light. The ATR-FTIR analysis was performed with light polarized either parallel (0°) to or normal (90°) to the plane of the deposited films. The transmission FTIR analysis was performed with unpolarized light.

The ATR-FTIR experiments on 5 layer, Z-type films indicated small but significant absorbance differences at frequencies (cm^{-1}) characteristic of aromatic ring modes (1611, 1569, 1518 cm^{-1}), C-O stretching (1183 cm^{-1}) and C-F$_n$ stretching modes (1100-1290 cm^{-1}). The absorbance differences were most significant for the 1518 cm^{-1} and the 1183 cm^{-1} bands, and the absorbance was higher (by ~5%) for the 0° polarization experiment in both cases. These results indicated that the lowest energy ring mode and the C-O and/or C-F stretches were slightly more aligned with the plane of the film rather than the normal direction. Surprisingly, no significant differences were observed at carbonyl (1715 cm^{-1}) and nitrile (2216 cm^{-1}) frequencies. This indicated that those stretching modes were likely oriented either randomly or at approximately 45° to the film plane normal. Unfortunately, we were unable to perform low-angle reflection experiments to determine the above orientations more precisely.

Second Harmonic Generation. The SHG experiments were performed using a Nd:YAG laser (λ = 1064 nm) having a 66 mJ/10 ns pulse at 10 Hz and a focused spot size of 0.15 cm (~370 MW/cm^2). All detectable second harmonic light (532 nm) was filtered from the fundamental beam. Second harmonic generation from each sample was detected with a photomultiplier tube or with an energy meter having a 1 pJ sensitivity. A boxcar and an averaging digital oscilloscope were used to improve the poor signal-to-noise ratio of the measurements. The background SHG from the glass surfaces was subtracted from the total SHG observed and the SHG from the LB films was referenced to the SHG from a Y-cut quartz crystal. The results are summarized below in Table III.

Although SHG was observed with all freshly prepared multilayer films, the SHG intensity was not stable over a period of days. The SHG did not increase in direct proportion to the square of the film thickness (number of layers) as would be predicted for stable, uniform noncentrosymmetric films. [More SHG experiments on freshly prepared films are needed to adequately determine the SHG as a function of thickness relationship.] In addition, the thinner films appeared to be more noncentrosymmetric than the thicker films. Surprisingly, The Y-type film also

Table III. Second Harmonic Generation from LB Films

Number of Layers	Type of Layers	SHG (% of Quartz)	Film Age (hours)
10	Z	0.02	72
18	Y	0.08	48
50	Z	0.07	12
100	Z	0.11	20

appeared to be more noncentrosymmetric than the Z-type films. These observations are consistent with the ellipsometry results which indicated a more spherical chromophore symmetry (depolarization factor approaching 1/3) for the thickest films and a more prolate ellipsoid shape for the chromophores in the Y-type films (q approaching 0). For the thicker films, perhaps the longer residence time of deposited layers in the subphase increased q (decreased noncentrosymmetry) through reduction of the T_g or aqueous acid (pH 5.6) catalyzed degradation. Unfortunately, thicker Y-type films (50 to 100 layers) were not available for SHG characterization. In light of the apparent temporal instability of the films and the large uncertainty of the SHG measurements (approximately 50%), we are hesitant to base much speculation regarding Y-type vs. Z-type film anisotropy on SHG measurements.

Conclusions

The preliminary results reported here indicate that syndioregic polymers can be appropriately designed and synthesized to produce polymeric monolayers at the air-water interface. These polymer molecules apparently participate in regular chain folding to produce films having a noncentrosymmetric arrangement of molecular dipoles when the films are composed of Y-type or Z-type multilayers. Both ellipsometry and SHG indicated that noncentrosymmetry was greatest for Y-type multilayers (FTIR and UV/visible spectroscopy were inconclusive in this regard). The molecular organization responsible for achieving noncentrosymmetry in Y-type multilayer films (as indicated by SHG) is not understood at this time, but it will be the focus of future studies using more quantitative characterization. Perhaps one type of deposition (X or Y) was converting to the other in order to achieve and/or diminish the apparent noncentrosymmetry, in addition to the apparent chemical instability that existed in films exposed to air and/or UV light. Lastly, these monolayer films can be readily and efficiently transferred to solid substrates to form novel, multilayered macromolecular assemblies.

Future Work

The chemical stability of α-cyanocinnamate chromophores within polyester and polyamide LB films is currently being investigated. We have also modified the syndioregic α-cyanocinnamate polymer structure by incorporating long (C_{16}) alkyl-amino substituents and by removing the main-chain (fluorocarbon) spacer which we suspect to be rigid and extended (from preliminary molecular modeling results). In addition, the carbon-fluorine bond polarizations have likely complicated the characterization of polarization anisotropy contributed by the α-cyanocinnamate chromophores. Monolayers of these modified molecules show dramatically increased collapse pressures (>40 mN/m) and significantly lower compressed areas per chromophore (~25 $Å^2$). Multilayer films of these modified accordion polymers will be prepared for study.

We also plan to use reflection SHG from Langmuir films to more accurately characterize chromophore orientation with respect to the film normal and study the assembly mechanics of these and other optically nonlinear accordion polymers.

Acknowledgments

Partial funding for this work was provided by the Office of Naval Research. The authors would also like to thank Dr. Kurt Baum of Fluorochem Inc. for providing the fluorinated diols used and Ms. Deborah Paull for thermal analysis and other technical assistance.

Literature Cited

1. Garito, A.F.; Wu, J.W.; Lipscomb, G.F. and Lytel, R. *Proc. MRS* **1990**, *173*, 467.
2. Green, G.D.; Weinschenk, J.I. III; Mulvaney, J.E. and Hall, H.K. Jr. *Macromolecules* **1987**, *20*, 722.
3. Willand, C.S. and Williams, D.J. *Ber. Bunsenges. Phys. Chem.* **1987**, *91*, 1304.
4. Mortazavi, M.A.; Knoesen, A.; Kowel, S.T.; Higgins, B.G. and Dienes, A. *J. Opt. Soc. Am.* **1989**, *6(4)*, 733.
5. Hayden, L.M.; Suater, G.F.; Ore, F.R.; Pasillas, P.L.; Hoover, J.M.; Lindsay, G.A. and Henry, R.A. *J. Appl. Phys.* **1990**, *68(2)*, 456.
6. Gaines, G.L. Jr. *Insoluble Monolayers at Liquid-Gas Interfaces*; Interscience: New York, NY, 1966.
7. Roberts, G.G. *Langmuir-Blodgett Films*, Plenum Press: New York, NY, 1990.
8. Ashwell, G.J. *Thin Solid Films* **1990**, *186*, 155.
9. Ashwell, G.J.; Dawnay, E.J.C.; Kuczynski, A.P.; Szablewski, M.; Sandy, I.M.; Bryce, M.R.; Grainger, A.M. and Hasan, M. *J. Chem. Soc. Faraday Trans.* **1990**, *86(7)*, 1117.
10. Girling, I.R.; Cade, N.A.; Kolinsky, P.V.; Jones, R.J.; Peterson, I.R.; Ahmad, M.M.; Neal, D.B.; Petty, M.C.; Roberts, G.G. and Feast, W.J. *J. Opt. Soc. Am. B* **1987**, *4(6)*, 950.
11. Kajzar, F. and Ledoux, I. *Thin Solid Films* **1989**, *179*, 359.
12. Ancelin, H.; Briody, G.; Yarwood, J.; Lloyd, J.P.; Petty, M.C.; Ahmad, M.M. and Feast, W.J. *Langmuir* **1990**, *6*, 172.
13. Young, M.C.J.; Lu, W.X.; Tredgold, R.H.; Hodge, P. and Abbasi, F. *Elec. Let.* **1990**, *26(14)*, 993.
14. Hall, R.C.; Lindsay, G.A.; Anderson, B.; Kowel, S.T.; Higgins, B.G. and Stroeve, P. *Proc. MRS* **1988**, *109*, 351.
15. Shen, Y.R. *The Principles of Nonlinear Optics*, John Wiley and Sons: New York, NY, 1984.
16. Stenger-Smith, J.D.; Fischer, J.W.; Henry, R.A.; Hoover, J.M. and Lindsay, G.A. *Accordion-Like Polymers for Nonlinear Optical Applications*, Navy Case No. 72950, 1991.
17. Lindsay, G.A.; Henry, R.A.; Hoover, J.M.; Kubin, R.F. and Stenger-Smith, J.D. *Proc. SPIE* **1991**, *1560*.
18. Lindsay, G.A.; Nee, S.F.; Hoover, J.M.; Stenger-Smith, J.D.; Henry, R.A.; Kubin, R.F. and Seltzer, M.D. *Proc. SPIE* **1991**, *1650(48)*.
19. Billmeyer, Jr., F.W. *J. Appl. Phys.* **1947**, *18*, 431.

RECEIVED September 24, 1991

Chapter 10

Langmuir–Blodgett Films of Stilbazolium Chloride Polyethers
Structural Studies

David D. Saperstein[1], John F. Rabolt[2], J. M. Hoover[3], and Pieter Stroeve[4]

[1]IBM Storage System Products Division, San Jose, CA 95193
[2]Almaden Research Center, IBM Research Division, San Jose, CA 95120
[3]Chemistry Division, Naval Weapons Center, China Lake, CA 93555–6001
[4]Department of Chemical Engineering and Organized Research Program on Polymeric Ultrathin Film Systems, University of California, Davis, CA 95616

>The orientation and order of Langmuir-Blodgett films of stilbazolium-epichlorohydrin polymer (PECH) interleaved with deuterated arachidic acid are studied with infrared radiation. Infrared absorption bands characteristics of the head and tail groups in both the dye and spacer are identified. Preliminary heating and aging studies show that the assembly undergoes a slow structural rearrangement at room temperature which is accelerated by heat. The observed rearrangement is consistent with previously measured SHG (second harmonic generation) signal loss in aged assemblies compared with freshly prepared samples.

Organic materials with large nonlinear susceptibilities can, in principal, generate significant amount of second harmonic light if they can be organized in films and maintained in a non-centrosymmetric geometry (1). One approach to organization has been to deposit Langmuir-Blodgett (LB) films of oriented organic materials with large nonlinear susceptibilities on transparent substrates (2). Some dyes such as the hemicyanines, which can be modified to permit LB deposition, show large second harmonic generation (SHG) from a single layer but show diminishing enhancement with the number of choromophore layers (3-4). Recent efforts have been devoted to constructing non-centrosymmetric multilayer LB films that show quadratic enhancement of the SHG signal with the number of dye layers (5-13).

It has been shown previously, through polarized IR spectroscopy and other techniques, that the orientation and order of LB films of a hemicyanine dye can be partially stabilized by interleaving the dye layers with a spacer such as cadmium arachidate (13). Stabilized multilayers show a nearly quadratic dependence of their second harmonic generation (SHG) signal with the number of dye layers. Although these interleaved assemblies maintain their orientation at room temperature, they are easily disrupted when heated mildly(13). In an attempt to improve the stability of the dye, Hall et al., have synthesized stilbazolium-polyepichlorohydrin (PECH) polymers (14). These stilbazolium-PECH polymers can also be deposited in layers using LB techniques to form non-centrosymmetric assemblies. Anderson et al., in 1989, showed quadratic behavior for multilayer assemblies with up to 10 active stilbazolium-PECH layers interleaved with calcium behenate (15). Anderson et al. now have shown that a multilayer assembly of stilbazolium-PECH layers interleaved with calcium behenate shows quadratic enhancement of its SHG out to 89 layers(16,17). To better understand these stilbazolium-PECH polymer assemblies, we have begun an evaluation of the orientation and order of a related assembly under mild heating conditions and during ambient aging over several months.

Experimental Summary

Figure 1 shows a schematic of the stilbazolium-PECH polymer where x is about 25; approximately 90% of the monomer units have stilbazolium groups attached, and the molecular weight of the polymer is about 12,800 Daltons (15).

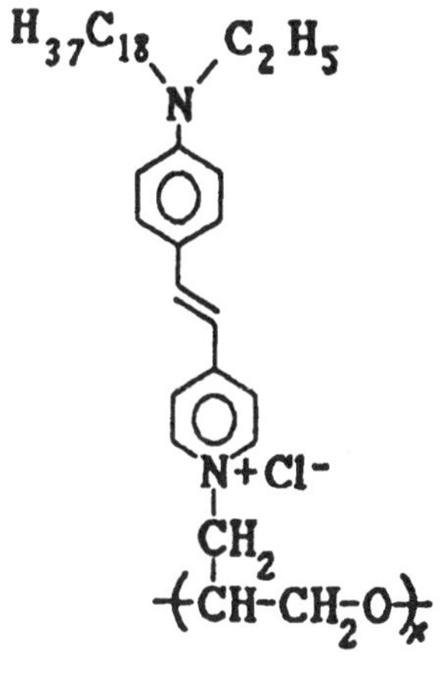

Fig 1. Structure of the repeat unit of the stilbazolium-PECH polymer

Stilbazolium-PECH and deuterated arachidic acid are spread at the air-water interface from 1 mg/ml solutions in chloroform. The stilbazolium-PECH LB layers are deposited at room temperature (23 °C) on the upstroke at 25 mN/m, and deuterated arachidic acid layers are deposited at 25 mN/m. (See ref. 9 for details of the preparation).

IR spectra are measured in a Bruker/IBM 32 Fourier Transform Spectrometer at 2 cm^{-1} resolution and 15,000 scans are co-added to improve signal-to-noise. The substrates are tilted slightly (7.5°) from the normal to minimize channeling. Sample measurements with and without the tilt show that the tilt has a negligible effect on the relative spectral intensities.

Results and Discussion

To learn more about the orientation of the stilbazolium-PECH without spectral interference from the spacer layer, the assemblies used in this study are constructed with deuterated arachidic acid so that the CD vibrations of the spacer can be distinguished from the CH vibrations of the polymer dye. Figure 2 shows the polarized infrared spectrum before and after heating of three layers of PECH interleaved with four layers of deuterated arachidic acid (d-AA). Figure 3 shows a 2nd sample before and after aging for two months at room temperature. The sequence of layers for both samples is, starting at the ZnSe surface, d-AA, d-AA, polymer, d-AA, polymer, d-AA and polymer. Hence, the dye polymer is separated from the surface by two spacer layers, and each polymer layer has one spacer layer separating it from the adjoining polymer layer(s).

Table I summarizes the identity of the principal vibrational bands shown in Figs. 2 and 3.

Heating Studies. Figure 2 shows a spectra of a sample of three polymer layers interleaved with deuterated arachidic acid on a zinc selenide substrate. Spectrum a (top) shows the unheated assembly; spectrum b shows the assembly after heating to 60 °C for 1 hour and then cooling to room temperature; spectrum c shows it after heating to 80 °C for 1 hour and then cooling to room temperature, and spectrum e shows it after heating to 120 °C for 1 hour and then cooling to room temperature. Several features in Figure 2a-e are noteworthy. Although the intensities of the CH_2 and CD_2 vibrations decrease as a function of increasing sample temperature, they show different trends. For example, the CH_2 vibrations lose 1/3 of their initial intensity after heating to 60 °C whereas the CD_2 vibrations are much less affected by the 60 °C temperature and show only a 10% intensity reduction. At 80 °C the CH_2 show even a greater intensity reduction compared to CD_2 bands. These intensity reductions are suggestive of a

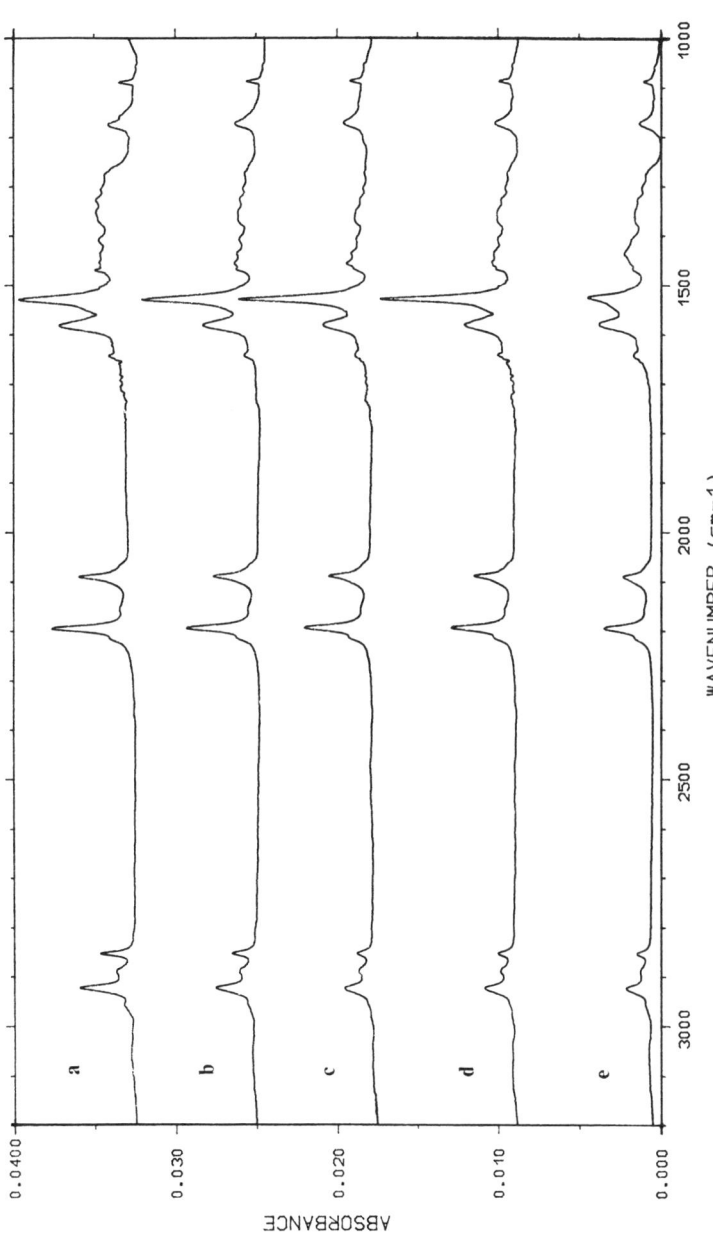

Fig 2. Transmission infrared spectra of a polymer assembly consisting of 3 layers of stilbazolium-PECH and 4 layers of deuterated arachidic acid (d-AA) on ZnSe. The ordering of the layers on the ZnSe surface is d-AA, polymer, d-AA, polymer, d-AA, polymer. All spectra are measured at room temperature, ca. 21 °C. Spectrum a (top) shows the unheated assembly; b shows the assembly after heating to 60 °C for 1 hour; c shows the assembly after heating to 80 °C for 1 hour; d shows the assembly after heating to 100 °C for 1 hour, and e shows the assembly after heating to 120° C for 1 hour.

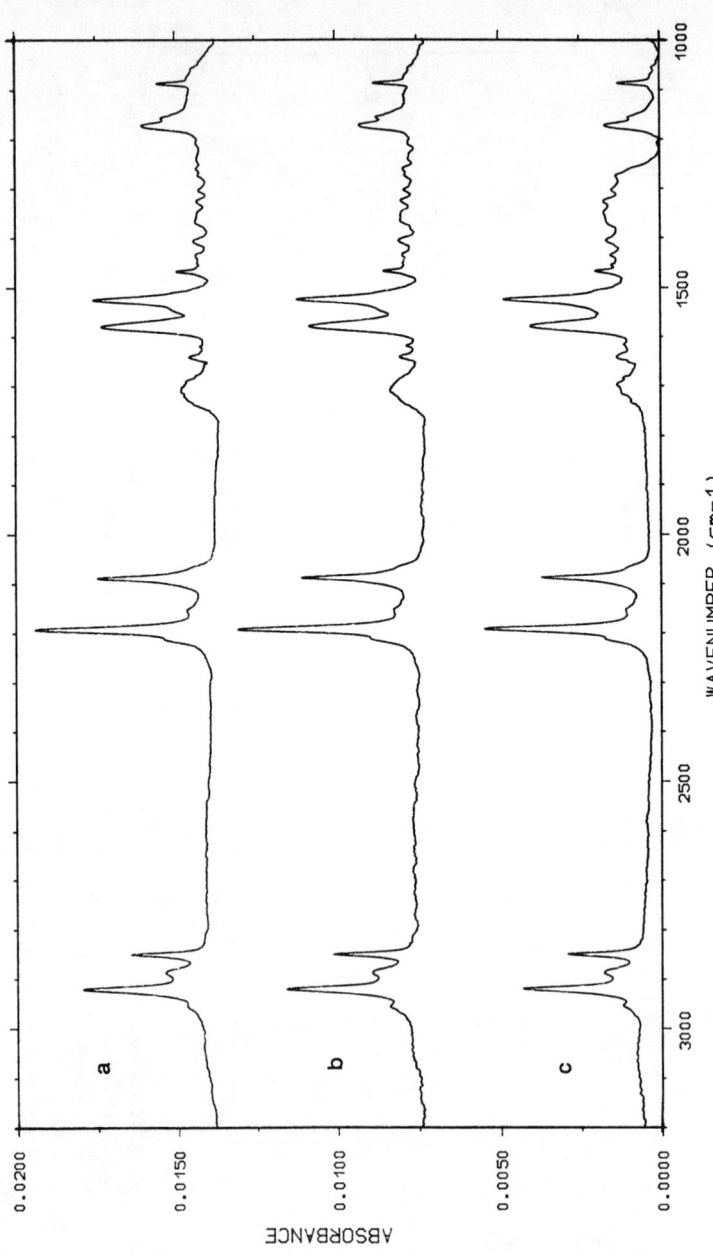

Fig 3. Transmission infrared spectra of a second polymer assembly consisting of 3 layers of stilbazolium-PECH and 4 layers of deuterated arachidic acid (d-AA) on ZnSe. The ordering of the layers is described in Fig 2. Spectrum a (top) shows the assembly shortly after preparation; b shows the assembly after aging at room temperature for 7 days, and c shows the assembly after aging at room temperature for 61 days.

TABLE I
SUMMARY OF PRINCIPAL INFRARED BANDS

Frequency	LB Layer	Substructure	Vibration
2923 cm^{-1}	polymer	CH$_2$ tail	a-sym CH str
2853	polymer	CH$_2$ tail	sym CH str
2194	spacer	CD$_2$ tail	a-sym CD str
2089	spacer	CD$_2$ tail	sym CD str
1720	spacer	COOH head	C = O str
1643	polymer	stilbazolium	C = C str
1582	polymer	stilbazolium	in-plane phenyl ϕ
1528	spacer (major)	COO$^-$	a-sym carboxylate str
	polymer (minor)	stilbazolium	in-plane phenyl
1469	polymer	CH$_2$ tail	CH$_2$ bend
1175	polymer	stilbazolium	C - ϕ str
~1130	polymer	backbone	C - O str
1089	spacer	CD$_2$ tail	CD$_2$ bend

loss of order and imply that the CH$_2$ is less stable than the CD$_2$ packing. The frequency of the CH$_2$ stretching vibrations, 2923 and 2853 cm^{-1}, Figure 2a-e, suggests that a number of the CH$_2$ groups are in a gauche conformation, and the CH$_2$ are not closest packed in the unheated polymer assembly. It is not surprising that heating disrupts the CH$_2$ conformation of the polymer more readily than the CD$_2$ packing of the arachidic acid spacer. However, instability at relatively low temperatures, e.g., 60 °C is not expected. At higher temperatures, Figure 2d-e, both the CH$_2$ and CD$_2$ bands show significant loss of intensity due to randomization of the packing and possible material evaporation.

A surprising feature of Fig 2a-e is that the intensity of the band at 1700 to 1720 cm^{-1} (COOH) is near zero whereas the 1528 cm^{-1} (COO$^-$) is sizable. It is expected that the COOH band would be quite distinct because Langmuir films of the acid form of the spacer were prepared at the air-water interface. The absence of a strong COOH band may arise from a combination of orientation effects (C = O normal mode aligned perpendicular to the substrate) and ZnSe contaminants reacting with the arachidic acid to make the salt. The effect of heat is dramatic upon the carboxylate band because it increases (sharpens) noticeably in Figure 2b-d compared to Figure 2a and then decreases dramatically in Figure 2e. The latter is probably indicative of the melting of the carboxylate groups. The appearance of the symmetric COO$^-$ stretch at 1431 cm^{-1} in Figure 2e supports this assertion. However, band sharpening in Figure 2b-d is unexpected and may arise from a subtle ordering (annealing) of the carboxylate groups. This ordering is observed in two independently prepared samples.

A more subtle feature in Figure 2a-e is the relative invariance of the intensity of the phenyl vibration(s) at 1582 cm^{-1} and the C = C stretch at 1643 cm^{-1}. These bands do not change by more than ± 10%. Since these bands show the orientation of the dye, it is tempting to speculate that the head group maintains a fixed orientation to the substrate under thermal stress. Unfortunately, the current studies only probe vibrations parallel to the substrate. Additional studies with reflecting substrates are needed to probe vibrations perpendicular to the substrate in order to more accurately assess the stability of the dye head group in the assembly.

Aging. Figure 3a-c shows a second sample of three layers of polymer interleaved with four layers of deuterated arachidic acid on ZnSe. This sample has not been heated and has been held in ambient conditions in the dark for several months. Spectrum a (top) shows the sample soon after preparation; spectrum b shows the sample seven days after spectrum a was measured, and spectrum c show the sample 61 days after spectrum a was measured. Note that the intensity of most bands remain near constant over the two month period. One apparent change is the increase of carboxylate at 1528 cm^{-1} relative to the in-plane phenyl vibration at 1582 cm^{-1}. Figure 3a-c shows that the carboxylate intensity at 1528 cm^{-1} increased by about 10% over the 60 day measurement period. Additional measurements show that after 150 days the increase in the carboxylate intensity, ca. 30%, causes the spectrum of the aged sample to resemble Figure 2a more than Figure 3a, i.e., the increase is reminiscent of the change observed when the polymer assembly is heated. Hence, it would appear that heating accelerates a reordering of the polymer assembly which occurs slowly even at room temperature. This reordering may be responsible for the sizable loss in SHG intensity between freshly made polymer assemblies and polymer assemblies which have been aged.

We have commented previously (13) on the instability of the dye assemblies to small changes in preparation. We now note similar effects in our present studies. Although the two samples shown in Figs. 2a and 3a were prepared simultaneously, their spectra are distinct. Figure 3a shows a sizable C=O band at 1718 cm^{-1} due to COOH and a carboxylate band at 1528 cm^{-1} approximately equal in size to the in-plane phenyl band at 1582 cm^{-1}. Figure 2a, as noted previously, shows only a very weak C=O band representing COOH whereas its carboxylate at 1528 cm^{-1} is 25% more intense than its in-plane phenyl band at 1582 cm^{-1}. Variability in preparation must be eliminated if successful devices are to be fabricated from these polymer dye assemblies.

Acknowledgements

Funding of this work was provided in part by I.B.M., in the form of the Paul J. Flory Sabbatical Fellowship to P. Stroeve, and a U.C. Faculty Research Grant to P. Stroeve.

Literature Cited

1. Williams,D.J. Angew. Chem. Int. Ed. Engl. 23, 690 (1984).
2. Neal, D.B., Petty, M.C., Roberts, G.G., Ahmad, M.M., Feast, W.J., Girling, I.R., Cade, N.A., Kolinsky,P.V., and Peterson, I.R. Electron. Lett. 22 460 (1986).
3. Girling, I.R., Cade, N.A., Kolinsky, P.V., Earls, J.D., Cross, G.H., and Peterson, I.R. Thin Solid Films 132 101 (1985).
4. Hayden, L.M., Kowel, S.T., and Srinivasan, M.P. Opt. Commun. 61 351 (1987).
5. Girling, I.R., Kolinsky, P.V., Cade, N.A., Earls, J.D., Cross, G.H., and Peterson, I.R. Opt. Commun. 55 289 (1985).
6. Geddes, N.J., Jurich, M.C., Swalen, J.D., Twieg, R. and Rabolt, J.F. J. Chem. Phys. 94 1603 (1991).
7. Cresswell, J.P., Tsibouklis, J., Petty, M.C., Feast, W.J., Carr, N., Goodwin, M., and Lvov, Y.M. Proc. SPIE 1337 358 (1990).
8. Era, M., Nakamura, K., Tsutsui, T., Saito, S., Niino, H., Takehara, K., Isomura, K., and Taniguchi, H. Japanese J. Appl. Phys. 29 L2261 (1990).
9. Draxler, S., Lippitsch, M.E., and Koller, E. Proc. SPIE 1319 88 (1990).

10. Travers, P.J., Miller, L.S., Sethi, R.S., and Goodwin, M.J. Chemtronics 4 239 (1989).
11. Ledoux, L. and Kajzar, F. Proc. SPIE 1127 137 (1989).
12. Hayden, L.M., Anderson, B.L., Lam, J.Y.S., Higgins, B.G., Stroeve, P., and Kowel, S.T. Thin Solid Films 160 379 (1988).
13. Stroeve, P., Saperstein, D.D., and Rabolt, J.F. J. Chem. Phys. 92 6958 (1990).
14. Hall, R.C., Lindsay, G.A., Kowel, S.T., Hayden, L.M., Anderson, B.L., Higgins, B.G., Stroeve, P., and Srinivasan, M.P. Proc. SPIE 824 121 (1987).
15. Anderson, B.L., Hoover, J.M., Lindsay, G.A., Higgins, B.G., Stroeve, P., and Kowel, S.T. Thin Solid Films 179 413 (1989).
16. Anderson, B.L., Higgins, B.G., Stroeve, P., Hoover, J.M., Lindsay, G.A., Mortazavi, M.A., Kowel, S.T., and Knoesen, A. in preparation.
17. Anderson, B.L., Ph.D. Thesis, Department of Chemical Engineering, University of California, Davis (1991).

RECEIVED September 24, 1991

Chapter 11

Static Secondary Ion Mass Spectrometric Analysis of Langmuir–Blodgett Film Multilayers

A Sampling Depth Study

Robert W. Johnson, Jr., Paula A. Cornelio-Clark[1], and Joseph A. Gardella, Jr.

Department of Chemistry, State University of New York at Buffalo, Buffalo, NY 14214

Alternating Langmuir-Blodgett multilayers of fatty acids and fatty acid salts were deposited on optically smooth germanium substrates. The multilayers consisted of one layer docosanoic and four layers eicosanoic acid, and the position of the docosanoic acid layer was varied. Static SIMS analysis was performed on these multilayers to determine the attenuation of the docosanoic acid molecular (or quasi-molecular) ion signal with increasing depth into the sample matrix. The results of this study suggest that the sampling depth is different for different molecular ions. It has been shown that sampling depth is affected by stability of the secondary molecular ion, substrate used for the LB films, and structural integrity of the LB films.

Secondary Ion Mass Spectrometry (SIMS) often gives information that is not available from any other surface analytical technique. In the analysis of synthetic polymers, SIMS can be used to make sensitive structural determinations. Gardella and Hercules used SIMS to distinguish between members of a homologous series of poly(alkyl methacrylates) based on length, chemical identity, or isomeric nature of the ester side chains. Electron Spectroscopy for Chemical Analysis (ESCA) could not differentiate between members of this series (1,2). These distinctions can also be made using attenuated total reflectance infrared spectroscopy (ATR-IR), however the greater sensitivity and shallower sampling depth (down to 10 A) of SIMS make it uniquely useful when information on the uppermost region of a sample surface is desired. SIMS has also been used to monitor surface sensitive reactions such as reactive modification and degradation of polymers (3,4). The extreme surface sensitivity of SIMS, as well as its ability to detect all elements, make it useful as a method for monitoring surface contamination (5).

[1]Current address: Air Products and Chemicals, Inc., 7201 Hamilton Boulevard, Allentown, PA 18195–1501

Static, or low damage, SIMS is normally used in polymer analysis. This technique has been used to characterize the surfaces of many different polymers and organic materials (6,7,8). In static SIMS, an inert gas primary ion impinges upon a solid surface causing sputtering (or desorption) of secondary ions. When SIMS is performed on a polymer surface, backbone bonds will often be broken and a secondary molecular ion (arising from the monomer repeat unit) will be formed (9). Molecular ions can result from protonation or deprotonation of the neutral molecule. Examples of ions that have a high yields are protonated amines (in the positive spectrum), and deprotonated acids (in the negative spectrum). If the molecular ion is not highly stable, then an 'ion pair' can form (4) consisting of a positive and a negative ion. The combined mass of these ions will be equal to that of the original monomer species. The characteristic fragmentation patterns formed by molecular and ion pair ions are the basis for structural identification.

Quadrupole mass filter (QMF) SIMS generally has the capability of detecting ions with mass up to 1000 daltons, although units are now available that can detect ions with mass up to 2000 daltons. Time of flight (TOF) SIMS can detect ions with mass up to 50,000 daltons (10,000 to 20,000 daltons for polymers) (10,11). Thus TOF-SIMS has the potential to give molecular weight distributions of the surface region of polymers. Success has been limited, primarily due to a lack of knowledge about the factors that control secondary ion formation. Thus one of the main goals of this study will be to obtain basic information about the mechanism of ion formation. This knowledge will then be used to help elucidate aspects of the SIMS applications that have already been discussed.

Since the results obtained from the SIMS analysis of polymers are often complex, these materials are not considered ideal sample matrices for the study of basic aspects of secondary ion formation. An ideal model system for such studies would consist of a highly ordered system of organic molecules. Langmuir-Blodgett films of fatty acids fulfill this requirement. The study of secondary ion formation from LB films should help to elucidate various aspects of the formation of secondary ions from polymer surfaces.

Studies have been done on various aspects of molecular secondary ion formation using LB films. Some of these studies have looked at the mechanism of proton transfer (12,13,14), the metal cationization mechanism (15-19), the mechanism of direct sputtering (9,19-21), film/substrate interactions (18), and quantitation of molecular ions (22,23). An important area that has not been extensively investigated is that of sampling depth of molecular ions in static SIMS. Sampling depth refers to the maximum distance below the sample surface from which secondary molecular ions, or quasi-molecular ions, can be detected. The type of quasi-molecular ion monitored in this study is formed when a neutral molecule combines with a metal ion to form a charged complex. A knowledge of sampling depth would aid in the analysis of polymer/substrate interactions, and in the quantitative analysis of thick polymer films. The maximum sampling depth of ions from a particular model system will give a general idea of the sampling depth from its polymer analogue. This sampling depth information will then be used to predict the maximum film thickness of polymer that can be quantitatively analyzed, and whose functional group/substrate interactions can be characterized. Cornelio and Gardella (24,25) have done a study of this type on alternating layers of eicosanoic (arachidic) (C20) and docosanoic (behenic) (C22) acids deposited on silver substrates. Alternating LB film layers of eicosanoic and docosanoic acid were laid down on the silver substrates. The signal intensity from the docosanoic acid layer was then monitored as a function of depth into the sample matrix. The level at which the signals from molecular, and quasi-molecular, ions disappeared was taken as a qualitative measure of the sampling depth for that particular ion. In this study they were able to determine the sampling depths of various molecular and quasi-molecular ions. They found that sampling depth was different for different ions, and was largely a function of secondary ion stability.

In the current study, we shall use an approach similar to that of Cornelio and Gardella for monitoring sampling depth. However, optically smooth germanium will be used as the substrate for the LB films. The previous study has shown that silver surfaces were roughened somewhat by the radio frequency glow discharge (RFGD) cleaning procedure, and that germanium surfaces were not (25). Results obtained on smooth germanium substrates should give an indication of the effect that substrate morphology has on sampling depth. LB films of the barium salts of these fatty acids will also be analyzed. Results obtained from the analysis of these films will allow us to determine the effect that structural integrity has on sampling depth. In addition, examination of these salt films will enable us to compare the signal intensity obtained from preformed ions (i.e. the acid salt films), versus that from neutral molecules (the acid films).

Experimental

Fatty acid LB samples were prepared by spreading films of eicosanoic (C20) and docosanoic (C22) acids from 1 mg/ml solutions (in benzene) on a pure triply distilled water subphase. Metal salts of fatty acid LB samples were prepared by spreading the same films on a $BaCl_2/KHCO_3$ subphase (26). After solvent evaporation (15-20 min.) the film was compressed to a surface tension of 25 dyne/cm corresponding a to close packed monolayer of 19.5 A^2/molecule (27). The films were transferred to a germanium prism at 4 mm/min. The deposition ratio and shape of the meniscus formed between the solid substrate and the insoluble monolayer were used to evaluate the quality of monolayer and multilayer transfer. The KSV Langmuir trough used for LB film preparations has been described elsewhere (21). To ensure a flat surface the germanium prisms were polished first with diamond paste (1 or 9 um, as required), and then with 0.3 um alumina. Surface roughness was evaluated using a Hitachi S-450 scanning electron microscope (SEM). SEM analysis indicated that the surfaces prepared using this polishing technique were only slightly rougher than the surfaces of unused, optically smooth, ATR prisms. The prisms were ultrasonically cleaned 3 times with soap and water and 3 times with hexane, and were then treated by radio frequency glow discharge to increase surface wettability and favor the attachment of the LB film (18). RFGD treatment of germanium removes hydrocarbon contamination and increases the thickness of the oxide layer, resulting in a high energy surface with increased wetting behavior (28,29). The prisms were stored under triply distilled water until ready for LB film deposition.

The SIMS experiments were carried out on a Leybold-Heraeus LHS10 Quadrupole mass filter (QMF) SIMS instrument which utilizes a 4 keV Xe^+ source. Typical base pressures of 5×10^{-10} mbar with operating pressures of 4×10^{-8} mbar Xe (differentially pumped) are achieved, which yield measured sample currents of 1 nA/cm^2 for Xe^+. The analysis time was approximately 10 minutes, which gave an ion dosage of approximately 2×10^{13} ions/cm^2.

Results and Discussion

The position of the docosanoic layer was varied from top to bottom, and the intensity of the molecular ion signals were monitored as a function of depth into the sample matrix. In the previous study of fatty acids on silver substrates (24), the signals from $(M-H)^-$ in the negative spectrum, and $(M+H)^+$ and $(M+Ag)^+$ in the positive spectrum were monitored. These signals were obtained for both eicosanoic and docosanoic acids. The sampling depth results of this study are shown in Table I. For the sake of simplicity, the table shows only the signals obtained from the docosanoic acid layer. The results suggest that the $(M-H)^-$ ion can be detected from a position that is much deeper in the sample matrix than either of the positive ions. The conclusion from this study was that the sampling depth was different for different ions and that it depends, at least partially, on secondary ion stability.

Table I SIMS Analysis of Fatty Acids on Silver

SAMPLE	$(M+H)^+$	$(M-H)^-$	$(M+Ag)^+$
4AA - 1BA/Ag	no	yes (w)	no
3AA - 1BA - 1AA/Ag	no	yes (w)	no
2AA - 1BA - 2AA/Ag	yes	yes (w)	no
1AA - 1BA - 3AA/Ag	no	yes	no
1BA - 4AA/Ag	yes	yes	yes

Key: w   - weak signal
      no - no signal obtained.
      yes - indicates that a signal of moderate intensity is obtained.

In the current study, this procedure was repeated using fatty acids on an optically smooth germanium substrate. The results from this study are shown in Table II. Note that the results indicate that the (M-H)$^-$ ion has a similar sampling depth for both germanium and silver substrates. However, the (M+H)$^+$ ion appears to have a much greater sampling depth on germanium than on silver. This is believed to be mainly a substrate effect. Previous studies by our group (*18,26*) have shown that silver is a very active metal and has a strong tendency to complex with LB films of fatty acids. This tendency to complex with the acid competes with protonation of the acid to form an (M+H)$^+$ ion. The net result is a less intense (M+H)$^+$ ion signal, and hence a shallower sampling depth. Germanium has been shown, by these same studies, to have a smaller tendency to complex with the acid. Thus the (M+H)$^+$ signal is much stronger and sampling depth is greater.

For studies on the fatty acid salts, we initially wanted to determine if barium from the KHCO$_3$/BaCl$_2$ subphase solution was adsorbing onto the germanium surface. To test this, clean germanium prisms were immersed in the subphase for approximately 3 hours and analyzed using SIMS. In addition, 1 and 5 LB film layers of barium arachidate were transferred to clean substrates and analyzed using SIMS. These SIMS spectra are shown in Figure 1. Figure 1a shows the absence of an intense peak in either the Ba$^+$ or BaO$^+$ regions, while Figure 1b has intense peaks in these regions. This suggests that little barium is adsorbed onto the surface from solution, and that the bulk of it is transferred to the surface with the LB film.

In Figure 1b we also see the presence of a signal of moderate intensity at 450 daltons. This signal has been assigned to the (M-H+Ba)$^+$ ion (*26*). In Figure 1c we also see this signal, as well as a signal of moderate intensity at approximately 314 daltons (expanded regions in upper right corner). This peak could result from an (M+H)$^+$ ion for eicosanoic acid. Its absence on the clean germanium, and its presence for both eicosanoic and docosanoic acids when multilayers are present, suggests that this is a strong possibility. Incomplete deprotonation of the acid groups in the LB film by the barium ions at the pH used (7.5 to 7.8) could be the reason that the (M+H)$^+$ ion appears. The positive and negative SIMS results from 1 and 5 LB film layers of barium arachidate on germanium are summarized in Table III. This table shows that, in the negative spectrum, the only molecular ion peak that appears is for the (M-H)$^-$ ion. It is interesting to note that the intensity of the (M-H)$^-$ peak from 1 layer of barium arachidate is approximately twice as intense as the peak resulting from five layers of barium arachidate.

It was necessary to collect this preliminary data on the fatty acid salts so that we could determine what signals were present, as well as their approximate intensity. Sampling depth data was obtained by repeating the sampling depth study described above using alternating layers of the eicosanoic and docosanoic acid barium salts. The results from this study are shown in Table IV. Note that in all cases the molecular, or quasi-molecular, ion signal from docosanoic acid disappears when it is buried deeper than 2 layers into the sample matrix. The general trend is much more rapid than when fatty acids are used. This result was not completely unexpected since it is well known that fatty acid salt films are more highly crystalline than LB films of fatty acids. Thus it appears that organic overlayers with fewer defects give greater attenuation of the molecular ion signal.

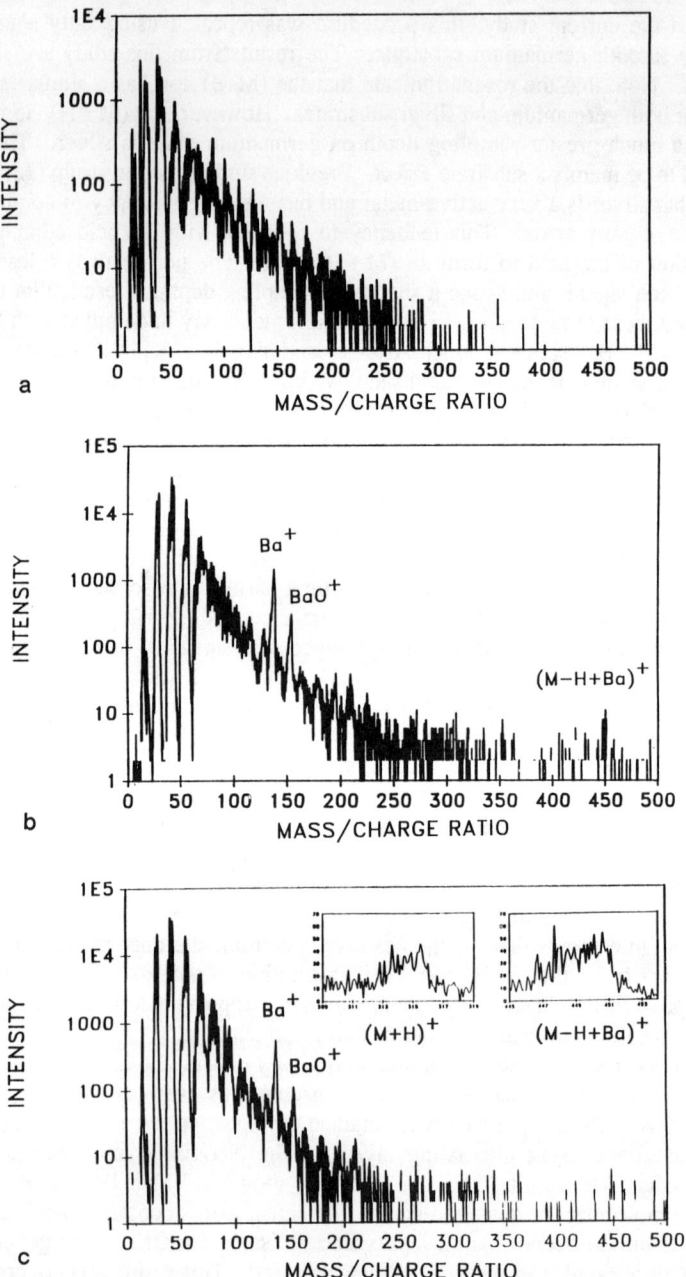

Figure 1. a) Positive SIMS spectrum of clean germanium after soaking in $KHCO_3/BaCl_2$ subphase for 3 hours, b) positive SIMS apectrum of 1 LB film layer of barium arachidate on germanium, c) positive SIMS spectrum of 5 LB film layers of barium arachidate on germanium.

Table II SIMS Analysis of Fatty Acids on Germanium

SAMPLE	$(M+H)^+$	$(M-H)^-$
4AA - 1BA/Ge	yes (vw)	yes (w)
3AA - 1BA - 1AA/Ge	yes (w)	yes (vw)
2AA - 1BA - 2AA/Ge	yes	yes (w)
1AA - 1BA - 3AA/Ge	yes	yes (w)
1BA - 4AA/Ge	yes	yes

Key: w   - weak signal
      vw - very weak signal

Table III SIMS Analysis of Arachidic Acid Salt Films on Germanium

SAMPLE	$(M+H)^+$	$(M-H)^-$	$(M-H+Ba)^+$
1 BaAA/Ge	no	yes	yes
5 BaAA/Ge	yes (?)	yes	yes

Key: (?) - uncertain signal assignment.

Conclusions

The main conclusion that we can draw from this work is that the sampling depth of molecular ions in SIMS varies greatly as the nature of the sample matrix changes. From the previous study we have seen that sampling depth varies from ion to ion and is related to secondary ion stability. In addition, the current study suggests that sampling depth for certain ions may be different when different substrates are present. Finally, it appears that sampling depth for all molecular and quasi-molecular ions decreases as the structural integrity of organic overlayers increases.

Table IV SIMS Analysis of Fatty Acid Salt Films on Germanium

SAMPLE	$(M+H)^+$	$(M-H)^-$	$(M-H+Ba)^+$
4AA - 1BA/Ge	no	no	no
3AA - 1BA - 1AA/Ge	no	no	no
2AA - 1BA - 2AA/Ge	no	no	no
1AA - 1BA - 3AA/Ge	yes (w) (?)	yes (w)	no
1Ba - 4AA/Ge	yes (w)	yes	yes (w)

Key: w - weak signal

Acknowledgements

We thank the National Science Foundation for support of this work (grant number DMR 8720650). We also thank Dr. Robert Baier and Dr. Anne Meyer for use of the LB film trough. In addition, we thank Peter Bush and Bob Barone for use of the scanning electron microscope.

Literature Cited

1. Gardella, J.; Hercules, D.; *Anal. Chem.*, **1980**, *52* (2), 226.
2. Gardella, J.; Hercules, D.; *Anal. Chem.*, **1981**, *53* (12), 1879.
3. Gardella, J.; Novak, F.; Hercules, D.; *Anal. Chem.*, **1984**, *56* (8), 1371.
4. Hook, K.; Hook, T.; Wandass, J.; Gardella, J.; *Appl. Surf. Sci.*, **1990**, *44*, 29.
5. Yamaguchi, N.; Suzuki, K.; Sato, K.; Tamura, H.; *Anal. Chem.*, **1979**, *51*, 695.
6. Briggs, D.; Wootton, A; *SIA, Surf. Interface Anal.*, **1982**, *4*, 109.
7. Briggs, D.; *SIA, Surf. Interface Anal.*, **1982**, *4*, 151.
8. Briggs, D.; Hearn, M.; Ratner, B.; *SIA, Surf. Interface Anal.*, **1984**, *6*, 184.
9. Benninghoven, A.; Rudenauer, R.G.; Werner, H.W. Eds.; "Secondary Ion Mass Spectrometry-Basic Concepts, Instrumental Aspects, Applications and Trends"; John *Wiley and Sons: New York, 1987.*
10. Blestos, I.; Hercules, D.; vanLeyen, D.; Benninghoven, A.; *Macromolecules*, **1987**, *20*, 407.
11. Blestos, I.; Hercules, D.; VanLeyen, D.; Niehuis, E.; Benninghoven, A.; *Proc. Phys. 9, Ion Formation from Organic Solids III*; Benninghoven, A., Ed.; Springer-Verlag: Berlin, 1986.

12. Benninghoven,A.; Lange, W.; Jirikowsky, M.; Holtkamp, D.; *Surf. Sci.*, **1982**, *123*, L721.
13. Sange, W.; Jirikowsky, M.; Benninghoven, A.; *Surf. Sci.*, **1975**, *79*, 549.
14. Parker, C.; Hercules, D.; *Anal. Chem.*, **1986**, *58*, (1), 25.
15. Wandass, J.H.; Schmitt, R.L.; Gardella, J.A., Jr.; *Appl. Surf. Sci.*, **1989**, *40*, 85.
16. Benninghoven, A.; Jaspers,D.; Sichtermann,W.; *Appl. Phys.*, **1976**, *11*, 35.
17. Grade, H.; Winograd, N.; Cooks, R.; *J. Am. Chem. Soc.*, **1977**, *99*, 7725.
18. Wandass, J.; Gardella, J.A., Jr.; *J. Am. Chem. Soc.*, **1985**, *107*, 6192.
19. Pachuta, S.J.; Cooks, R.G.; *Chem. Rev.*, **1987**, *87*, 647.
20. Clark, M.B., Jr.; Ph.D. thesis, University at Buffalo, SUNY, Sept. 1989.
21. Hook, K.; Gardella, J.A., Jr.; *J. Vac. Sci. Technol.*, **1989**, *A7*(3), 1795.
22. Clark, M.B., Jr.; Gardella, J.A., Jr.; *Anal. Chem.*, **1990**, *62*, 870.
23. Cornelio, P.; Gardella, J.; *J. Vac. Sci. Technol. A*, **1990**, *3*, 2283.
24. Cornelio-Clark, P.; Gardella, J.; *Langmuir*, in press.
25. Cornelio, P.A.; Ph.D. thesis, University at Buffalo, SUNY, July 1990.
26. Blodgett, K.B.; Langmuir, I.; *Phys. Rev.*, **1937**, *51*, 964.
27. Gaines, G.L.; *"Insoluble Monolayers at Liquid-Gas Interfaces"*; Wiley Interscience: New York, 1966.
28. Wandass, J.H.; Ph.D. thesis, University at Buffalo, SUNY, August 1986.
29. Baier, R.; DePalma, V.; Calspan Corp. Report no. 176, Buffalo, NY, 1970.

RECEIVED November 1, 1991

Chapter 12

Langmuir–Blodgett Affinity Surfaces
Targeted Binding of Avidin to Biotin-Doped Langmuir–Blodgett Films at the Tip of an Optical Fiber Sensor

Shulei Zhao and W. M. Reichert

Center for Emerging Cardiovascular Technologies, Department of Biomedical Engineering, Duke University, Durham, NC 27706

Here we demonstrate the targeted binding of protein ligands to receptor-containing LB affinity layers at the tip of optical fiber sensor. Solutions of fluorescein-labelled avidin exposed to a series of biotin-doped LB films deposited at the sensor tip all produced fluorescence intensities that increased monotonically to a saturation asymptote. Fitting the fluorescence data to a bimolecular reaction model revealed that the fluorescence at target saturation was linear with biotin surface density as long as the number of targets per surface-bound avidin did not exceeded unity. The linear portion of these data, assuming a 1:1 molar correspondence of target surface density to surface-bound avidin, allowed us to calculate the avidin surface density at the sensor tip as well as the sensor's molecular sensitivity and detection limit.

The binding of ligands to receptor containing Langmuir-Blodgett (LB) films and supported bilayer membranes (BLMs) have long been proposed as chemically selective layers with potential applications in sensing.[1,2] Optical fibers have proven to be attractive in remote sensing applications,[3] and several sensors based on distally located, fluorophore-doped, chemically selective LB films have been proposed.[4,5] An alternative, but similar, approach would be to fabricate LB films or BLMs that contain properly oriented antibodies, antibody fragments, whole antigens, or haptens that bind the analyte ligand in which the analyte concentration would be determined subsequently by conventional sandwich or competitive fluoroimmunoassay. Although several fiber optic affinity-type biosensors have appeared in the last ten years,[6] and the deposition of LB films onto fiber optics have been reported,[7,8] to these authors' knowledge no LB or BLM affinity binding configurations have been demonstrated with fiber optic excitation and collection.

The past decade has also evidenced a rapid increase in the use of total internal reflection fluorescence (TIRF) spectroscopy[9] and microscopy[10] in the monitoring of protein adsorption and specific ligand-receptor interactions at the solid/liquid interface - primarily owing to the surface sensitivity of evanescent wave excitation. The first use of TIRF in the detection of targeted protein binding to LB films at the solid/liquid interface involved the binding of the lectin concanavalin A to glycolipid-derivatized, polymerized, diacetylene lipid monolayers.[11] However, the ability to modulate the level of targeted binding of proteins to hapten-doped lipid bilayers was not adequately demonstrated at the solid-liquid interface until the recent TIRF microscopy work of Tamm and co-workers,[12] who observed that the level of binding of fluorescently labeled anti-dinitrophenol antibodies to supported lipid bilayers increased with the mol % doping of the bilayer film with dinitrophenol lipid hapten.

In this paper we demonstrate the targeted binding of a protein to an LB affinity surface deposited at the tip of an optical fiber sensor. The key features of the sensing configuration used in this study include evanescent excitation and collection of surface-bound fluorescence in the well known 'dip stick' configuration,[13] and the deposition of a head-group-out, hapten-doped LB film at the surface of the distal sensing tip (Fig. 1). The ligand-receptor model chosen for this study was the highly characterized, and commercially available avidin-biotin complex noted for its extremely high binding affinity with a K_a on the order of 10^{15} L/mol.[14] In all of our binding experiments the fluorescence intensity of the surface bound fluorescein-labelled avidin increased monotonically with time to an asymptote characteristic of saturation binding to the biotin-doped sensing surface. In addition, the fluorescence intensity at target saturation increased linearly with target density in the LB film as long as there appeared to be less than one target per surface-bound protein molecule. This observation allowed us to calculate the molecular avidin surface density at the sensor tip as well as the sensor's molecular sensitivity and detection limit.

Materials and methods.

Sensor fabrication. Two, step-index, quartz optical fibers with silicone gel cladding (Quartz Products Corp.) were used to make the sensor illustrated in Figure 1. Core diameters of these two fibers were 200 µm and 600 µm with lengths of approximately half a meter and one meter, respectively. The plastic coating and cladding were stripped from a 2 cm terminal segment of the 200 µm fiber and the exposed fiber core was cleaned in chromic acid followed by rinsing with distilled-deionized water. The coating and cladding of a 4 cm segment in the middle of 600 µm fiber were stripped and the exposed core was cleaned with ethanol soaked Kimwipes. The stripped sections of the two fibers were then laid parallel and heated with a butane-oxygen torch. A low temperature flame was used initially to physically attach the tip of the 200 µm fiber to the core of the 600 µm fiber (Fig. 2a). The attached fibers were then 'feathered' with flames of sequentially increasing temperature taking care to heat the fiber junction on both sides (Fig. 2b). The temperature of the flame was increased until the quartz fibers became "white hot" and the two fibers spontaneously fused together. After allowing the

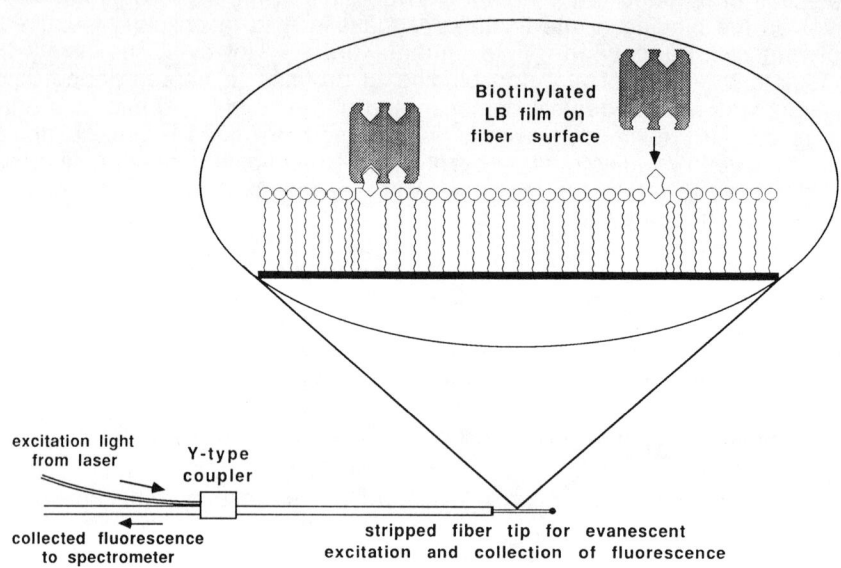

Figure 1. Schematic illustration of the 'dip stick' type evanescent fiber optic sensor showing the head-group-out, biotin-doped LB film at the sensing tip.

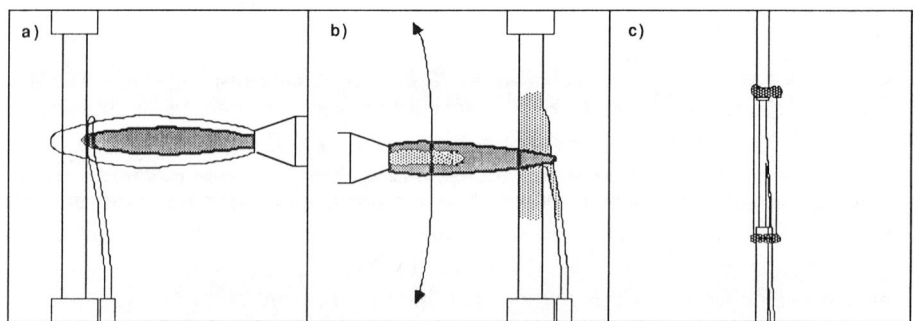

Figure 2. Fusion of Y-type fiber optic coupler with oxygen-butane torch: a) attachment of 200μm fiber to 600μm fiber using low heat flame; b) fusing of fibers by 'feathering' junction with flame of increased heat; c) encasement of fused coupler in tube.

fused fibers to cool, the fragile fiber junction was then covered with a small plastic tube and both ends of the tube were sealed with silicone rubber cement to prevent contamination (Fig. 2c). The light coupling efficiency of the coupler was measured to be about 50% which is comparable to that of the half-silvered mirror.

End surfaces of the fibers were polished with 3μm and then 1μm lapping films (Amphenol Fiber Optic Products). The distal sensing tip was made by stripping the coating and cladding from a 3 cm segment at the end of the 600μm fiber. The exposed tip was cleaned in hot (80°C) chromic acid for 30 minutes and rinsed with copious amounts of distilled-deionized water, and then dried in air. The majority of the exposed sensor tip was wrapped with a small piece of lens paper (Clay Adams), leaving an approximately 3 mm long segment at tip still exposed. The exposed tip was capped with opaque epoxy (#320, Epoxy Technology, Inc.) and cured at room temperature overnight and then in an oven (70°C) for two days. After curing, the epoxy capped fiber tip was unwrapped and immersed in chromic acid at room temperature for 10 minutes, followed by rinsing with distilled water and then air drying. Finally, the epoxy capped fiber tip was immersed in a 10% (v/v) solution of dimethyldichlorosilane (DDS, Sigma) in toluene (Aldrich) for 10 min. and then rinsed sequentially with methanol, distilled water, and again methanol. This final step was necessary to hydrophobize the fiber surface so that a head-group-out oriented LB monolayer could be deposited at the sensor tip (Fig. 1).

Deposition of biotinylated LB monolayers. 1.2 mg/mL and 1.0 mg/mL stock solutions of arachidic acid (Kodak) and the biotinylated phospholipid N-(biotinoyl)dipalmitoyl-L-α-phophatidylethanolamine triethylammonium salt (Molecular Probes), respectively, were made in chloroform (Fischer). Stock solutions were combined in precise amounts to produce spreading solutions containing biotinylated phospholipid and arachidic acid in molar ratios of 1/600, 1/350, 1/250, 1/200, 1/158, 1/130, and 1/100. 30μL of spreading solution were deposited drop-wise on a subphase of distilled-deionized water in a PTFE-lined film balance (NIMA). After allowing 15 min. for equilibration, the fatty acid film was compressed to a pressure of 35 mN/m. With the silane treated fiber tip poised above the compressed film, a single monolayer was deposited onto the silane treated sensing tip on the down stroke. Although the degree of silane coupling was not quantified, a silane treated sensing tip was readily LB deposited with fatty acid on the downstroke, while no fatty acid was deposited on the downstroke with an untreated fiber tip. Calculations of transfer ratio of arachidic acid onto the fiber tip (0.9) and estimates of the deposited film quality using this configuration are discussed in detail elsewhere.[8] Assuming no mass loss during LB deposition, the arachidic acid stock solution and the seven biotin containing spreading solutions produced LB films of precisely known biotinylated phospholipid doping densities ranging from 0-1 mol %. Unfortunately, pulling the coated fiber tip back through the air/water interface either deposits an improperly oriented second LB layer or strips off the initial layer, both of which are undesirable. This problem was averted by submerging a glass beaker in the LB trough that received the fiber tip upon the deposition downstroke (Fig. 3a-b). The submerged beaker then provided a vehicle for transferring the LB coated fiber tip from the trough to an adjacent immersion tank (Fig. 3c) where the fiber tip was assembled into the flow cell (Fig. 3d). The sensor tip/flow cell assembly was then removed from the tank for the binding measurements. The flow cell itself was a simple device consisting of a cleaved Pasture pipette with a glass blown inlet port, silicone rubber tubing, and a stopper for immobilizing the sensor tip and preventing leakage through the top orifice (Fig. 3d). Total volume of the flow cell was approximately 1 mL.

Avidin binding experiments. Fluorescein isothiocyanate (FITC) labelled avidin (67,000 daltons, Sigma) with an average of 3.9 labels per protein molecule was dissolved without further purification to a concentration of 0.01 mg/mL in 0.01 M phosphate buffered saline (PBS, pH 7.5) with 0.5 M NaCl concentration. Protein solutions were stored at 4°C in the dark until used. High ionic strength PBS (0.5 M NaCl) was used to reduce nonspecific adsorption of avidin to the sensor surface.[15] Using previously described instrumental[16] and fiber optic sensor[17] configurations, fluorescence was excited with the 488 nm line of an air cooled Ar ion laser and collected from 490nm to 600nm at 5nm intervals. After the fiber tip was assembled into the flow cell, the residual water in the flow cell was displaced with PBS followed by collection of 10 spectra to provide an average background. Next, 4 mL of protein solution was infused into the flow cell to displace the buffer. Starting with the initiation of protein infusion (t=0), fluorescence intensity was collected at increasing time intervals for a maximum of 60 min. The difference between the collected fluorescence intensity (S_1) and the PBS background (B_1) was the fluorescence intensity attributed to the fluorescently-labelled avidin (F_1).

$$F_1 = S_1 - B_1 \tag{1}$$

Relative fluorescence intensity. At the end of each binding experiment the flow cell was disassembled, the fiber tip was cleaned with Kimwipes and ethanol, and then the tip was immersed in a vial containing distilled-deionized water and 10 spectra were collected to give another average background (B_2). The fiber tip was then immersed into a vial containing 0.4 mg/mL water-soluble fluorescein (Sigma) and a final fluorescence signal (S_2) was collected. Fluorescence attributed to the water soluble fluorescein molecules was the difference between the fluorescein solution signal and the water background

$$F_2 = S_2 - B_2 \tag{2}$$

In all experiments, the background intensity B_2 did not differ significantly from B_1 indicating that the optics system had not changed during disassembling the fiber tip from the flow cell and that the bound FITC-avidin was essentially cleared from the fiber surface. Finally, since each binding experiment necessitated an individually prepared sensing tip, and since several different fiber probes were employed, the avidin signal intensity collected with each sensor F_1 was normalized with the corresponding value of F_2 determined at the end of each experiment. This normalization yielded the relative fluorescence intensity that was independent of individual sensor effects.

$$F_r = F_1/F_2 \tag{3}$$

Bimolecular model of ligand-receptor binding.

A reasonable model for the specific binding of the protein ligand avidin (A) to a surface doped with a given population of biotin binding sites (B) is the reversible bimolecular reaction[18]

$$A + B \underset{k_r}{\overset{k_f}{\rightleftarrows}} A\text{-}B \tag{4}$$

where k_f and k_r are the forward and reverse rate constants and A-B is the surface bound avidin-biotin complex. The rate of appearance of the A-B complex is described by the differential equation

$$d[A\text{-}B]/dt = k_f[A][B] - k_r[A\text{-}B] \tag{5}$$

where [A] is the solution avidin concentration (e.g. molecules/cm^3), while [B] and [A-B] are surface densities (e.g. molecules/cm^2) of the vacant and avidin-bound biotin targets, respectively. Solving Equation. 5 with the boundary condition [A-B]=0 at t=0, and assuming that [A] remains constant, yields an exponential expression for the time-dependent increase in the surface density of the avidin-biotin complex

$$[A\text{-}B](t) = [A\text{-}B]_{max}[1-\exp(-\beta t)] \tag{6}$$

where $[A\text{-}B]_{max}$ is the maximum surface density of the avidin-biotin complex and ß is the affinity time constant, both of which are defined functions of k_f, k_r, and the initial concentrations $[A]_o$ and $[B]_o$. If one assumes that the relative fluorescence intensity collected with the sensor is linearly proportional to the surface density of bound avidin, then Equation 6 becomes

$$F_r(t) = F_{max}[1-\exp(-\beta t)] \tag{7}$$

where

$$F_{max} = \alpha[A\text{-}B]_{max} + F_o, \tag{8}$$

α is the sensitivity of the sensor with units of intensity per unit molecular surface density (e.g. intensity/molecules/cm^2) and F_O is the background fluorescence intensity corresponding to $[A\text{-}B]_{max}=0$.

Results.

Figure 4 is displays the time course of a typical avidin binding experiment. The monotonic increase in fluorescence to a saturation asymptote in a

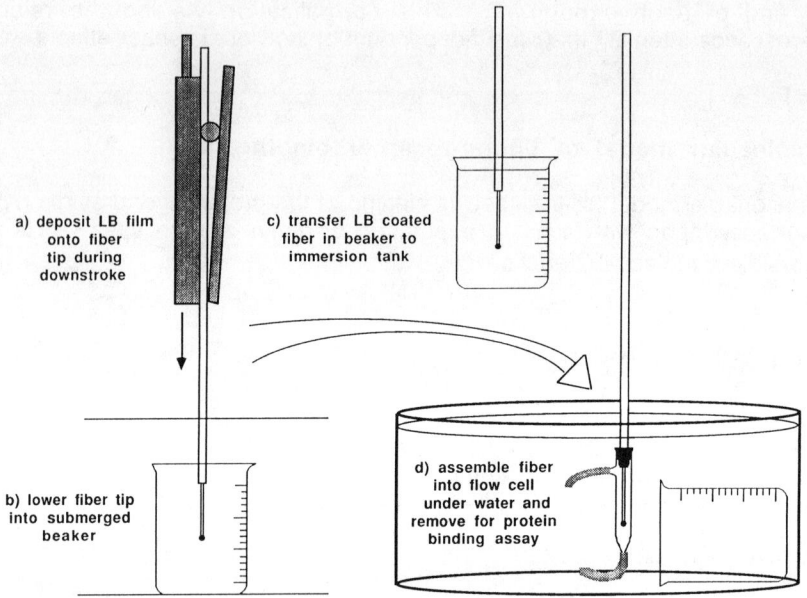

Figure 3. Fiber transfer technique used to maintain head-group-out orientation at fiber tip.

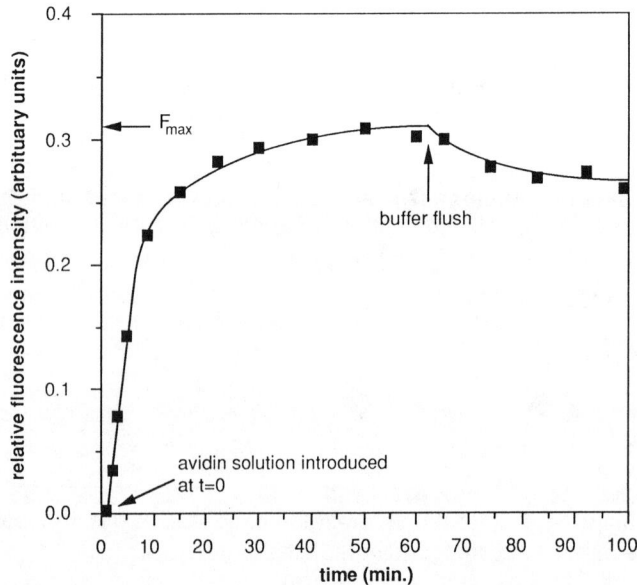

Figure 4. Time course of FTIC-labelled avidin binding to a 1 mol % biotin-doped LB film. Note the monotonic fluorescence intensity increase to a saturation asymptote (F_{max}) after introduction of protein solution and the small fluorescence decrease after buffer flush.

manner characteristic of Equation 7 suggests that avidin continued to bind with the vacant biotin targets until all of the available targets were occupied. After buffer flush a small drop in fluorescence intensity was observed resulting from a combination of 1) flushing the flow cell of fluorescent solution, thus eliminating the bulk contribution to the evanescently excited signal, 2) desorption of non-specifically adsorbed avidin, and 3) avidin-biotin dissociation from the sensor surface. In this particular case (1 mol % biotin doping), the fluorescence intensity dropped by 13.5% after 40 min. exposure to buffer. In all experiments the protein solution in the flow cell was static with a concentration of 0.01mg/mL (8.99×10^{14} molecules/cm^3).

The relative fluorescence intensity vs. time data (for t≤60 min.) collected from the eight binding trials were fit to Equation. 7 using a nonlinear regression routine.[19] The best fits of the parameters F_{max} and ß to these data are listed in Table I as a function of mol % biotin doping of the LB film. Although the molecular area of phospholipids and arachidic acid are 40-60 and 18Å^2/molecule, respectively,[20] the highest doping level in our experiments was only 1 mol %. Therefore, the net area per molecule in the LB films differed from the molecular area of arachidic acid by less than 0.5Å^2/molecule (i.e. for 1 mol % doping: 0.99(18)+0.01(40-60)=18.22-18.42Å^2/molecule) and all of the values of target surface density in Table I were estimated by simply using a molecular area 18Å^2/molecule in the LB films.

The left ordinate of Figure 5 plots the values of F_{max} in Table I against the corresponding target surface density in the control and biotin-doped LB films. Least squares regression of the linear portion of these data (filled squares) produced a slope of $\alpha = 9.58 \times 10^{-14}$ units of relative fluorescence intensity collected per unit target molecular surface density (e.g intensity/molecules/cm^2), with a y intercept of $F_o = -3.10 \times 10^{-2}$ and a correlation coefficient of 0.995. If one assumes a 1:1 molar correspondence of target surface density to surface-bound avidin in the linear region of Figure 5, then inserting the above slope and intercept values into Equation 8 yields

$$F_{max} = 9.58 \times 10^{-14} [A-B]_{max} - 3.10 \times 10^{-2} \qquad (9)$$

which becomes upon rearrangement

$$[A-B]_{max} = 1.04 \times 10^{13} F_{max} + 3.24 \times 10^{11} \qquad (10)$$

where 3.24×10^{11} is the avidin surface density in molecules/cm^2 that produces no fluorescence intensity ($F_{max}=0$). The values of avidin surface density on the right ordinate of Figure 5 were calculated directly from Equation 10. The assumptions inherent in this calculation are given below.

Discussion.

Given the reported affinity of avidin for biotin, one would expect that avidin in solution would bind in a 1:1 molar ratio with the surface density of biotin

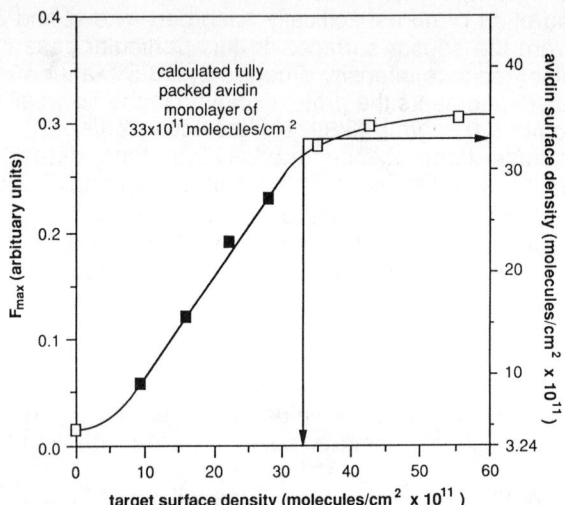

Figure 5. Best fit values of F_{max} listed in Table I (left ordinate) plotted as a function of target surface density for the eight LB affinity surfaces tested. The point-slope formula derived from the linear portion of these data (filled squares) was used to calculate avidin surface density at sensor tip (right ordinate). Note that the deviation of the data from linearity and the maximum observed avidin surface density both occur near the value of 33×10^{11} molecules/cm^2 calculated for a fully packed avidin monolayer.

Table I. Fitted parameters from avidin-biotin binding experiments

mol % biotin doping of LB film	target density in molecules/cm^2 x10^{11}	$F_{max} \pm \sigma_D \times 10^{-2}$	ß (sec^{-1}) $\pm \sigma_D \times 10^{-3}$
0	0	1.42±0.17	3.03±0.92
0.167	9.27	5.71±0.26	2.97±0.71
0.286	15.9	11.9±0.66	1.49±0.26
0.400	22.2	19.0±0.71	0.97±0.10
0.500	27.8	23.1±0.51	1.19±0.08
0.633	35.2	27.9±0.78	1.53±0.16
0.769	42.7	29.8±0.76	1.23±0.11
1.000	55.6	30.7±0.91	1.95±0.21

σ_D = standard deviation

binding sites at the sensor tip *as long as there is less than or equal to one biotin binding site per surface-bound avidin molecule.* The upper limit of target density beyond which a 1:1 binding of avidin to biotin is no longer possible can be estimated from the surface density of a fully packed monolayer of surface bound avidin. Specifically, according to a review by Green,[14] avidin has a somewhat rectangular shape with molecular dimensions of 55Åx55Åx41Å. The four biotin binding sites on each avidin molecule are located on the two 55Åx55Å faces of the molecule, two binding sites per face. Consequently, each surface-bound avidin molecule requires a minimum surface area of $(55Å \times 55Å) \times (10^{-16} cm^2/Å^2) = 3.03 \times 10^{-13}$ cm^2/molecule. Therefore, assuming a fully packed avidin monolayer, the maximum number of targets per unit area that would still allow a 1:1 molar correspondence of biotin binding sites to surface-bound avidin is simply $(3.03 \times 10^{-13} cm^2/molecule)^{-1} = 33 \times 10^{11}$ molecules/cm^2.

Now, consider the following observations that: 1) the most highly biotin-doped LB film in the linear region of Figure 5 has a surface density of 27.8×10^{11} molecules/cm^2; and 2) the collected fluorescence intensities in Figure 5 break from linearity and begin to plateau at a target density just below 33×10^{11} molecules/cm^2. Combined, these suggest that avidin binding is stoichiometric with target surface density in the linear region of Figure 5, but a target surface density that approaches or exceeds 33×10^{11} molecules/cm^2 becomes crowded with avidin and further increases in target density cannot significantly increase the level of avidin binding. Invoking the above observations not only allowed us to justify using Equations 9 and 10 to calculate avidin surface density, but it also provides a means for determining the molecular sensitivity and detection limit of the sensor directly from the data in Figure 5 and Table I. However, this analysis does not take into consideration the possibility of target aggregation at these low doping levels.

Using well known conventions,[21] the sensitivity of the sensor is simply the slope the of the linear portion of Figure 5, and thus the value of α in Equation 9, while for the 95% confidence level, the detection limit is equal to three times the standard deviation of the measurement. In the experiments conducted here, the sensor displayed a surface sensitivity of 9.58×10^{-14} intensity units for every avidin molecule/cm^2, or 0.06 intensity units/pmol avidin/cm^2. Three times the average standard deviation of the F_{max} data in Table I suggests that changes in avidin surface density can be detected reliably in increments of 0.02 intensity units which, from the sensitivity, corresponds to $0.02/9.58 \times 10^{-14} = 2.09 \times 10^{11}$ molecules avidin/cm^2, or 0.33pmol avidin/cm^2. These values, along with the surface density of avidin non-specifically adsorbed to the control LB film, are listed in Table II.

Conclusions.

Here we demonstrated the targeted binding of protein ligands to LB affinity layers at the tip of optical fiber sensors using the avidin-biotin model system. Analysis of the collected fluorescence at target saturation allowed us to calculate the surface density of avidin bound to the sensor surface (Fig. 5), as well as the sensor's sensitivity and detection limit (Table II). Results suggest

Table II. Sensor sensitivity, detection limit, and nonspecific adsorption

sensitivity		detection limit			nonspecific adsorption to control LB film	
$\dfrac{\text{intensity}}{\text{molecule/cm}^2}$	$\dfrac{\text{intensity}}{\text{pmol/cm}^2}$	a intensity	$\dfrac{\text{molecules}^b}{\text{cm}^2}$	$\dfrac{\text{pmols}^c}{\text{cm}^2}$	$\dfrac{\text{molecules}^d}{\text{cm}^2}$	$\dfrac{\text{pmols}^e}{\text{cm}^2}$
9.58×10^{-14}	0.06	0.02	2.09×10^{11}	0.33	4.72×10^{11}	0.78

a: calculated from F_{max} in Table I via $3[\Sigma_i(\sigma_{Di})^2/n]^{1/2}$; b: $0.02/9.58 \times 10^{-14}$; c: $0.02/0.06$; d: calculated from F_{max} in Table I via Eq. 10; e: $1.42 \times 10^{-2}/6.02 \times 10^{11}$ molecules/pmol that avidin binding was stoichiometric with target surface density as long as there was less than one biotin target per surface bound avidin molecule. The calculated detection limit and Equation 10 suggests that this fiber optic system can reliably detect the binding of avidin at target densities of 2-3×10^{11} molecules/cm^2, while Pasarchick and Thompson report that total internal reflection fluorescence microscopy is capable of detecting antibody binding to target densities of 6×10^{11} molecules/cm^2.[22] However, the degree of resolution is specific to the model system studied and varies with the degree of fluorescent labelling, the fluorophore quantum efficiency, the surface area occupied by the analyte, and the affinity of the ligand for the target.

Acknowledgements.

This work was funded by a biomedical research grant from the Whitaker Foundation and a graduate student fellowship from the Center for Emerging Cardiovascular Technologies (S.Z.). These authors thank S.S. Saavedra and G.A. Truskey of Duke University, H. Ringsdorf of the University of Mainz, and F.S. Ligler of the Naval Research Labs for helpful discussions, for providing the nonlinear regression software (G.A.T.), and for suggesting the avidin-biotin model system (H.R.).

Literature cited.

1. W.M. Reichert, C.J. Brucker, and J. Joseph, Thin Solid Films, 152 (1987) 345-376.
2. T. Moriizumi, Thin Solid Films, 160 (1988) 413-429.
3. O.S. Wolfbeis, in S.J. Schulman Ed., Molecular Luminescence Spectroscopy: Methods and Applications- Part II, John Wiley and Sons, New York (1988), 129-281.
4. B.P.H. Schaffar, O.S. Wolfbeis, and A. Leitner, Analyst, 113 (1988) 693-697.
5. M. Aizawa, M. Matsuzawa, and H. Shinohara, Thin Solid Films, 160 (1988) 477-481.
6. D.L. Wise and L.B. Wingard Ed., Biosensors with Fiber Optics, Humana Press, Clifton, NJ, 1991.
7. R.H. Selfridge, S.T. Kowel, P. Stroeve, J.Y.S. Lam, and B.G. Higgins, Thin Solid Films, 160 (1988) 471-476.
8. S. Zhao and W.M. Reichert, Thin Solid Films, 200 (1991) 363-373.
9. W.M. Reichert, Crit. Rev. Biocompat., 5 (1989) 173-205.
10. D. Axelrod, N.L. Thompson, and T.P. Burghart, J. Microsc., 129 (1982) 19-28.
11. H. Bader, R. VanWagenen, J.D. Andrade, and H. Ringsdorf, J. Colloid Interface Sci., 101 (1984) 246-249.
12. E. Kalb, J. Engel, and L.K. Tamm, Biochemistry, 29 (1990) 1607-1613.
13. J.D. Andrade, R.A. VanWagenen, D.E. Gregonis, K. Newby, and J.-N. Lin, IEEE Trans. Electron. Dev. ED-32 (1985) 1175-1179.
14. M. Green, Avidin, Adv. Protein Chem. 29 (1975) 85-133.
15. R.C. Duhamel and J.S. Whitehead, Meth. Enzymol., 184 (1990) 201-207.

16. J.P. Bromberg, Physical Chemistry, Allyn and Bacon, Inc., Boston, 1980.
17. D.W. Marquadt, J. Soc. Ind. Appl. Math, 11 (9163) 431-441.
18. W.M. Reichert, C.J. Bruckner, and S.R. Wan, Appl. Spectro., 42 (1988) 605-608.
19. H.A. Strobel and W.R. Heineman, Chemical Instrumentation: a Systematic Approach, John Wiley and Sons, New York, 1989.
20. G. Roberts, Langmuir-Blodgett Films, Plenum Press, New York, NY, 1991.
21. H.A. Strobel and W.R. Heineman, Chemical Instrumentation: a Systematic Approach, John Wiley and Sons, New York, 1989.
22. M.L. Pisarchick and N.L. Thompson, Biophys. J., 58, 1235-1249 (1990).

RECEIVED September 24, 1991

Chapter 13

Mixed Monolayers of Lecithin and Bile Acids at the Air–Aqueous Solution Interface
Effect of Temperature and Subphase pH

M. J. Gálvez-Ruiz and M. A. Cabrerizo-Vilchez

Department of Applied Physics, Biocolloid and Fluid Physics Group, University of Granada, Granada 18071, Spain

A study of mixed monolayers at the air-aqueous solution interface has been carried out. These kind of systems could be considered the most simple model to get molecular information and hence, their study is the first step to undertand the properties and the behavior of macromolecular assemblies.
Surface pressure-molecular area isotherms of mixed monolayers of lecithin with chenodeoxycholic, deoxycholic and cholic acid spreaded at aqueous solution/air interface have been recorded over a wide range of pH (2 to 12) and different temperatures (25-40°C).
The monolayer properties of bile acids are influenced by the subphase pH: important changes occur in both the electric characteristics and the stability of the monolayers. Also, the behavior of these films is modified with the temperature. These changes are attributed to a decrease in the hydrophobic forces between the molecules forming the monolayer and a progressive dehydration of the polar groups as temperature increases.
The addition of lecithin to a bile acid film produces molecular condensation. Consequently the mixed monolayers are more stable than the simple bile acid films. This molecular association is mainly due to the attractive forces of the van der Waals type between the hydrophobic regions. The different behavior observed is largely influenced by subphase pH, temperature and surface pressure values and, therefore, by the orientation of the molecules with respect to the liquid surface.

The understanding of the behavior of mixed monolayers at the aqueous solution-air interface is a step more towards a general theory for the cholelithiasis processes.

Investigating the properties of mixed monolayers is of great interest, because it enables to gain knowledge about the interactions between the monolayer components. These interactions play an important role in some biological processes.

The interaction between amphiphilic molecules at the air-aqueous solution interface is controlled by three different forces: electrostatic, hydrophobic and hydration forces (1). The electrostatic and hydration contributions depend on subphase pH, whereas the hydrophobic and hydration forces are dependent on temperature.

The behavior of the simple monolayers formed by lecithin or bile acids as a function of pH and temperature has been studied in previous papers (2, M.J.Gálvez-Ruiz and Cabrerizo-Vílchez, *Colloids and Surfaces*, in press). The behavior of the simple monolayers of bile acids studied as a function of the subphase pH is to be expanded when the pH values are shifted away from the acid pK_a value (M.J. Gálvez-Ruiz and M. A. Cabrerizo-Vílchez, *Colloids and Surfaces*, in press) and the lecithin simple monolayers become more expanded when the pH increases (2) due, in both cases, to the increase of repulsive electrostatic forces between the polar groups.

Also, when the temperature is increased all the simple monolayers become more expanded.

In this work we study the behavior of mixed monolayers formed by lecithin and bile acid (chenodeoxycholic, deoxycholic and cholic acids) as a function of pH and temperature. Those compounds are involved in the cholelithiasis processes. An attempt has been made for explaining the mechanisms involved in gallstone formation on the basis of the interactions in mixed monolayers.

Materials and Methods

Analytical grade chenodeoxycholic and deoxycholic acids were from Serva and the other monolayer components (L-α-phosphatidylcholine and cholic acid) were from Sigma. The spreading solvent was a mixture of n-hexane/ethanol 4:1 (v/v) (Merck A.R. grade), and 0.05% amylalcohol was added to improve spreading (3). A Britton-Robinson buffer was used in all the experiments since it is suitable for the preparation of solutions in a wide pH range. This buffer consists of a solution containing acetic, phosphoric and boric acids, and its pH can be adjusted between 2 and 12 by addition of adequate amounts of sodium hydroxide. All these compounds were supplied by Merck with analytical quality.

Surface pressure-molecular area isotherms were

performed using the previously described (2) Langmuir method with a computer-controlled "Lauda Filmwaage" balance.

Results and Discussion

Mixed Monolayers Characteristics of Chenodeoxycholic Acid and Lecithin. In Figure 1 we observe that all mixed monolayers formed by different proportions of lecithin and chenodeoxycholic acid show an intermediate behavior between those and the simple monolayers. This behavior is due to the important condensing effect of lecithin over chenodexycholic acid (more expanded than the lecithin monolayers) on the whole pH (2-12) and temperature (25-40°C) range. Even, when the parameters increase, practically all isotherms, obtained by compressing of the mixed monolayers, show the liquid expanded (LE)-liquid condensed (LC) phase transition, in opposition with the behavior of the bile acid simple monolayers at high pH and temperature values. Then, this condensing effect is responsible for the increase of the mixed monolayer stability at the aqueous solution-air interface.

The interaction between amphiphilic molecules at the air-aqueous solution interface will be controlled by the contribution of electrostatic, hydrophobic and hydration forces.

For this mixed system, we examine the influence of these forces, analyzing the effect of the subphase pH and temperature.

When the pH is 2 or 6, and the surface pressure value is 5 mNm^{-1}, we observe (in Figure 2) positive deviations from the law of ideal mixing. Such positive deviations indicate that the polar groups of the molecules interact more strongly by repulsive electrostatic forces with each other than with the more hydrophobic part of the molecules. Also, the hydration forces can result in the formation of H-bonding due to the presence of carboxyl and hydroxyl groups in the molecules.

If the pH is 4 and the surface pressure is 5 mNm^{-1}, the deviations are negative. Under those experimental conditions the predominant forces are hydrophobic. This could be explained because of the chenodeoxycholic acid pK$_a$ value is around pH 4 (4) and at this pH value the molecules are not deprotonate.

When the pH is 8, at the same surface pressure value, we have found negative deviations at high molar fraction of chenodeoxycholic acid and positive deviations for the monolayers containing higher proportion of lecithin.

At basic pHs (10-12) the deviations from the law of ideal mixing are negative. These results could be explained by the progressive disappearance of the hydration forces as the pH increases since, under these pH values the molecules are ionized and the deviations should

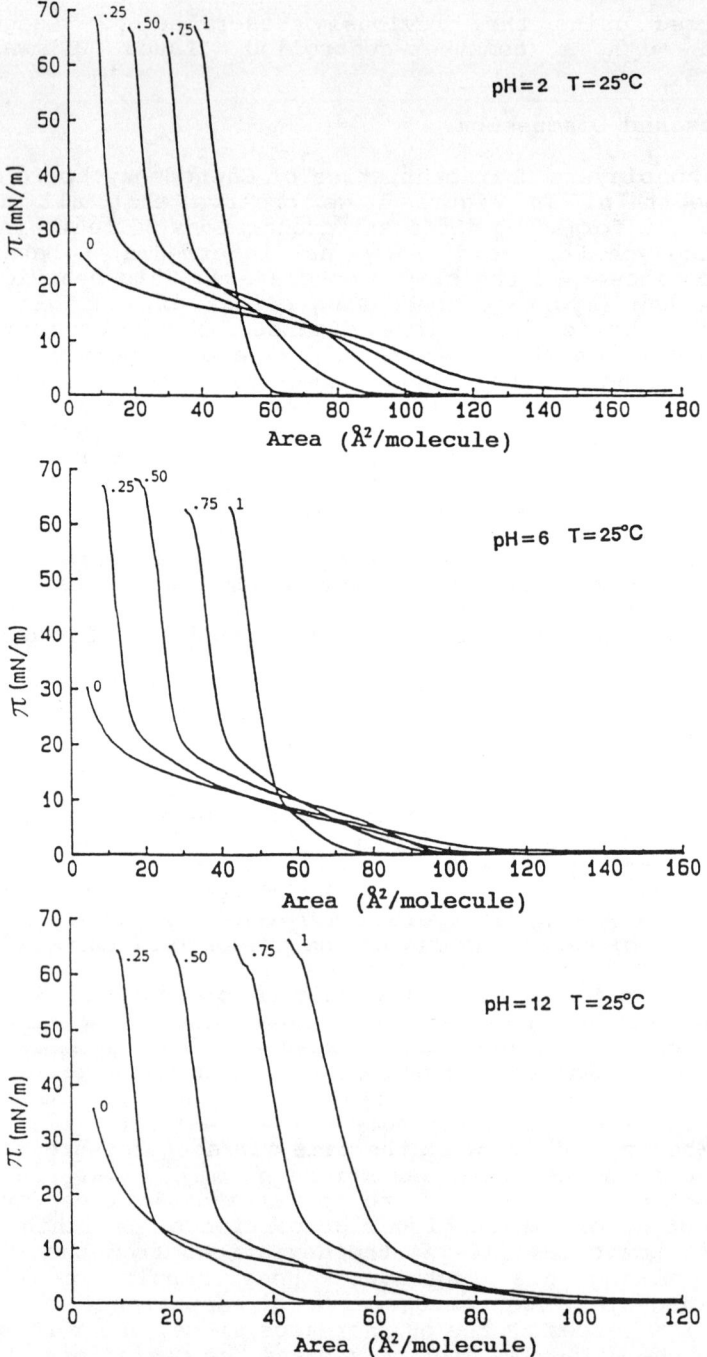

Figure 1. Surface pressure-area isotherms for monolayers containing lecithin and chenodeoxycholic acid in different proportions. Lecithin mole fraction is indicated. *Continued on next page*

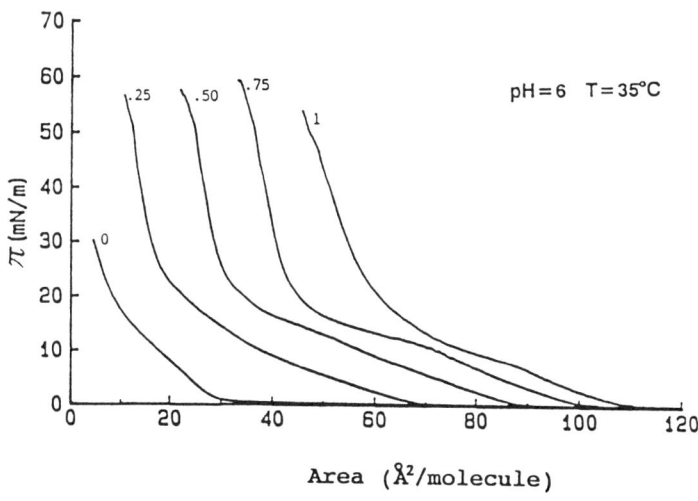

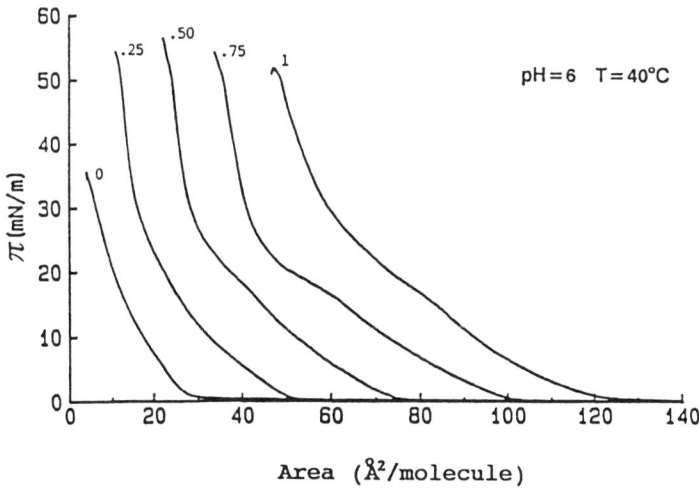

Figure 1. Continued

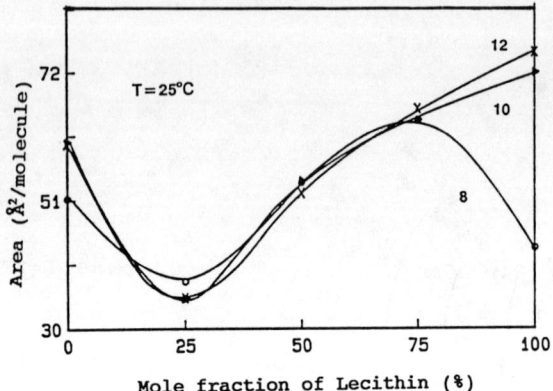

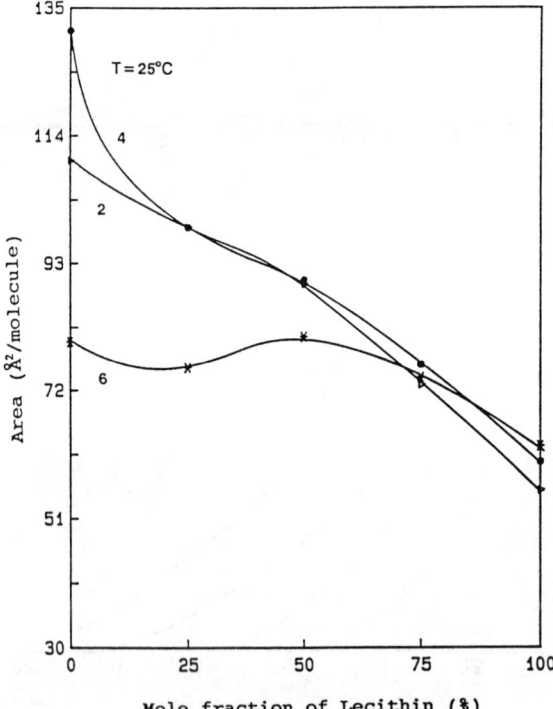

Figure 2. Molecular area as a function of the lecithin mole fraction in mixed lecithin-chenodeoxycholic acid monolayers, at π=5 mNm^{-1}. Varying pH at fixed temperature or varying temperature at fixed pH, as indicated.

Continued on next page

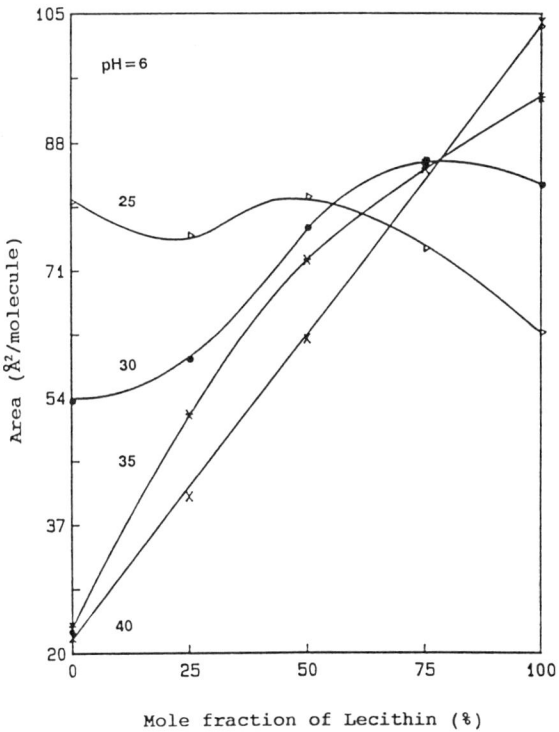

Figure 2. Continued

be positive. Nevertheless, we have to take in account the possibility of solubilization of the chenodeoxycholic acid molecules in the subphase as basic pH values (8-12). In order to study this, we have obtained the limiting molecular areas, A_0, by extrapolation of the steep linear portions of the π vs. A curves to $\pi = 0$. In Table I these values for three mixed monolayers, at 25°C and at different pH values are shown.

Table I. Limiting Molecular Area Values, A_o ($Å^2$/molecule), for Mixed Monolayers of Lecthin and Bile Acids at 25°C

Mixed Monolayer/pH		2.0	4.0	6.0	8.0	10.0	12.0
25% Lecithin- 75% Bile acid	Chenodeoxy.	14.5	19.9	18.9	18.4	17.1	16.5
	Deoxychol.	15.1	16.0	17.1	17.7	17.0	17.0
	Cholic	15.2	15.0	15.7	15.1	16.9	19.6
50% Lecithin- 50% Bile acid	Chenodeoxy.	27.3	28.1	31.8	30.9	31.6	30.3
	Deoxychol.	28.2	29.7	30.7	30.9	32.2	31.0
	Cholic	28.0	29.0	29.1	28.7	30.3	29.6
75% Lecithin- 25% Bile acid	Chenodeoxy.	39.4	40.0	44.0	47.7	47.0	46.9
	Deoxychol.	40.2	43.2	43.8	45.0	45.5	44.6
	Cholic	40.6	41.5	42.1	41.7	42.4	43.4

We observe that the isotherms are shifted toward lower A_0 values with increasing pH above the 8 value. This effect is more pronounced as the proportion of bile acid increases. It should be noted that behavior characterized by contraction of the areas as a function of pH has been verified by Tomoaia-Cotişel et al. (5) for other compounds (fatty acids) and this is attributed to the dissolution of the ionized acids in the subphase.

Studying the effect of temperature, we observe (in Figure 2), at surface pressure value 5 mNm^{-1}, and 25, 30 and 35°C, positive deviations from the law of ideal mixing. These deviations disappear at 40°C. Also, in Table II, the limiting molecular area values, A_0, at pH = 6.00 and as a function of temperature are shown.

A temperature rise favours the following effects: (i) an increase in the mobility of the hydrophobic parts and, therefore, a decrease in the attractive energies with the increased area that accompanies the temperature increase (6), (ii) a partial or total dehydration of the polar groups with a decrease in the number of water molecules bound to the monolayers and the resulting condensation of the surface phase (6), and (iii) the dissolution of the ionized molecules in the subphase.

Then, for these systems, when the temperature increases to 35°C the most important effect is that the attractive hydrophobic forces decrease which gives rise to a more expanded character of mixed monolayers. At 40°C, the more noticeable effects will be the dehydration of the polar groups in the monolayers and the solubilization of the molecules in the subphase.

When the compression of the monolayers is high, at 30 mNm^{-1}, we cannot observe any deviations from the law of ideal mixing on the whole pH and temperature range. In the condensed state, attractive and repulsive electrostatic intermolecular forces are compensated.

Table II. Limiting Molecular Area Values, A_o (Å^2/molecule), for Mixed Monolayers of Lecithin and Bile Acids at pH = 6.0

Mixed Monolayers/T(°C)		25.0	30.0	35.0	40.0
25% Lecithin- 75% Bile acid	Chendeoxy. Deoxychol. Cholic	18.9 17.1 15.7	20.3 18.2 22.9	22.3 17.8 20.6	20.1 25.3 27.6
50% Lecithin- 50% Bile acid	Chendeoxy. Deoxychol. Cholic	31.8 30.7 29.1	32.5 33.2 32.7	34.4 35.7 33.7	31.9 41.1 44.4
75% Lecithin- 25% Bile acid	Chenodeoxy. Deoxychol. Cholic	44.0 43.8 44.1	45.4 46.2 48.7	46.9 49.3 47.3	47.9 47.9 59.2

Mixed Monolayers Characteristics of Deoxycholic Acid and Lecithin. In Figure 3, the isotherms obtained by compressing mixed monolayers formed of deoxycholic acid and lecithin, are shown. These are similar to those found for the last system (chenodeoxycholic acid-lecithin) on the whole pH and temperature range. We are interested in comparing the behavior of this system with the last one, because the only difference is that the molecules of chenodeoxycholic and deoxycholic acids have a hydroxyl group in a different position (trans- and cis-, respectively, with respect to the other hydroxyl group).

At acidic pH values, we observe the same type of deviations from the law of ideal mixing that we found previously for the chenodeoxycholic acid-lecithin system.

However, at basic pH values (8-12) all the deviations are positive. In this case, the presence of two hydroxyl groups in cis-position provokes a increase of the repulsive electrostatic forces and also, the hydration forces appear to be present resulting in the formation of H-bonding in opposition with the chenodeoxycholic acid molecules, where the hydroxyl groups are in trans-position and the formation of H-bonding is less favoured.

The temperature effect on the deoxycholic acid-lecithin mixed monolayers is the same as that previously

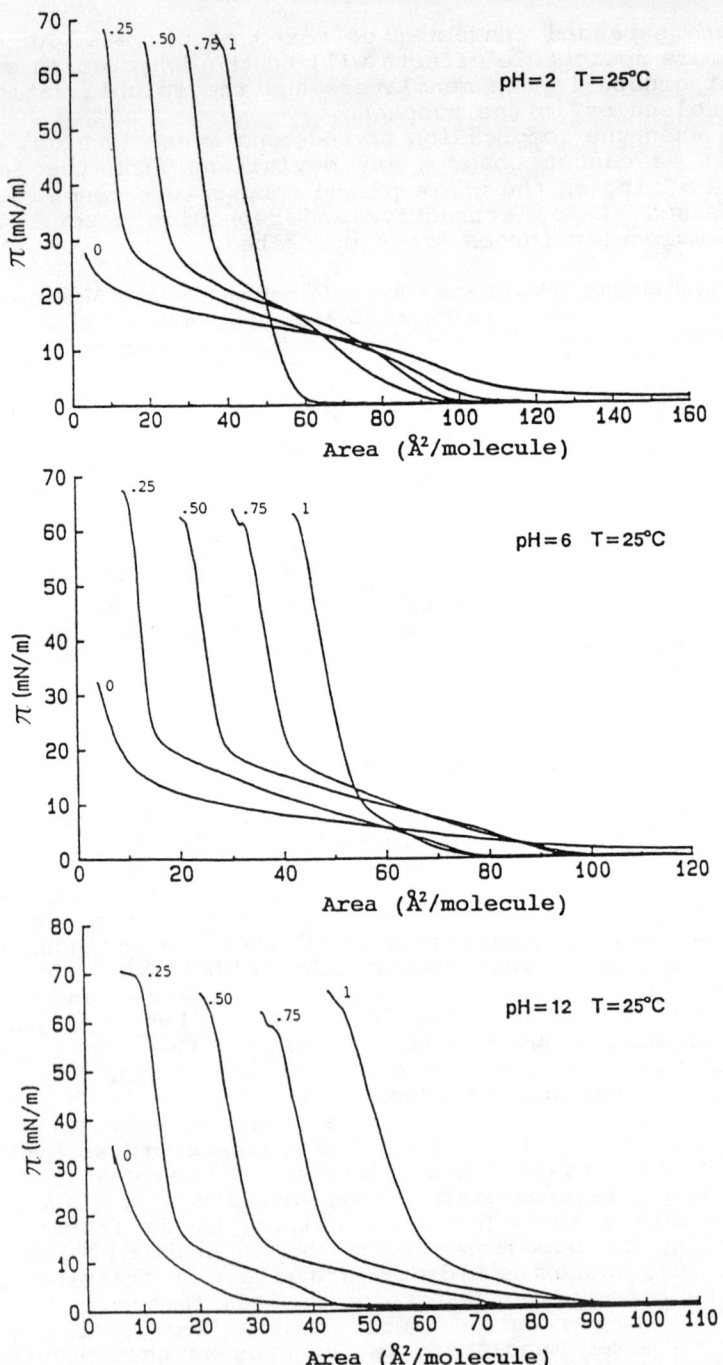

Figure 3. Surface pressure-area isotherms for monolayers containing lecithin and deoxycholic acid in different proportions. Lecithin mole fraction is indicated. *Continued on next page*

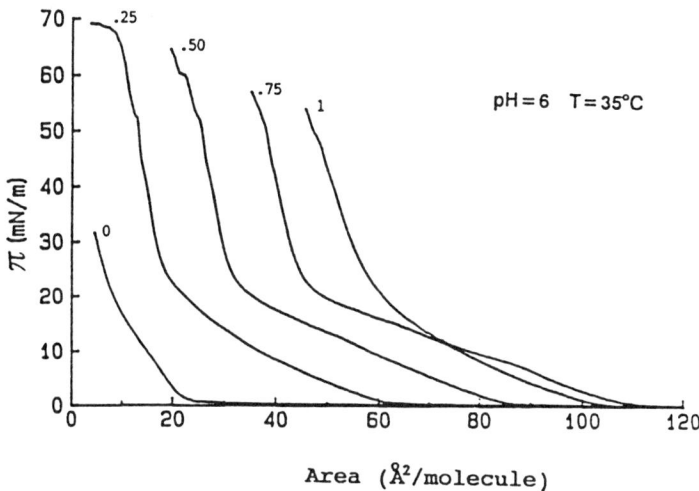

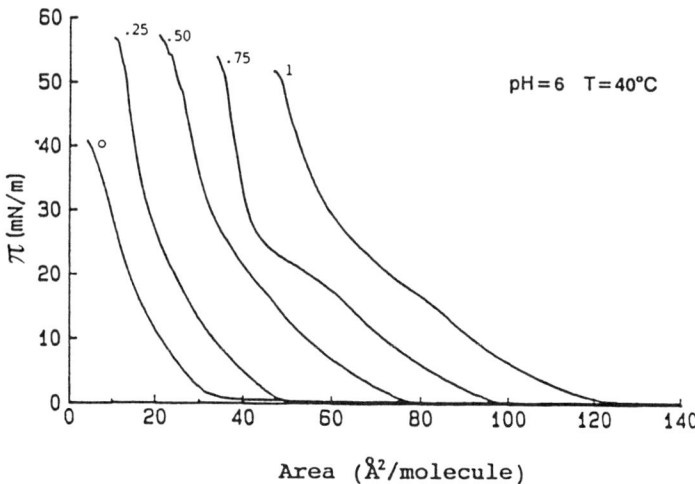

Figure 3. Continued

described for the chenodeoxycholic acid-lecithin system.

Then, we can conclude that this different position of the hydroxyl groups, practically has no influence on the possible interactions with the lecithin molecules except at high pH values. The condensing effect of the lecithin over deoxycholic acid can be seen in Figure 4.

Mixed Monolayers Characteristics of Cholic Acid and Lecithin. In Figure 5, the isotherms obtained by compressing monolayers formed by cholic acid and lecithin mixed in different proportions are shown. It can be observed that mixed monolayers of these compounds show a somewhat different behavior compared with that shown by the previous systems with other bile acids.

The main difference is that at high pH and temperature values and when the monolayers have a high proportion of cholic acid, the monolayers are more condensed and the EL-CL phase transition disappears. Previously, we discussed (M.J. Gálvez-Ruiz and M.A. Cabrerizo-Vílchez, *Colloids and Surfaces*, in press) that under those experimental conditions the cholic acid molecules are dissolved in the subphase. This is due to the presence of three hydroxyl groups in these molecules. However, the presence of a more hydroxyl group in the molecules does not have an important influence on the interactions with the lecithin molecules, because studying the law of ideal mixing, we have found the same deviations as for the previous systems. At lower pH values (2-8) the deviations are positive and at basic pH values (10-12) the deviations are negative (Figure 6). These deviations are explained in the same terms as before. The only difference is, that the positive deviations are more pronounced, even at pH 4 these do not disappear where the molecules are not deprotonate. Consequently, the hydration forces can be more significant in this case, due again to the presence of a third hydroxyl group in the molecules of the cholic acid.

In Figure 6 we observe a light decrease of the positive deviations when the temperature increase. Unlike the previous system, these deviations no disappear at 40°C due to the presence of a third hydroxyl group in the cholic acid molecules. Again, the possibility of molecules' solubilization in the subphase must be taken into account.

Comparing the behavior of the three mixed systems, the most noticeable result is, that the same character for the interactions between molecules of lecithin and bile acid has been found. When the pH increases from 2 to 12, the positive deviations become negative and an increase of temperature between 25° and 40°C brings about the progressive disappearance of the positive deviations. Then, the condensing effect of the lecithin over bile acid molecules is very similar for the three acids studied. The

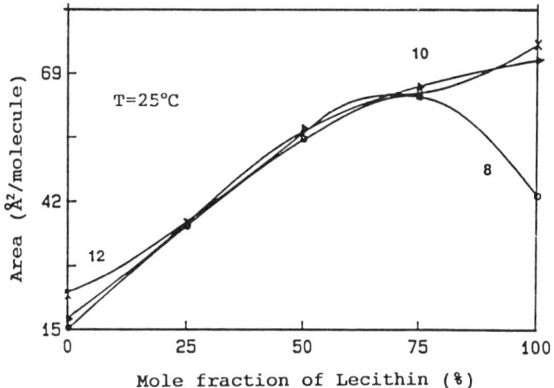

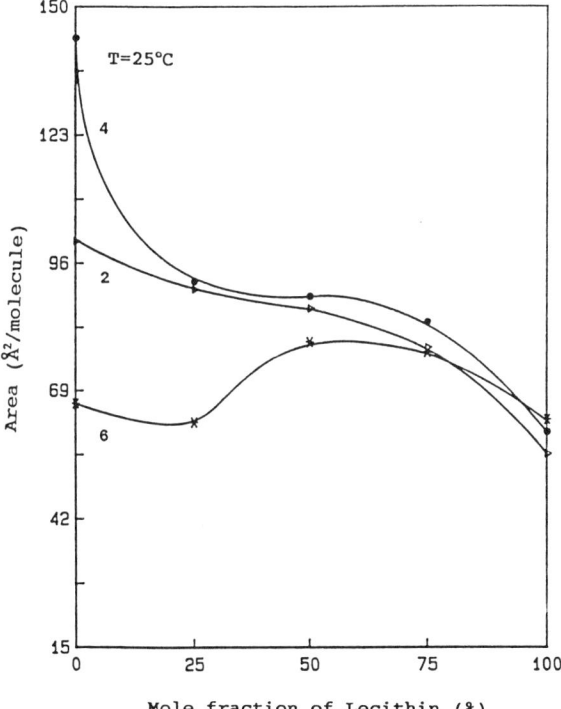

Figure 4. Molecular area as a function of the lecithin mole fraction in mixed lecithin-deoxycholic acid monolayers, at π=5 mNm^{-1}. Varying pH at fixed temperature or varying temperature at fixed pH, as indicated.

Continued on next page

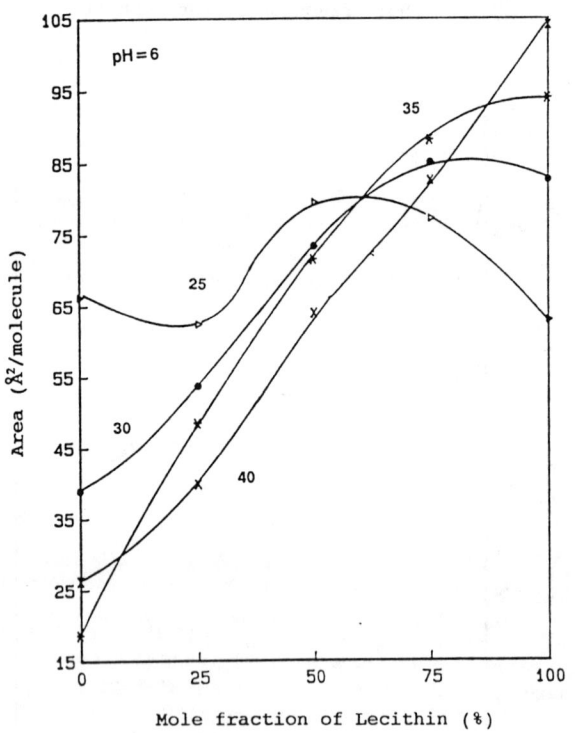

Figure 4. Continued

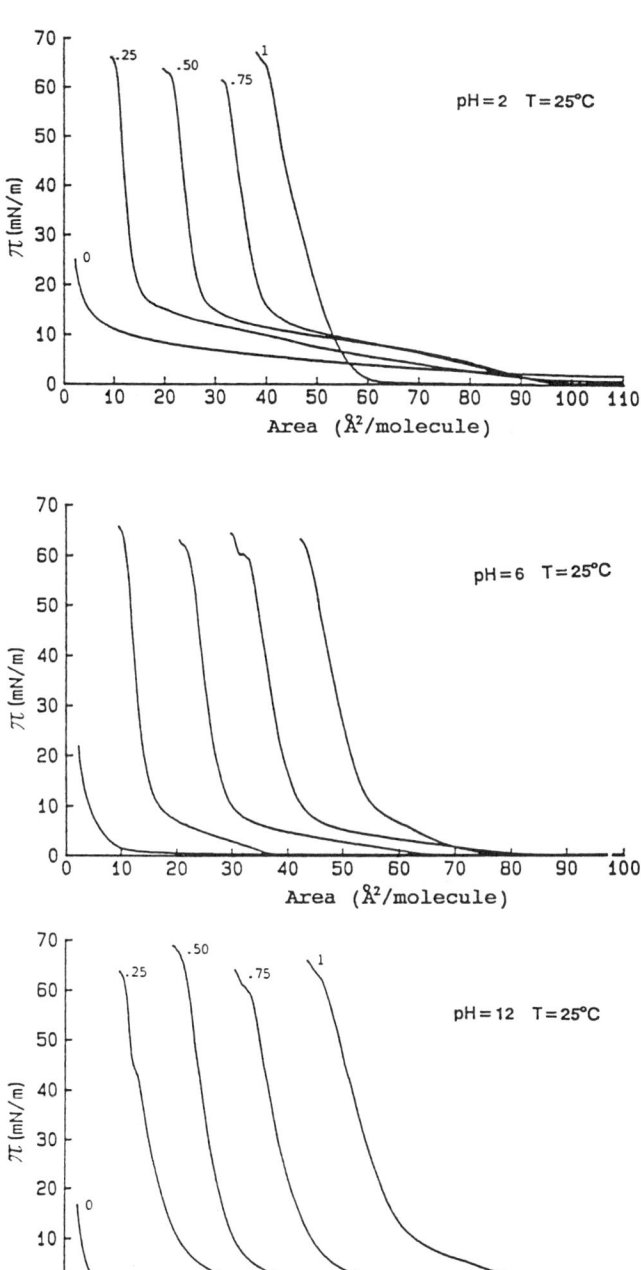

Figure 5. Surface pressure-area isotherms for monolayers containing lecithin and cholic acid in different proportions. Lecithin mole fraction is indicated.

Continued on next page

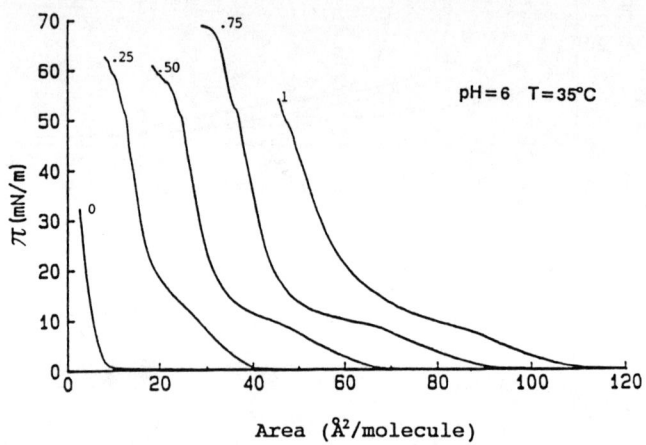

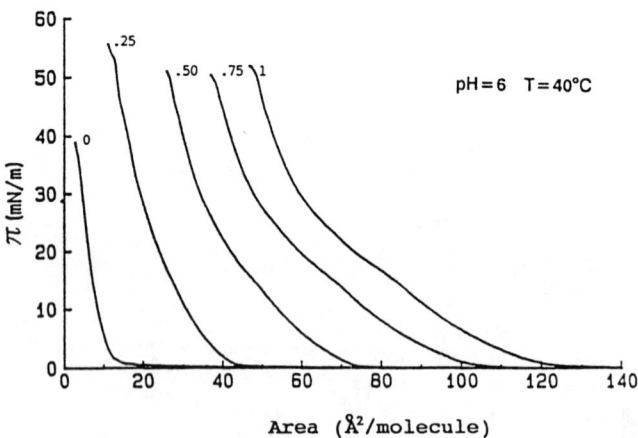

Figure 5. Continued

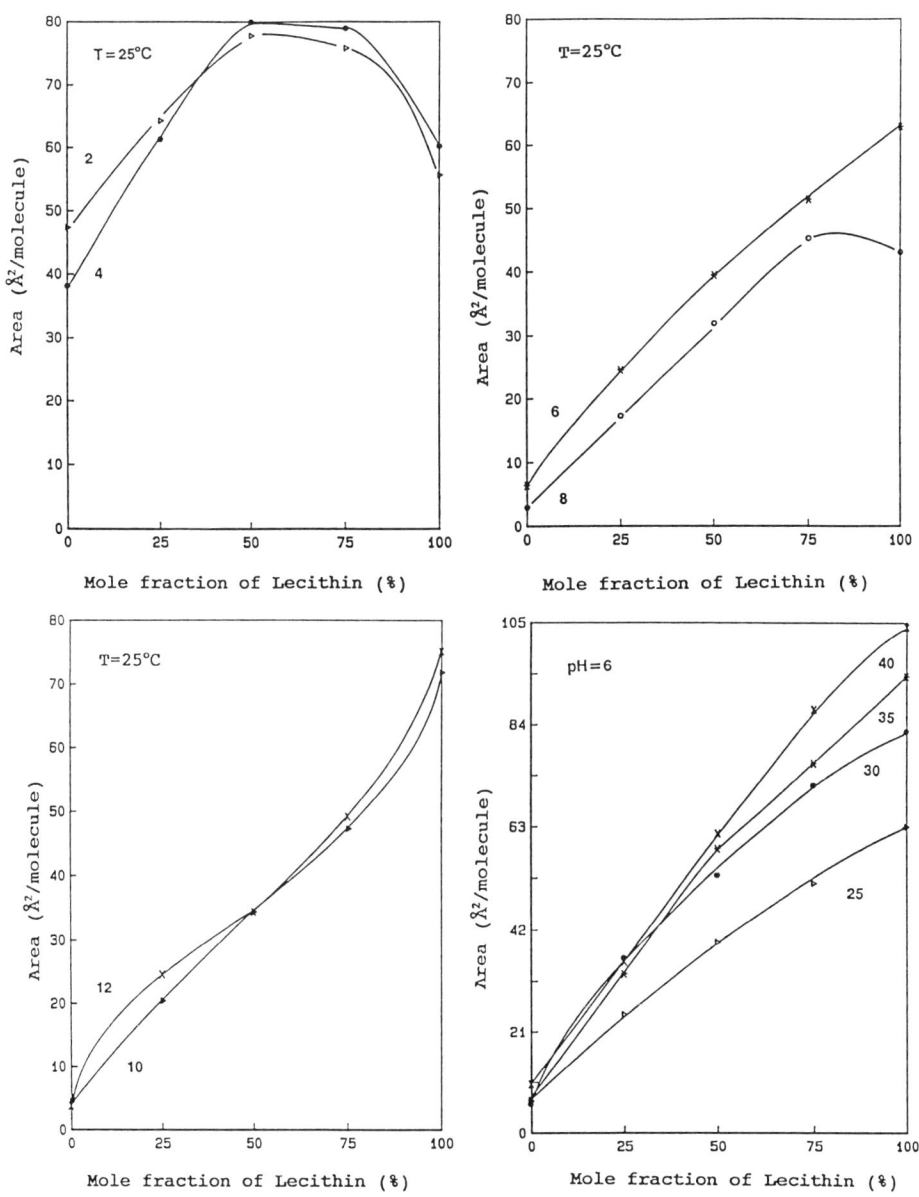

Figure 6. Molecular area as a function of the lecithin mole fraction in mixed lecithin-cholic acid monolayers, at π=5 mNm^{-1}. Varying pH at fixed temperature or varying temperature at fixed pH, as indicated.

presence of different number of hydroxyl groups and/or their different positions in the bile acid molecules, have no influence on the interactions with lecithin molecules. Also, in Tables I and II we observe that the limiting molecular area values for the same proportion of lecithin are very similar.

Conclusions

All mixed monolayers formed by different proportions of lecithin and bile acid show an intermediate behavior between those of the simple monolayers. This is due to the important condensing effect of lecithin over bile acids on the whole pH (2-12) and temperature (25-40°C) range.

The mixed monolayers are more stable than the simple bile acid films.

The interactions between lecithin and bile acid molecules, at the aqueous solution-air interface, depend on electrostatic, hydrophobic and hydration forces.

The hydration and repulsive electrostatic forces seem to be dominant. This conclusion is supported by the deviations, from the ideal mixing, obtained at different pH and temperatures.

Solubilization of the bile acid molecules in the subphase has to taken into account, at high pH and temperature values, mainly for the cholic acid molecules.

Attractive and repulsive electrostatic intermolecular forces are compensated, at high surface-pressure values (LC state), for all monolayers.

The presence of different number of hydroxyl groups and their different positions in the bile acid molecules exert practically no influence on the interactions with lecithin molecules, but we observe an increase of these when the cholic acid molecules are present in the films at low pH.

Literature Cited

(1) A.D. Sorokina, N.D. Yanopolskaya and G.A. Deborin, *Bioelectrochemistry and Bioenergetics*, 23 (1990) 271 A Section of *J. Electroanal. Chem.*, 298 (1990)
(2) M.J. Gálvez-Ruiz and M.A. Cabrerizo-Vílchez, *Colloid and Polym. Sci.*, 269 (1991) 77
(3) J. Miñones-Trillo, S. García-Fernández and P. Sanz-Pedrero, *J. Colloid Interf. Sci.*, 26(4) (1968) 518
(4) A. Fini, A. Roda and P. De María, *Eur. J. Med. Chem.-Chim. Ther.* 17(5) (1982) 465
(5) M. Tomoaia-Cotişel, J. Zsakó, A. Mocamo, M. Lupea and E. Chifu, *J. Colloid Interf. Sci.* 117(2) (1987) 464
(6) C. Gabrielli, M. Puggelli, E. Ferroni, G. Carubia and L. Pedocchi, *Colloids and Surfaces*, 41 (1989) 1

RECEIVED September 24, 1991

THREE-DIMENSIONAL SYSTEMS

Macromolecular systems in three dimensions include structures that form spontaneously in solution, such as micelles, liposomes, tubules, and microspheres. The medium of interest is often aqueous, particularly when dealing with biological systems. Biological macromolecular assemblies are extremely complex because they are multi-component systems. For example, the bilayer membrane found in the cell membrane can contain phospholipids, cholesterol, enzymes, and proteins. To understand the metabolism of cells it is necessary to understand the mechanisms of mass transport through the cell membrane. Constituents in the cell membrane that protrude from the cell membrane surface play an important role in surface recognition of molecules and other cells. Within the cell other complex structures are present. Even in the red blood cell, for example, where normally internal structures are not present, a generic difference in the human hemoglobin molecule can cause the hemoglobin molecules to reversibly polymerize into macromolecular structures. The consequence to oxygen transport from the cell and the rheological behavior of a red blood cell suspension are drastic when reversible polymerization of hemoglobin takes place. One is faced with the same problem as with ultrathin films: What are the properties of the collective if the properties of the individual molecules are known?

Chapter 14

Photoinduced Morphological Changes in Plasmalogen Liposomes Using Visible Light

Valerie C. Anderson and David H. Thompson[1]

Biomolecular Materials Research Center, Department of Chemical and Biological Sciences, Oregon Graduate Institute of Science and Technology, Beaverton, OR 97006

Morphology changes of semi-synthetic plasmalogen liposomes have been triggered using a hydrophilic phthalocyanine dye, aluminum chloro phthalocyanine tetrasulfonate, AlClPcS, as sensitizer in the presence of visible light and oxygen. Irradiation ($\lambda > 630$ nm; 37 °C) of unilamellar, air-saturated liposomes composed of 1-alk-1'-enyl-2-palmitoyl-sn-3-phosphocholine/dihydrocholesterol (DHC; 30 mol%) results in release of encapsulated glucose which exceed 100% after 90 minutes of photolysis. TLC indicates formation of new photo-products tentatively identified as oxidized lipid species. Electron microscopy confirms significant photoinduced changes in liposome size and lamellarity. The potential application of this system as an in vivo drug delivery agent is discussed.

Targeted drug delivery methods based on triggered release via an external stimulus are currently undergoing intensive research and development efforts. Polymeric matrices (1) and vesicles, comprised of natural or synthetic phospholipids both represent important classes of delivery vehicles for controlled release of drugs. The ease of vesicle preparation and selection of particle size (2), high encapsulation efficiencies (3), and compatibility with living tissues makes them particularly well suited for use in targeted drug delivery. In addition, recent studies have been shown that incorporation of therapeutic agents such as anthracycline drugs (4) within liposomes can reduce their side effects and improve the pharmacokinetics of delivery to the target site (5). Thus, use of liposomes to encapsulate drugs can also provide a means of buffering the effects of the therapeutic agent itself.

Although encapsulation of pharmaceuticals in liposomes can produce

[1]Corresponding author

dramatic improvements in patient response, a serious limitation to their widespread use is the lack of a rapid triggering mechanism by which the drug is completely released at a predetermined site. Phototriggered delivery systems can, in principle, provide a level of pharmacological control that is not easily obtained with many biochemical or physical targeting methods (6) due to the specificity of light absorption and the ability to modulate intensity, pulse frequency, and duration of directional optical sources. While several reports of vesicle contents release using UV or visible light have appeared (7), few of these systems have used both biologically compatible lipids and excitation by light frequencies having tissue penetration depths exceeding 1 mm (8).

We have been exploring ways to trigger release of materials from liposomes utilizing visible/near IR light as an extended application of photodynamic therapy (PDT). Photodynamic therapy of diseased sites typically involves selective retention of a tumor-localizing photosensitizer (such as hematoporphyrin derivative or Photofrin II) followed by visible irradiation of the affected tissue (9, 10). Absorption of light by the photosensitizer and subsequent reactions with dissolved oxygen results in cell death and tissue necrosis (11). The effectiveness of PDT is limited by several factors. First, the photochemotherapeutic agent must interact with diseased tissue such that selective localization in affected tissue at concentrations which are clinically relevant is possible without causing additional adverse effects (12). Secondly, delivery of light of appropriate wavelength to the diseased tissue must be possible while minimizing intensity losses due to scattering or preferential absorption by other cellular components (8). High efficiency coupling to optical fibers now makes it possible to deliver a therapeutically relevant intensity of visible/near IR laser light to many previously inaccessible sites.

Our approach to targeted drug delivery combines the advantages of liposomal encapsulation of pharmaceuticals and biocompatibility with PDT. The triggering principle involves absorption of near IR or visible light by a sensitizer to effect lipid decomposition thereby causing a breakdown of the liposome bilayer and release of encapsulated drugs. We anticipated that photoinduced plasmalogen vesicle leakage, based upon cleavage/modification of the labile sn-1 vinyl ether linkage, would be well suited as a triggering mechanism for therapeutic applications based on earlier reports of the photooxidative instability of plasmalogen lipids (13,14). This report describes the use of AlClPcS, a water-soluble phthalocyanine localizing in the aqueous core of the vesicle, to produce photomodifications of plasmalogen liposomes via photodynamic type mechanisms. The phthalocyanines were chosen as sensitizers due to their favorable absorption and photooxidative stability (15). The structures of the semi-synthetic plasmalogen and photosensitizer, AlClPcS, used in this study are shown below.

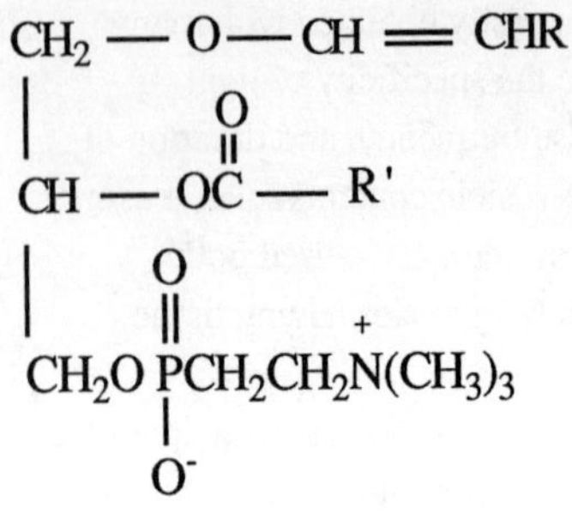

PlasPPC

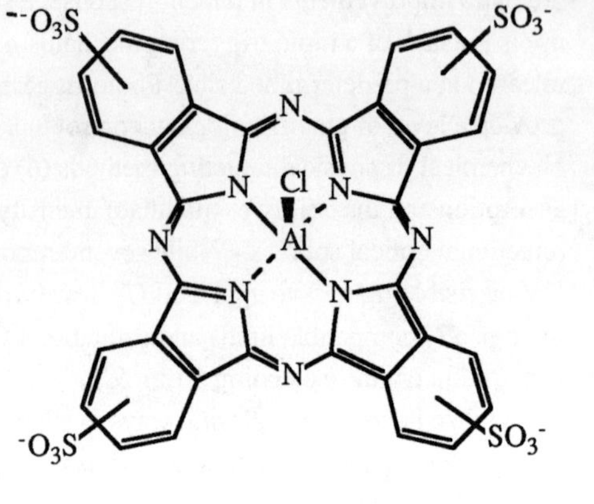

AlClPcS

EXPERIMENTAL

Melting points were measured on a Mel-Temp II capillary melting point apparatus and are corrected. AlClPcS (Midcentury, Chicago, IL), and palmitic anhydride (American Tokyo Kasei, Portland, OR) were used as received. NADP, ATP, hexokinase, Bakers yeast glucose-6-phosphate dehydrogenase, and D-glucose were obtained from Sigma (St. Louis, MO). Dihydrocholesterol (DHC; Aldrich (Milwaukee, WI), m.p. 144-5 °C (lit. (16) m.p. 140-142 °C)), 1,2-dipalmitoyl-sn-glycero-3-phosphocholine (DPPC; Avanti Polar Lipids, Alabaster, AL) and bovine heart phosphocholine (BHPC; Avanti) were single spots by silica TLC (DHC: 90:10 CH_2Cl_2/MeOH; DPPC and BHPC: 65:35:6 $CHCl_3$/MeOH/NH_4OH) and were used without further purification. 1-alk-1'-enyl-2-palmitoyl-sn-glycero-3-phosphocholine (PlasPPC) was prepared by alkylation of 1-alk-1'-enyl-sn glycero-3-phosphocholine following alkaline deacylation of BHPC (17). Lipids were stored at -70 °C under argon until use. $CHCl_3$ was dried over P_2O_5 and freshly distilled prior to use. Silica for column chromatography was pre-washed in 1:1 $CHCl_3$/MeOH and dried at 110 °C.

Steady-state absorption measurements were recorded on a Hewlett-Packard 8452A UV-Vis diode array spectrophotometer. Differential scanning calorimetry measurements were performed on a Perkin Elmer DSC7 equipped with Intercooler for subambient operation. Samples were typically heated at 2-5 °C/min over the appropriate temperature range. Samples for electron microscopy were prepared by coating copper grids with extruded liposomes and staining with 2% ammonium molybdate in 20 mM Tris buffer, pH 8. Samples were vacuum dried for 3 hours and then visualized by transmission electron microscopy.

Preparation of PlasPPC liposomes. 10 mM liposome suspensions were prepared by dissolving appropriate amounts of PlasPPC and either DHC or DPPC in 1-2 mL $CHCl_3$ and evaporating to a dry film using a stream of Ar. Following complete solvent removal under high vacuum for 1-2 hr, 3 ml of 20 mM Tris/100 mM NaCl buffer, pH 8, containing 0.3 M glucose and 200-250 µL of a 40 µM stock AlClPcS solution was added. This mixture was extensively vortexed at 55-60 °C, resulting in a pale blue suspension that was hydrated fully via five freeze-thaw cycles. The emulsion was then extruded ten times through stacked 1000 and 800 A Nucleopore filters at 48 °C using a thermostatted Lipex Biomembranes Extruder. Unentrapped sensitizer and glucose were removed via gel filtration on a buffer-equilibrated Sephadex G-25 column. Concentration of AlClPcS in the final suspension varied from 3-7 µM.

Experiments requiring the use of NaN_3 or D_2O within the liposomes were prepared in the same manner. pD of the buffer was adjusted to 8.4 using an Orion SA 720 pH meter (pD= pH + 0.4 (*18*)) and DCl. All manipulations of sensitizer-incorporated liposomes were conducted in the dark or under reduced lighting conditions.

Photoinduced Glucose Release. Stirred, thermostated solutions of liposomes were irradiated using a 100 W tungsten lamp whose output was first filtered through a 10 cm cell of water, followed by a Corning OG-630 cut-off filter, and then focused to fill a 1.5x0.5 cm^2 area of a standard 3 mL cuvette. Light intensity at the cuvette surface was measured with a Coherent 210 power meter and found to be 20 ($\pm$ 5) mW/ cm^2. The temperature inside the photolysis cell was monitored periodically with a calibrated Fluke 51 K/J thermocouple and found to be constant to within $\pm$ 0.2 °C.

Contents leakage at 37 °C was monitored by withdrawing 25 µL of liposomal solution at various times during the course of the photolysis. Released glucose was quantitated using the enzymatic glucose oxidase assay (*19*) which involves monitoring production of NADPH (λ_{max} 340 nm) that occurs by reduction of NADP during the enzymatic phosphorylation of released glucose. The total amount of entrapped glucose was determined following disruption of liposomes with Triton X-100.

Thin Layer Chromatography. Liposomes (75 µL) were added to 150 µL of 2:1 $CHCl_3$/MeOH and vortexed to mix. The organic layer was removed, spotted on Baker silica 1B-F plates, and developed in 65:35:6 (v/v/v) $CHCl_3$/MeOH/NH_4OH.

Spots were visualized after charring plates dipped in 20% H_2SO_4.

RESULTS

Differential Scanning Calorimetry. Table I gives the main gel-to-liquid-crystalline transition temperature, T_c, of DPPC, PlasPPC, and mixed PlasPPC/DPPC or PlasPPC/DHC liposomes. Several aspects of the data are notable. First, pure PlasPPC liposomes show a T_c slightly lower than that observed for the diacyl analogue, DPPC. This result contrasts with previous DSC studies of ethanolamine plasmalogens (20) where varying chain lengths and unsaturation sites yielded T_c's that were indistinguishable from those of their diacyl counterparts. A comparable study of phosphatidylcholine plasmalogens has not appeared. It is possible that the slight transition temperature depression observed here is a result of different packing and molecular conformations of the PCs induced by the vinyl ether linkage of the sn-1 chain. Such differences between plasmenylcholines and PCs have been reported recently(21).

Secondly, addition of DPPC (12 mol%) has little effect on T_c of PlasPPC liposomes. Apparently, the structural similarity of DPPC and PlasPPC precludes any calorimetrically discernible regions of immiscibility (i.e., phase-separated domains) of the two lipids. In contrast, 30 mol% DHC increases the transition temperature by 7 °C. This result is consistent with reported increases in T_c for DPPC: 30 mol% cholesterol mixtures of 5-10 °C (22,23). The effect of steroids on vesicle T_c's has been ascribed to a perturbation of the packing of the alkyl chains induced by more solid-like cholesterol regions within the bilayer (24).

Mixed plasmalogen liposome thermograms indicate that relative to pure DPPC, the enthalpy of the transition is affected by both PlasPPC and, more severely, by 30 mol% DHC incorporation (data not shown). Increases in the mole ratio of DPPC within PlasPPC:DPPC mixtures resulted in an increase in peak width at half-height of from 2 to >5 °C. The broad, non-cooperative nature of the transition may result from the polydispersity of molecular species which comprise the semi-synthetic plasmalogen liposomes (25). (PlasPPC prepared by the method described above has been shown to contain a mixture of 1-alk-1'-enyl lipid chains of which 96% are saturated, with 68% of these being $C_{16:0}$ (26).) Incorporation of 30 mol% DHC reduced the magnitude and broadened the PlasPPC transition even more, making the transition unobservable in some samples. It has been postulated that in diacyl PCs this effect may be due to non-cooperative interactions of lipid and cholesterol-rich domains (23).

Finally, it should be noted that incorporation of glucose and AlClPcS at the concentrations used here did not result in significant variations in T_c from that measured in unloaded vesicles. It is reasonable to assume, therefore, that macroscopic molecular packing is not appreciably disturbed by solubilization of glucose and sensitizer within the core.

Photoinduced Glucose Release. Encapsulation of AlClPcS in 800 A diameter PlasPPC liposomes resulted in a small shift of the absorption maximum from 674 nm (20 mM Tris, pH 8; lit. maximum 675 nm (27)) to 672 nm. Absence of a large absorption maximum shift in liposomal solution is consistent with localization of AlClPcS predominantly in the interior aqueous core of the vesicles where interactions with the hydrocarbon bilayer of the liposome are minimal. Dye photobleaching was checked by monitoring the optical density at λ_{max} throughout the course of the irradiation and/or measuring Triton X-100-disrupted aliquots of the

TABLE I. Transition Temperatures of Plasmalogen Liposomes

Composition[a]	T_c(°C)[b]
PlasPPC	39.4
DPPC	41.5[c]
PlasPPC/DPPC[d]	38.9
PlasPPC/DHC	46.5[e]

[a] 10 mM lipid in 20 mM Tris buffer, pH 8; < 800 A diameter
[b] ±1 °C
[c] cf 41.4 °C (13)
[d] 10 mol% DPPC
[e] 30 mol% DHC

liposomal solution. No bleaching of the AlClPcS was observed under the experimental conditions used.

Figure 1a shows the photoinduced release kinetics of glucose from PlasPPC/DPPC (10 mol%) liposomes at 37 °C loaded with AlClPcS ([lipid]/[AlClPcS] ca. 10^3). While release of entrapped glucose appears essentially complete within 90 min of irradiation, permeation of glucose due to thermal (dark) processes results in release of 60% of the entrapped glucose during the same time interval. Control experiments performed in the absence of sensitizer resulted in leakage rate profiles similar to those observed in dark experiments.

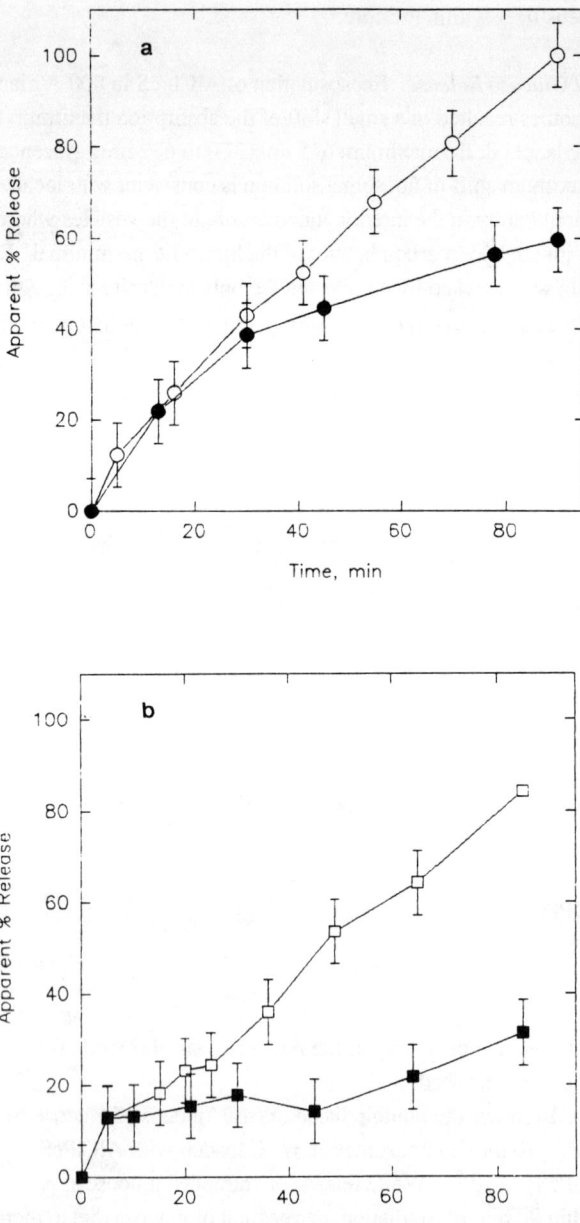

Figure 1. (a) AlClPcS-sensitized release of glucose from PlasPPC/DPPC at 37 °C. $\lambda > 630$ nm (○); and dark (●). (b) Apparent AlClPcS-sensitized release of glucose from PlasPPC/DHC at 37 °C. (□) $\lambda > 630$ nm ; (■) dark .

PlasPPC liposomes containing DHC as co-additive were prepared in an attempt to reduce the dark leakage rate of glucose in PlasPPC/DPPC vesicles. Glucose release from PlasPPC/DHC (30 mol %) vesicles is plotted in Figure 1b. Given minor variations in lamp intensity and AlClPcS concentration, the results suggest that photoinduced release rates from PlasPPC/DHC are within experimental error of those observed in PlasPPC/DPPC liposomes. However, after 90 min of dark permeation only 25% of the entrapped glucose has been released from the PlasPPC/DHC liposome core, a 2.4-fold decrease from the dark leakage observed in PlasPPC/DPPC vesicles.

Release rates monitored in the presence of 100 mM sodium azide, a known quencher of $^1O_2$ (28), are shown in Figure 2. Azide-containing liposomes exhibited a pronounced induction period of 25-30 minutes during which time little or no photoinduced leakage of glucose was measured. In comparison, after 25 minutes of irradiation of azide-free liposomes (Fig 1b), ca. 10-15% of encapsulated glucose had been released in addition to that leaked as a result of dark permeation processes. This result is consistent with azide quenching of a pathway responsible for short-time modification of liposome properties (see below). UV/Vis analysis before and after 90 minutes of photolysis indicated a slight increase in optical density at 672 nm from 0.454 to 0.567. It appears, therefore, that while processes responsible for damage of liposomal integrity are quenched at early times, 90 minutes of irradiation does result in some changes in the light scattering properties of liposomes which contain the $^1O_2$ quencher.

Photoinduced contents leakage was also measured in vesicles prepared with Tris/D_2O/NaCl buffer (Figure 3). Following a short induction period (ca. 10 min), the apparent release rate increases dramatically to a final value of more than 200% after 55 min of irradiation. This large apparent release of glucose is consistent with liposome aggregation or fusion as discussed below.

In order to probe the purely light-induced changes in liposome contents leakage, dark permeation rates were subtracted from those of irradiated liposomes (Figures 1b, 2, and 3) and replotted in Figure 4. In all cases (i.e., Tris/NaCl, Tris/NaN$_3$, Tris/D_2O/NaCl), a distinct induction period is observed, ranging from 10 to 30 minutes. While fluctuations in photon density make precise quantitative interpretations difficult, the induction time appears to be shortest in vesicle solutions containing D_2O, followed by PlasPPC/DHC liposomes prepared in Tris and finally those containing Tris/NaN$_3$. Similar studies of glucose leakage accompanying sensitization by a membrane-bound phthalocyanine (Anderson and Thompson, submitted), showed no such lag-time.

TLC analysis of liposomes after irradiation indicated the formation of several new products each more polar than starting PlasPPC. Table II gives the R_fs in 65:35:6 CHCl$_3$/MeOH/NH$_4$OH of products from photolysis (90 minutes) of

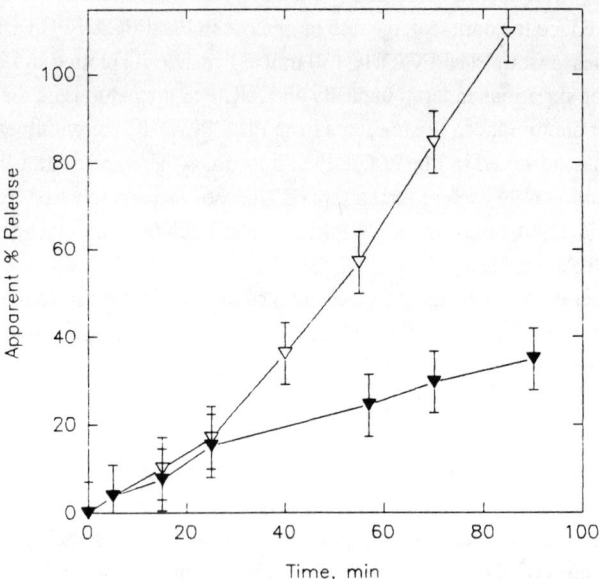

Figure 2. Apparent glucose release from PlasPPC/DHC (30 mol%) liposomes in 20 mM Tris/100 mM NaN$_3$, pH 8 at 37 °C. (▽) λ > 630 nm; (▼) dark.

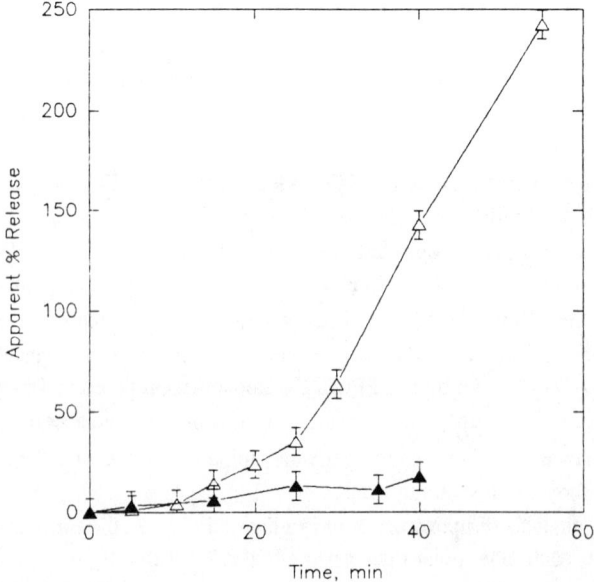

Figure 3. Apparent glucose release from PlasPPC/DHC (30 mol%) liposomes in 20 mM Tris/D$_2$O/100 mM NaCl, pD 8.4, 37 °C. (△) λ > 630 nm; (▲) dark.

PlasPPC/DHC/Tris/NaCl liposomes as well as those from Tris/azide and Tris/D$_2$O/NaCl experiments. Authentic samples of D-glucose, DHC, PlasPPC and 1-palmitoyl-sn-glycero-3-phosphocholine (LysoPC) were also spotted. In each case, an aliquot from the associated dark control sample was also spotted. All dark samples showed only a single spot with R_f 0.54 corresponding to that of PlasPPC.

Photoinduced Morphology Changes. As described above, photolysis of liposomes in Tris/D$_2$O/NaCl resulted in a large increase in solution turbidity. The UV/Vis absorption spectra of liposomes before and after photolysis are given in Figure 5. Clearly, a large increase in light scattering has resulted from irradiation. Such an

TABLE II. TLC of Photolysis Products of PlasPPC/DHC/AlClPcS Liposomes

Compound	R_f			
PlasPPC	0.54			
DHC	1.0			
Glucose	0.09			
LysoPC	0.27			
PlasPPC/DHC Liposomes:				
Tris/ NaCl	0.23	0.55		
Tris/ NaN$_3$	0.26	0.55		
Tris- D$_2$O/ NaCl	0.16	0.22	0.32	0.50

effect could be induced by aggregation and/or fusion of the liposomes. No comparable change in the scattering properties was evidenced in the absorption spectrum of liposomes used in the dark control.

Electron micrographs of PlasPPC/DHC liposomes in 20 mM Tris/D$_2$O/NaCl before and after irradiation, Figures 6a and b, respectively, indicate a much larger population of large, multilamellar vesicles following photolysis. Before irradiation, the liposomes (extruded through stacked 0.05 μm Nucleopore filters for this experiment) were unilamellar and exhibited a narrow size distribution with median diameter of 600 A. In contrast, irradiated vesicles were predominantly multilamellar and statistically much larger, with a median diameter of 1100- 1200 A.

DISCUSSION

Photodynamic generation of reactive oxygen species (ROS), e.g., $^1O_2$, superoxide

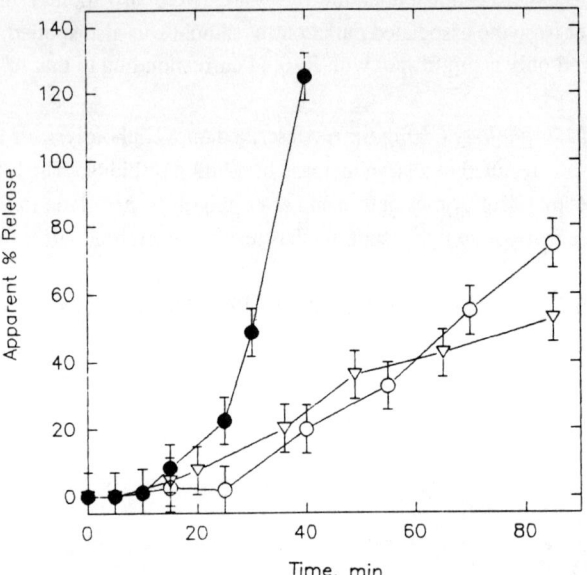

Figure 4. Dark-subtracted apparent glucose release kinetics from irradiated ($\lambda >$ 630 nm) PlasPPC/DHC liposomes at 37 °C. ($\triangledown$) 20 mM Tris/H_2O/100 mM NaCl, pH 8; (○) 20 mM Tris/100 mM NaN_3, pH 8; (●) 20 mM Tris/D_2O/100 mM NaCl, pD 8.4.

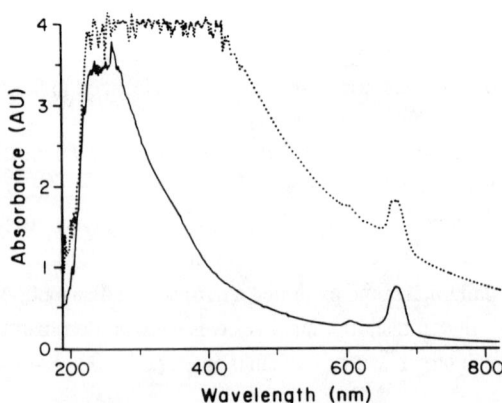

Figure 5. Absorption spectra of PlasPPC/DHC (30 mol%) liposomes in 20 mM Tris/D_2O/100 mM NaCl before (——) and after (•••) 90 minutes of photolysis ($\lambda > 630$ nm).

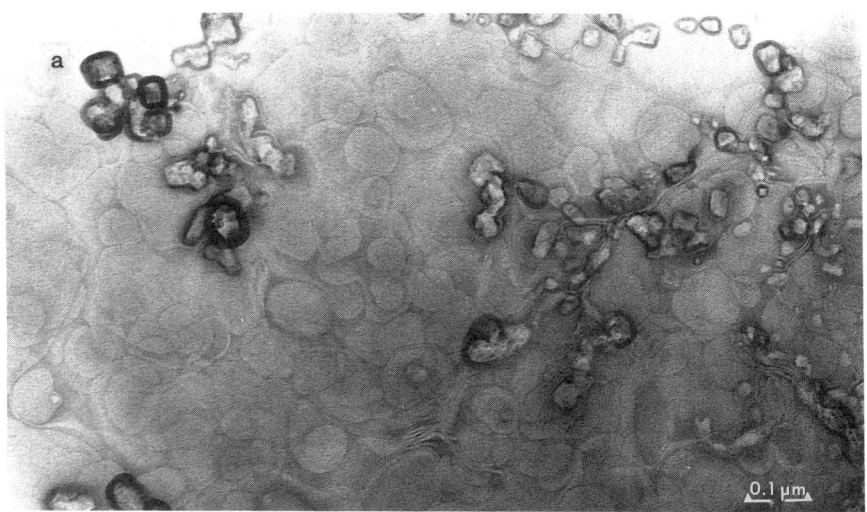

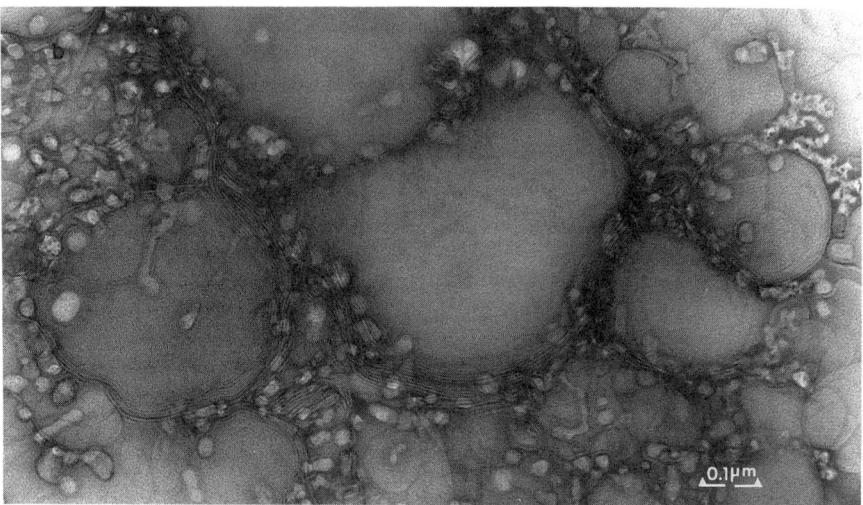

Figure 6. Negative stain (2% ammonium molybdate, pH 8 Tris) transmission electron micrographs of PlasPPC/DHC (30 mol%) liposomes before (a) and after (b) 100 minutes of photolysis ($\lambda > 630$ nm).

anion, and hydroxyl radical in membranes is known to result in significant modification of lipid and proteinaceous components (*29*). While the relative importance of ROS/protein versus ROS/lipid reactions with respect to alteration of cellular function is an unsettled issue, photodynamic reactions of lipid membranes (both naturally occuring and synthetic) generally result in peroxidation of polyunsaturated fatty acids (*30-34*). However, the exact mechanism of subsequent reactions depends upon medium pH, sensitizer composition and concentration, light intensity, oxygen concentration, and other parameters (*29*), and photodynamic interactions are known to be responsible for a wide variety of membrane modifications.

Previous studies of membrane-bound zinc phthalocyanine and PlasPPC/DPPC liposomes (Anderson and Thompson, submitted) showed that photostimulated release of hydrophilic reagents from plasmalogen liposomes is possible. In those experiments, liposomes were irradiated at or near T_c. After 60 minutes, greater than 60% of entrapped glucose was released. No evidence of photoinduced liposomal structural changes was observed.

In the present study, we have investigated the potential of a water-soluble phthalocyanine, AlClPcS, to produce ROS which stimulate release of entrapped glucose. These studies differ from our previous work in several significant ways. First, in the present study the sensitizer is localized in the hydrophilic core of the liposome, away from the vinyl ether linkage. Lipid peroxidation requires diffusion of the ROS across the bilayer/aqueous interface. Although quantitative measures vary substantially (*35-37*), the penetration efficiency of $^1O_2$ through the phosphatidylcholine/ water bilayer interface seems to be substantially less than that observed in either isotropic solution or in micelles. Secondly, DSC data indicates that in contrast to PlasPPC/DPPC liposomes, PlasPPC/DHC vesicles are in the gel phase at 37 °C . Measurements of O_2 solubility in 1,2-dimyristoyl-sn-3-phosphatidylcholine liposomes above and below the phase transition temperature demonstrate that O_2 is three to four times less soluble in the gel phase than in the liquid crystalline state (*38*). Finally, in the present study, we have used DHC to stabilize the bilayer structure and reduce thermal permeation of glucose. Addition of steroids like cholesterol and DHC is known to result in increased viscosity of bilayer membranes (*39*). As a result, diffusion of ROS to the vinyl ether linkage may be more difficult in PlasPPC/DHC liposomes than the PlasPPC/DPPC vesicles studied previously.

The data in Figure 4 clearly demonstrate that PlasPPC/DHC liposomes are substantially affected by irradiation in the presence of oxygen. The observation of a glucose release greater than 100% suggests that some photoinduced morphological changes have caused an increase in light scattering of the liposomes independent of any increases due to reduction of NADP. Changes in liposome size and structure are confirmed by electron microscopy which shows a substantial increase in the average liposome size after visible irradiation. The abundance of large diameter (> 1200 A) multilamellar structures in the micrographs of photolysed sample (Fig. 6b) is most likely responsible for the increased tailing observed throughout the absorption spectrum (Figure 5). Finally, the abundance of large, multilamellar structures in the micrographs suggests (but does not require) that a significant portion of glucose was released by photolysis. Additional experiments are underway to clarify this issue.

TLC results indicate the formation of new products in the photolysed samples not found in the dark controls. The observation that the major product formed during photolysis exhibits an R_f very similar to that of LysoPC suggests that the predominant reaction induced by AlClPcS sensitization is cleavage of a lipid chain. The saturated nature of the sn-2 chain in PlasPPC makes it more likely that such a reaction would take place at the vinyl ether linkage of the sn-1 chain. Scherrer and Gross (14) have demonstrated that photooxidation of plasmenylcholine by methylene blue in $CHCl_3$/MeOH solution results in formation of a product consistent with oxidation of the vinyl ether linkage. Studies aimed at identifying the phthalocyanine-sensitized PlasPPC/DHC photooxidation products are currently underway.

CONCLUSION

PlasPPC forms unilamellar vesicles when combined with DHC at concentrations of 30 mol%. Passive permeation at 37 °C of encapsulated glucose is significantly reduced from that observed in PlasPPC/DPPC vesicles. Visible irradiation of PlasPPC/DHC liposomes in the presence of AlClPcS entrapped within the aqueous core results in significant changes in size and morphology of the vesicles as well as formation of new photoproducts. Observation of an induction time in apparent glucose release plots is consistent with an ROS- dependent mechanism for photoinduced morphology changes.

ACKNOWLEDGEMENTS

The authors gratefully acknowledge Mike Webb (Oregon Regional Primate Center) for his assistance with the electron microscopy and the financial support of the American Cancer Society, Oregon Division.

LITERATURE CITED

1) a) *Pulsed and Self-Regulated Drug Delivery*; Kost, J., Ed.;CRC Press: Boca

Raton, FL, 1990. b) *Macromolecules as Drugs and as Carriers for Biologically Active Materials*; Tirrell, D.A.; Donaruma, L.G.; Turek, A.B., Eds.; New York Academy of Science; NY Acad. Sci.: New York, NY, 1985, Vol. 146.

2) a) Nayar, R.; Hope, M.J.; Cullis, P.R. *Biochim. Biophys. Acta* **1989**, *898*, 200-206. b) Hope, M.J.; Bally, M.B.; Mayer, L.D.; Janoff, A.S.; Cullis, P.R. *Chem. Phys. Lipids* **1986**, *40*, 89-107. c) Lichtenberg, D. *Meth. Biochem. Anal.* **1986**, *33*, 337-460.

3) Mayer, L.D.; Bally, M.B.; Hope, M.J.; Cullis, P.R. *Chem. Phys. Lipids* **1986**, *40*, 333-45.

4) Mayer, L.D.; Tai, L.C.; Ko, D.S.; Masin, D.; Ginsberg, R.S.; Cullis, P.R.; Bally, M.B. *Cancer Res.* **1989**, *49*, 5922-5930.

5) a) Gabizon, A.T.; Peretz, T.; Sulkes, A.; Amselm, S.; Ben-Yosef, R.; Ben-Baruch, N.; Catane, R.; Biran, S.; Barenholz, Y. *Eur. J. Cancer Clin. Oncol.* **1989**, *25*, 1795-1803. b) Ostro, M.J. *Liposomes: From Biophysics to Therapeutics*; Marcel Dekker: New York, NY, 1987. c) Goren, D.; Gabizon, A.; Barenholz, Y. *Biochim. Biophys. Acta* **1990**, 1029, 285-294. d) Rahman, A.; White, G.; More, N.; Schein, P.S. *Cancer Res.* **1985**, *45*, 796-803.

6) Wilson, B.C. In *Photosensitizing Compounds: Their Chemistry, Biology and Clinical Use*; CIBA Foundation Symposium 146; Wiley: New York, NY, 1989; pp. 60-77.

7) a) O'Brien, D.F.; Zumbulyadis, N.; Michaels, F.; Ott, R. *Proc. Natl. Acad. Sci. U.S.A.* **1977**, *74*, 5222-5226. b) Kano, K.Y.; Tanaka, Y.; Ogawa, T.; Shimomura, M.; Kunitake, T. *Photochem. Photobiol.* **1981**, *34*, 323-329. c) Pidgeon, C,; Hunt, C. *Photochem. Photobiol.* **1983**, *37*, 491-494. d) Yemul, S.; Berger, C.; Estabrook, A.; Edelson, R.; Bayley, H. *Ann. N.Y. Acad. Sci.* **1985**, *446*, 403-414. e) Frankel, D.A.; Lamparski, H.; Liman, U.; O'Brien, D.F. *J. Am. Chem. Soc.* **1989**, *111*, 9262-9263. f) Kusumi, A.; Nakahama, S.; Yamaguchi, K. *Chem. Lett.* **1989**, 433-436.

8) a) Wilson, B.C.; Jeeves, W.P.; Lowe, D.M. *Photochem. Photobiol.* **1985**, *42*, 153-162. b) Bolin, F.P.; Preuss, L.E.; Cain, B.W. In *Porphrin Localization and Treatment of Tumors;* Doiron, D.R.; Gomer, C.J. , Eds.; Alan R. Liss: New York, 1984; pp 211-225.

9) Kato, H.; Kawate, N,; Kinoshita, K.; Yamamoto, H.; Furukawa, K,; Hayata, Y. In *Photosensitizing Compounds: Their Chemistry, Biology and Clinical Use*; CIBA Foundation Symposium 146; Wiley-Interscience: New York, NY, 1989.

10) a) Dougherty, T.J. *Photochem. Photobiol.* **1987**, *45*, 879-89. b) Star, W.M.; Marijnissen, H.P.A.; Jansen, H.; Keijzer, M.; van Gemert, M.J.C. *Photochem. Photobiol.* **1987**, *46*, 619-624.

11) a) *Photosensitizing Compounds: Their Chemistry, Biology and Clinical Use*; CIBA Foundation Symposium 146; Wiley-Interscience: New York, NY, 1989.

b) Dougherty, T.J. In *Medical Radiology- Innovations in Radiation Oncology;* Springer-Verlag: Berlin, 1988; pp. 175-88.

12) Henderson, B.W.; Bellview, D.A. In *Photosensitizing Compounds: Their Chemistry, Biology and Clinical Use*; CIBA Foundation Symposium 146; Wiley: New York, NY, 1989; pp. 95-111.

13) Morand, O.; Zoeller, R.A.; Raetz, C.R.H. *J. Biol. Chem.* **1988**, *263*, 11597-11606.

14) Scherrer, L.; Gross, R.W. *Mol. Cell. Biol.* **1989**, *88*, 97-105.

15) *Phthalocyanines: Properties and Applications*; Leznoff, C.C.; Lever, A.B.P., Eds.; VCH Publishers: New York, NY, 1989.

16) Hellinger, O. *Biophysik* **1969**, *6*, 63-8.

17) a) Bergelson, L.D. *Lipid Biochemical Preparations*; Elsevier/North Holland Biomedical: New York, NY, 1980; pp. 141-2. b) Guisvisdalsky, P.N.; Bittman, R. *J. Org. Chem.* **1989**, *54*, 4643-48.

18) Glascoe, O.; Long, F.A. *J. Phys. Chem.* **1960**, *64*, 188-90.

19) a) Demel, R.A.; Kinsky, C.B.; Van Deenen, L.L. *Biochim. Biophys. Acta* **1968**, *150*, 655-665. b) Kinsky, S.C.; Haxby, J.; Kinsky, C.B.; Demel, R.A.; Van Denen, L.L.H. *Biochim. Biophys. Acta.* **1968**, *152*, 174-185.

20) Boggs, J.M.; Stamp, D.; Hughes, D.W.; Deber, C.M. *Biochemistry* **1981**, *20*, 5436-5735.

21) Han, X.; Gross, R.W. *Biochemistry* **1990**, *29*, 4992-4996.

22) a) Snyder, B.; Freire, E. *Proc. Natl. Acad. Sci. USA*, **1980**, *77*, 4055-4059. b) Mabrey, S.; Sturtevant, J.M. *Proc. Natl. Acad. Sci. USA* **1976**, *73*, 3862.

23) Estep, T.N.; Mountcastle, D.B.; Biltonen, R.L.; Thompson, T.E. *Biochemistry* **1978**, *17*, 1984-89.

24) Blume, A. In *Physical Properties of Biological Membranes*; Hidalgo, C., Ed.; Plenum Press: NY, 1988; pp 71-121.

25) Cullis, P.R.; Hope, M. In *Biochemistry of Lipids and Membranes*; Vance, D.E.; Vance, J.E., Eds.;Benjamin/Cummings: Menlo Park, CA, 1985; pp. 25-73.

26) Creer, M.H.; Gross, R.W. *J. Chromatogr.* **1985**, *338*, 61-69.

27) Darwent, J.R.; Douglas, P, Harriman, A.; Porter, G.; Richoux, M.-C. *Coord. Chem. Rev.* **1982**, *44*, 83-126.

28) Pooler, J.P. *Photochem Photobiol.* **1989**, *50*, 55-68.

29) Valenzeno, D.P.; Tarr, M. In *Photochemistry and Photophysics*; Rabek, J.F., Ed.; CRC Press: Boca Raton, FL, 1991, Vol. III; pp. 137-191.

30) Muller-Dunkel, R.; Blais, J.; Grossweiner, L.I. *Photochem. Photobiol.* **1981**,*33*, 683-687.

31) Kanofsky, J.R. *Photochem. Photobiol.* **1991**, *53*, 93-99.

32) Valenzeno, D.P. *Photochem. Photobiol.* **1987**, *46*, 147-160.

33) Sikurova, L.; Haban, I.; Frankova, R. *Stud. Biophys*.**1988**, *128*, 163-168.
34) Pooler, J.P. *Photochem. Photobiol*.**1989**, *50*, 55-68.
35) Gorman, A.A.; Lovering, G.; Rodgers, M.A.J. *Photochem. Photobiol*. **1976**, *23*, 399.
36) Matheson, I.B.C.; Lee, J.; King, A.D. *Chem. Phys. Lett*. **1978**, *55*, 49.
37) Rodgers, M.A.J.; Bates, A.L. *Photochem. Photobiol*. **1982**, *35*, 473-477.
38) Smotkin, E.S.; Moy, F.T.; Plachy, W.Z. *Biochim. Biophys. Acta* **1991**, *1061*, 33-38.
39) Bloch, K. In *Biochemistry of Lipids and Membranes*; Vance, D.E.; Vance, J.E., Eds.; Benjamin/Cummings: Menlo Park, CA, 1985; pp. 1-24.

RECEIVED September 24, 1991

Chapter 15

Percolation Process and Structural Study in Docusate Sodium Reverse Micelles Containing Cytochrome c

J. P. Huruguen[1,2], T. Zemb[2], C. Petit[1,2], and M. P. Pileni[1,2]

[1]Laboratoire SRI, Batiment de Chimie-Physique, Université Pierre et Marie Curie, 11 Rue Pierre et Marie Curie, Paris 75005, France
[2]CEN Saclay, DRECAM, SCM, Gif sur Yvette 91191, France

We report changes in the percolation threshold of AOT water droplets by solubilizing cytochrome c. The changes in the critical percolation factors (volume fraction, temperature, water content) are attributed to the increase in the attractive interactions between droplets by solubilizing cytochrome c in water pools. The critical exponents determined from percolation theories are compared in the presence and in the absence of cytochrome c. By SAXS before and at the percolation onset the scattering by the clusters is characteristic of cylinders. The average distance between the cylinder is determined from a geometrical model and from X-ray scattering experiments using the "correlation hole" model.

It is well known that the main structural pattern of biological membranes is the flat bilayer of lipid molecules. However, the notion of the lipid bilayer as the only possible way of organization of membrane lipids, which represents the essence of the widely accepted fluid mosaic model of biological membranes, does not agree with established facts of structural rearrangements of lipids, for example, from the bilayer to the hexagonal phase. Further investigations of the structure of lipid membranes resulted in the discovery of other types of non-bilayer lipid stuctures, in particular, so-called lipidic particles, representing reversed lipid micelles sandwiched between monolayers of the lipid bilayer. The concept of non-bilayer structures in lipid membranes made a basis for a new "metamorphic mosaic" model of biomembranes which explains elegantly many processes occurring in the living cell, such as fusion and compartmentalization of membranes, exo- and endocytosis, lipid flip-flop, etc. From the point of view of our discussion, the ability of certain proteins to induce the

NOTE: Docusate sodium is also known as AOT and aerosol OT.

formation of non-bilayer structures upon incorporation into model and biological membranes is of particular significance. It is quite possible that the formation of non-bilayer structures represents a general mechanism for the protein incorporation into membranes and for the regulation of their activity.

Taken together, the above data allow one to conclude that model studies of enzymes and enzymatic catalysis in systems of reverse surfactant micelles in organic solvents are of great importance in understanding of the enzyme functioning in natural lipid systems. By the present time many reversed micellar systems containing dozens of different enzymes have been studied *(1)*.

The dissolution of sodium sulfosuccinate usually called Aerosol OT or AOT in isooctane induces the formation of spheroidal aggregates called reverse micelles *(1)*, microemulsions or water droplets. Water is readily solubilized in the polar core, forming a so called "water pool", characterized by w, the ratio of water concentration over surfactant concentration, (w=[H_2O]/[AOT]). For AOT in isooctan, above w=15, the water pool radius, r_w, is found linearly dependent on the water content *(2)*. So as the size of the droplet increases, the concentration of discrete micelles decreases. To demonstrate that the presence of some proteins induces a change in the percolation threshold in AOT-water-isooctane microaggregates we choose to solubilize cytochrome c in these microemulsions. This protein is a water soluble hemoprotein with a small molecular weight (12,400) which is responsible for several electron transfer reactions across membranes. Previously we have developed a geometrical model tested by SAXS *(3)* and a kinetic model *(4)* to determine the average location of low molecular weight proteins or enzymes at low volume fraction of water in reverse micellar systems. We demonstrated that cytochrome c is located at the interface and its interfacial contribution increases with the water content *(3,4)*.

In this paper we report an analysis of low frequency permittivity measurements of AOT microemulsion W/O in the presence of cytochrome c. It is shown that the solubilization of small amounts of cytochrome c (10^{-4} - 10^{-3} M) in reverse micelles favours a percolation process at lower temperature and volume fraction of water values than that observed with protein-free water droplets system. The structural study that we performed through X-ray scattering measurements before and at the percolation threshold confirms the formation of highly extended aggregates.

Abstract of the results of percolation phenomena obtained in AOT-water-isooctan solution

Percolation threshold. The interactions between water droplets (AOT-water-isooctan), largely studied at relatively high AOT concentrations *(5-8)*, favour the formation of clusters of water droplets: dimers, trimers,... If the water-in-oil microemulsion polar volume fraction is large enough, an aggregate of macroscopic size appears. Several groups *(9-11)* showed that a divergence of the static dielectric permittivity and a sharp increase in the conductivity are attributed to a percolation transition. When an electrical field is applied, the existence of water droplets clusters induces a critical permittivity behaviour close to the percolation threshold by a capacity effect between different clusters *(5-11)* : van Dijk et al. explained the appearance of this divergence as originating from a significant additive capacity effect due to the vicinity of the polarized clusters. Hence the percolation threshold is attributed to the formation of an infinite cluster of water droplets allowing the charge carriers to percolate through the system. The percolation threshold corresponds to the onset of the conductivity and to the maximum of the static permittivity coming from the cross-over of its critical behavior appearing on the either sides of the percolation threshold *(12)*. These features have been observed in reverse AOT-water-isooctan micelles *(9,10)*. Hence temperature, T_p, volume fraction of water, ϕ_p, and water content, w_p, percolation thresholds can be determined. Such percolation transition is observed by increasing factors such as water content, temperature or volume fraction of water and keeping the other factors constant. The volume fraction percolation threshold, ϕ_p, decreases by increasing temperature: At w=27 it is found *(11)* ϕ_p equal to 13,8%, 18,3% and 27,5% at T_p equal to 52°C, 47°C and 35°C respectively. These two factors (T_p, ϕ_p) decrease by increasing the water content, w.

Critical exponents. Percolation theories have been mainly developed on the basis of lattice models *(13)*. In this case, the critical behaviours of some physical properties such as dc conductivity, $\kappa(\phi)$, and static permittivity, $\varepsilon_s(\phi)$, are described through critical exponents t and s close to the percolation threshold ϕ_p where an infinite aggregate exists *(12-14)*:

$$\kappa(\phi) \sim (\phi-\phi_p)^{-s} \text{ for } \phi<\phi_p \qquad (1)$$

$$\kappa(\phi) \sim (\phi-\phi_p)^{t} \text{ for } \phi>\phi_p \qquad (2)$$

$$\varepsilon_s(p) \sim |\phi-\phi p|^{-s} \text{ for } \phi>\phi_p \text{ and } \phi<\phi_p \qquad (3)$$

The value found for s (static permittivity) and t are of the same order than the theoretical predictions: t ~ 2 and s ~ 0.6 - 0.8. The critical exponent s describing the critical behaviour of conductivity below the percolation threshold has been found different and attributed to the dynamic character of the droplet clusters *(11)*. The same critical exponents t for conductivity and s for static permittivity have been found with temperature variation at fixed polar volume fraction. This has been explained as an increase of the attractive interactions when temperature increases *(15)*. In such case, the theorical studies predict no changes of the critical exponents *(13)*. At the percolation threshold, the critical behaviour of κ and ε versus the frequency ω is described by the following *(16)*:

$$\kappa(\omega) \sim \omega^x , \text{ and } \varepsilon(\omega) \sim \omega^{-y} \qquad (4)$$

This behaviour is attributed to the fractal structure of the aggregates. A good agreement between the experimental data and theoritical predictions is obtained for y. A relationship between s, t and y is deduced *(16)*:

$$y = s/(s+t) \qquad (5)$$

However in the case of x, the experimental data and the theoritical values differ. This is attibuted to the dynamic character of the aggregates.

Experimental section

AOT was obtained from Sigma, cytochrome c and isooctan from Fluka and they were used without further purification.

Let us define the number of water droplets, n, by:

$$n = V_{aq.}/V_M \qquad (6)$$

where $V_{aq.}$ is the total volume of water inthe sample and V_M is the average water droplets volume. If V_T is the total volume of the sample, the number of water droplets per unit volume is given by:

$$n/V_T = 3.\phi_w/4.\pi.r_w \qquad (7)$$

where ϕ_w is the volume fraction of water and r_w is the average radius of the water droplets. For r_w given in angströms, one obtains:

$$[RM] = 3 \times 10^3 \, \phi_w /4 \, N\pi \, r_w^3 \quad (M/l) \qquad (8)$$

where N is the Avogadro's number.

The experimental investigations were carried out either at various water contents, w, or at a constant w value (w=40) and at a given cytochrome c number per micelle. The cytochrome c concentrations are equal to [RM], 2[RM] or 4[RM] respectively, corresponding to an average cytochrome c number equal to 1, 2 and 4. At w=40, [AOT]=0.1M, [RM] is equal to 1.23×10^{-4}M. We choose the ratio, w', of AOT concentration over cytochrome c concentration, i.e. the average number of AOT molecules per cytochrome c molecule. For 1, 2 and 4 cytochrome c per water droplet, we have respectively w' equal to 900, 450 and 225. The samples used are optically isotropic in all the volume fractions of water and at the various temperatures under investigation.

The conductivity measurements have been made with a Tacussel CD 810 instrument. The dielectric measurements have been described previously (17).

The small-angle X-ray scattering was done at L.U.R.E. (Orsay) on the D22 diffractometer. The wave vertor range, q, under the present investigation was:

$$4.10^{-3} Å^{-1} < q < 0.11 Å^{-1}.$$

Results and discussion

Conductivity and permittivity measurements. At room temperature and at fixed w (w=40), the conductivity of AOT in isooctane reverse micelles is very low and increases with the volume fraction of water (Figure 1). The conductivity onset is found to be equal to 25%. In the presence of cytochrome c, initially the conductivity was very low which increases abruptly with increase in volume fraction of water. Figure 1 shows that this sharp increase in the conductivity occurs at lower volume fraction of water by increasing the average number of cytochrome c per droplet.

The static permittivity and the conductivity was measured at various temperatures, at fixed volume fraction, (ϕ_w=7.4%) and water content (w=40), in the absence and in the presence of cytochrome c ([cyt] = 4[RM]) (18). The divergences in static dielectric permittivity with temperature are shown in figure 2.

In the absence of protein (Figure 2A) a divergence in the static permittivity and in the conductivity is observed. In the presence of cytochrome c, the static permittivity reaches a maximum which is associated to the conductivity onset (Figure 2B). Such behaviour, characteristic of a percolation transition (9-11), occurs with or without proteins. The divergence in the permittivity and in the conductivity occurs at lower temperature in the presence of cytochrome c. Figure 2 shows that, at a given volume fraction (ϕ_w=7.4%), the temperature percolation threshold is lower (T_p =34°C) in the presence than in the absence of cytochrome (T_p~45°C). This indicates that addition of

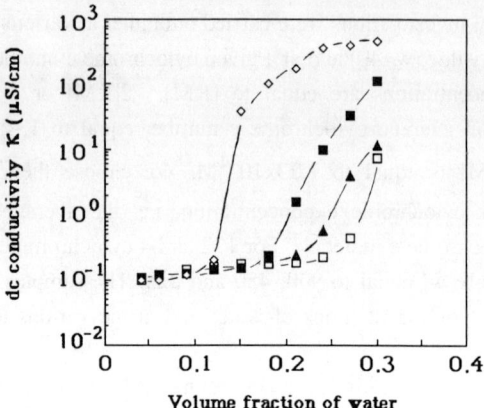

Figure 1: Variation of the conductivity with the volume fraction of water in AOT-isooctan-water solution in the absence and for different values of the average number of AOT molecules per cytochrome c molecule:
[cyt.c]=0 (□); w'=900 (▲); w'=450 (■); w'=225 (◇).

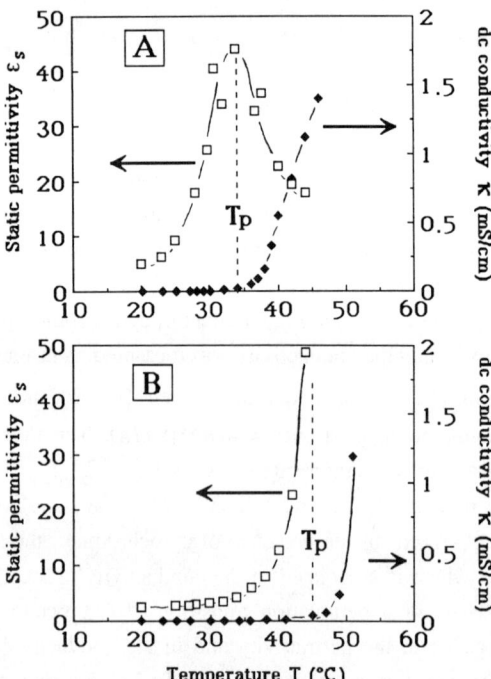

Figure 2: A-Variation of the static permittivity (□) and conductivity (◆) with temperature, w=40, ϕ_w =7.4% and w'=225.

B-Variation of the static permittivity (□) and conductivity (◆) with temperature, w=40, ϕ_w =7.4% and [cyt.c]=0.

cytochrome c favours the percolation process, with a decrease in the percolation threshold by protein addition.

Similar behavior is observed from the measurement of the static dielectric constant at various polar volume fractions, at fixed temperature (T=20°C) and water content (w=40), in the absence and in the presence of various cytochrome c concentrations: the divergence in the static permittivity occurs at lower volume fraction in presence of cytochrome c. This confirms the data given above from which it has been deduced that the percolation threshold is lower in the presence of the protein. This decrease in the percolation onset is more important as the cytochrome c concentration increases.

From the slope of the linear behaviour of log κ and log ε versus log ω, the critical exponents (described in equations (2) and (3)) are calculated. Table I shows that the experimental values obtained with protein-free micelles are similar to those previously published *(11)*. In the presence of cytochrome c, the t value is not changed and the value of s is in the same order of magnitude to those obtained with protein-containing micelles. From the critical exponent s and t, the y exponent is also calculated (Table I).

Figure 3 shows the variation of the the log-log plots of permittivity versus frequency. A linear behavior is observed around the percolation threshold at 45°C with ptoyein-free micelles (Figure 3A), and at 34°C with protein-containing micelles (4[RM]) (Figure 3B). These results are in good agreement to those estimated from the percolation threshold in Figure 2. From extrapolation of the slope at the percolation temperature values, the critical exponent y is deduced (Table I).

In the absence of cytochrome c, the y value obtained by Huang et al. *(11)* is in good agreement to that obtained in our experimental conditions (Table I). Table I shows an unchanged value of y in the presence of protein. This indicates that the percolation mechanism is similar with or without cytochrome c.

Structural study by small angle X-ray scattering. SAXS has been performed at room temperature, x=4 and at low volume fraction of water (7,4%; 13,6%). From Figure 1 it can be observed that 7,4% corresponds to a sample below the percolation threshold (low conductivity). The water droplets aggregates are probably highly extended, but an infinite aggregate doesn't exist yet. For 13,6%, the sample is at the percolation threshold, where the conductivity begins to increase: the infinite aggregate exists allowing the charge carriers to percolate through the system.

Figures 4 A and B show the scattered intensity I(q) as a function of the wave vector q in a log-log plot. An asymptotic behavior close to -1 on a large range of q values is osberved. This behaviour is characteristic of a cylindrical microscopic structure whose length is much important than its radius R_c, with a long persistance length. The scattered intensity by a homogeneous cylindrical structure is given by *(19)* :

Table I: critical exponents obtained in water-in-oil microemulsions of AOT/isooctane/water at various concentrations of cytochrome c determined from the increase in the conductivity and the divergence of the static permittivity at room temperature. The critical exponents deduced from the behaviour with frequency are also reported

[cyt]	0	w'=225
T_p (°C)	45	34
t	2 2 *(7)*	2
s	0.83 0.6-0.8 *(7)*	0.56
y = s/(s+t)	0.27 0.29 *(5)*	0.25

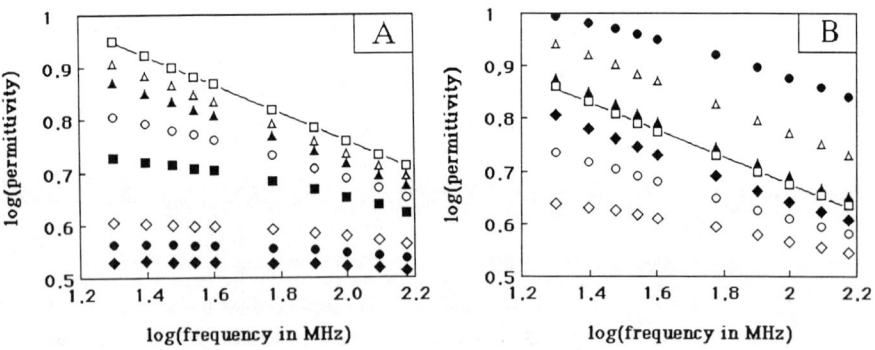

Figure 3: Variation of the permittivity with the frequency in a log-log plot for different temperatures, at w=40.

A-[cyt.c]=0: 30°C (●); 32°C (◇); 36°C (■); 38°C (O); 40°C (▲); 42°C (Δ); 45°C (□).

B-w'=225: 25°C (◇); 29°C (O); 31°C (◆); 34°C (□); 36°C (▲); 38°C (Δ); 40°C (●).

$$I(q) \sim q^{-1} \cdot [J_1(qR)/qR]^2 \tag{9}$$

$J_1(x)$ is the first order Bessel's function. Figure 4A and 4B shows that the cylinder simulated curve and the experimental data are in good aggreement with each other. The normalisation has been made by invariant. The radii determined from the slope of the linear relationship by plotting $Ln[q.I(q)]$ versus q^2, from the Porod plot and from the cylinder simulation are in good agreement (Table II). From these data it can be concluded that the scattering behaviour of cylinder is observed at low volume fraction before and at the percolation threshold. This could indicate the formation of polymers made up with water droplets connected by cytochrome c.

Using the "correlation hole" model *(20)* developed by de Gennes to describe the semidilute regime in polymer solutions, the average mesh size ξ can be determined. The intensity can be expressed as following:

$$I(q) = S(q) \cdot I_c(q) \tag{10}$$

$I_c(q)$ is the intensity scattered by a single chain, and $S(q)$ is the structure factor given by *(21)*:

$$S(q) \sim q^2\xi^2 / (1+q^2\xi^2) \tag{11}$$

Assuming that the behavior of the AOT-Isooctane-water-cytochrome c solutions is similar to that of polymers solution in semidilute regime, the previous model has been tested by SAXS. From simulation of the structure factor, the average distance between particles is calculated. Table II shows a decrease of the distance, denoted ξ_{exp}, when volume fraction of water increases. (See also Figure 5.)

These distances can be determined using a geometrical model that assumes the aggregate to be a homogeneous cylinder. Its surface, S_c, and its volume, V_c, are given respectively by:

$$S_c = 2.\pi.R_c\, l_v \quad \text{and} \quad V_c = \pi.R_c^2\, l_v \tag{12}$$

R_c is the radius and l_v is the length of cylinder per unit volume. The total volume, V_T, and the total surface, S_T, of cylinders are equal to nV_c and nS_c respectively, where n is the total number of cylinders in the solution. Then the ratio V_T/S_T is equal to $R_c / 2$. Assuming that the total surface of the cylinder is due to the AOT molecules, the specific interface, Σ, expressed per volume unit, is given by:

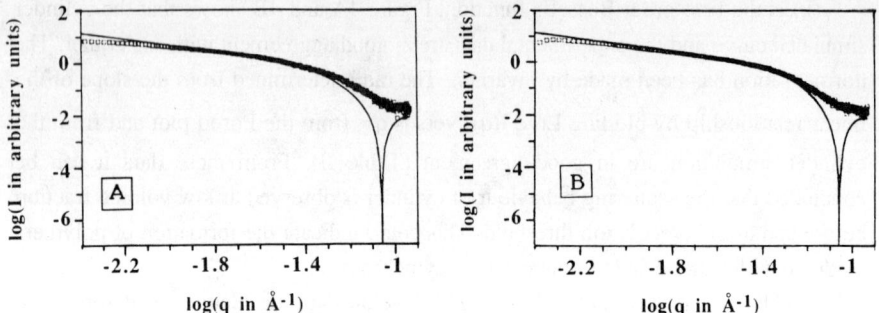

Figure 4: Scattered intensity variation versus wave vector in a log-log plot for a volume fraction of water equal to 7.4% (A) and equal to 13.8% (B). The full line represents the simulated curve of the scattering by a homogeneous cylinder (R_c=45Å). w'=225.

Table II: Radii obtained from the various treatment of the X-rays scattering data

ϕ_w	Porod	Ln I(q).q versus q^2	simulation	ξ_{geo}	ξ_{exp}
7.4%	45Å	40Å	45Å	326Å	310Å
13.8%	44Å	36Å	45Å	236Å	260Å

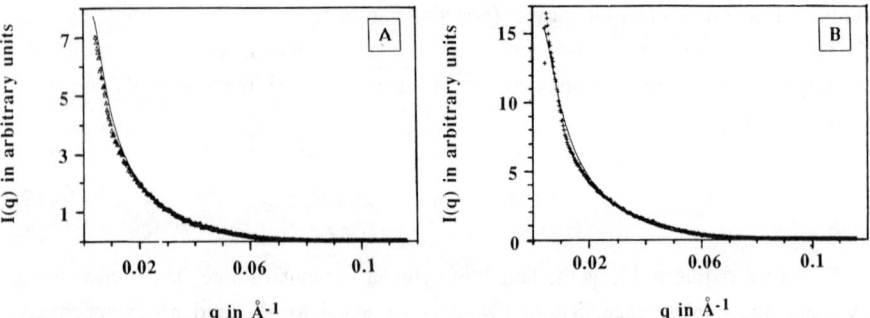

Figure 5: Scattering behavior with wave vector q for a volume fraction of water equal to 7.4% (A) and equal to 13.8% (B). In full line is the fit using the "correlation hole" model, with ξ_{exp} equal to 310Å for 7.4% and 260Å for 13.8%. w'=225.

$$\Sigma(A^2/\text{Å}^3) = [AOT].N \cdot \sigma_{AOT} \cdot 10^{-27} \quad (13)$$

σ_{AOT} is the surface area per polar head group and found to be equal to 60Å² *(21)*. For a given volume of solution, V, the total interface is : $S_T = V\Sigma$. Assuming that the total volume of the cylinders, V_T, is due to the volume of the water molecules, V_{aq} it is deduced :

$$V_T/S_T = R_c/2 = V_{aq}/(V\Sigma) = \phi_w/\Sigma \quad (14)$$

From the surface of the cylinder, S_c, is equal to Σ, and it is deduced:

$$l_v = \Sigma^2 / 4.\pi.\phi_w \quad (15)$$

The average mesh size, ξ and denoted ξ_{geo} for this geometric model, is the radius of the correlation tube containing an homogeneous cylinder from which the other cylinders are strongly repelled. From this model *(18)*, a simple geometric calculation leads to:

$$\pi.(\xi_{geo}/2)^2.l_v = 1 \quad (16)$$

And then:
$$\xi_{geo} = 2 (\phi_w)^{0.5}/\Sigma \quad (17)$$

Table II shows a decrease in the distance with increasing the volume fraction which is in good agreement to the data obtained from simulation of the structure factor from the "correlation hole" model.

Conclusion

Because of its location at the interface, the addition of cytochrome c to water droplets induces an increase in the attractive interactions between the microemulsion aggregates which promote a percolation process. This usually takes place, using protein-free reverse micelles, at larger volume fraction values and higher temperature. The percolation volume fraction in our experimental conditions (T=20°C, w=40) is close to 10% and to 20% in the presence (4[RM]) and in the absence of cytochrome c respectively. The percolation temperature, at w=40, ϕ_w = 7.4% is equal to 34°C and to 45°C with and without cytochrome c respectively. Such changes in the percolation threshold is all the more so important as the cytochrome c concentration is high and are due to strong attractive interactions which promote a phase transition with two optically clear phases. The similarity in the critical exponents obtained in the absence and in the presence of protein shows that the percolation process observed with protein-containing reverse micelles is similar to that previously observed with protein-free reverse micelles: in presence of cytochrome c the behaviour of this system can always be described through an aggregation process.

As we have demonstrated, the solubilization of a macromolecule can induce a percolation process in dilute solution. This phenomenon is followed by a phase transition which could be very important for carrying out, in reverse micelles, chemical reactions catalyzed by enzymes. Assuming that the reactants (A, B) and the product (C) are mainly solubilized in the bulk hydrocarbon phase and that the enzyme is located into the micellar core, the chemical reaction could occur in the water pool. At the end of the reaction, the product C is in the bulk phase. Addition of water to the micellar solution induces the percolation and the appearance of a biphasic system. The upper phase would contain product C and all the other components (water, AOT, enzyme) were in the lower phase. The upper phase could then be removed and be replaced by pure isooctane. Addition of AOT and the reactants (A and B) to this system would favor the reformation of reverse micelles containing the enzyme and then gives the enzyme the ability to play its catalytic role and favors the regeneration of the reaction.

Literature cited

-1-*Structure and reactivity in reverse micelles;* Pileni M.P., Ed.; Studies in physical and theorical chemestry 65; Elsevier: Amsterdam, 1989.
-2-Pileni M.P.; Zemb T.; Petit C. *Chem. Phys. Lett.* **1985**, *118*, 414.
-3-Brochette P.; Petit C.; Pileni M.P. *J.Phys.Chem.* **1988**, *92*, 3505.
-4-Petit C.; Brochette P.; Pileni M.P. *J.Phys.Chem.* **1986**, *90*, 6517.
-5-Van Dijk M.A.; Boog C.C.; Casteleijn G.; Levine Y.K. *Chem. Phys. Lett .* **1984**, *111*, 571.
-6-Van Dijk M.A.; Broekman E.; Joosten J.G.H.; Bedeaux D. *J. Phys.* **1986**, *47*, 727.
-7-Safran S.A.; Webman I.; Grest G.S. *Phys.Rev.* **1985**, *32*, 506.
-8-Koppel. D.E. *J. Chem. phys.* **1972**, *57*, 4814.
-9-Van Dijk M.A., *Phys. Rev. Lett.* **1985**, *55*, 1003.
-10-Van Dijk M.A.; Casteleijn G.; Joosten J.G.; Levine Y.K. *J. Chem. Phys.,* **1986**, *85*, 626.

-11-Bhattacharya S.; Stokes J.P.; Kim M.W.; Huang J.S. *Phys. Rev. Lett.* **1985**, *55*, 1884.

-12-Efros A.; Shkolvskii B.I. *Phys. Stat. Sol. B* **1976**, *76*, 475.

-13-Stauffer D., in *Physics Reports; Rev. Sec. Phys. Lett.*, North-Holl. Pub. Comp., **1979**, Vol. 54, N°1, pp 1-79.

-14-Kirkpatrick S. in *Hill-Condensed Matter, Proceeding of the Les Houches Summer School, session XXXI;* R.Balian, Ed.; North-Holl. Pub. Comp., 1979.

-15-Kotlarchyk M.; Chen S-H.; Kim M.W. *Phys. Rev. A* **1984**, *29*, 4, 2054.

-16-Bergman D.J.; Imry Y. *Phys. Rev. Lett.* **1977**, *39*, 19.

-17-Huruguen J.P.; M.Authier M.; Greffe J.L.; Pileni M.P. *Langmuir* **1991**, 7, 243.

-18-Huruguen J.P.; Authier M.; Greffe J.L.; Pileni M.P. *J. Phys.: Cond. Matt.*, **1991**, *3*, 865.

-19-Guinier A.; Founet G. in *Small-Angle Scattering of X-rays;* Wiley J., Ed.; Wiley and Sons: N.Y., 1955.

-20-de Gennes P.G. *Scaling Concept in Polymer Physics;* Cornell University Press: London, 1979.

-21-Hayter J.; Janninck G.; Brochard-Wyart F.; de Gennes P.G. *J. Phys. Lett.* **1980**, *41*, L-451.

RECEIVED September 24, 1991

Chapter 16

Polymerization and Phase Transitions in Deoxy Sickle Cell Hemoglobin

Muriel S. Prouty

Department of Chemistry, University of the District of Columbia, Washington, DC 20008

The polymerization and gelation of deoxygenated sickle cell hemoglobin (HbS) is reviewed here as a classic example of biopolymer self assembly. Under near physiological conditions deoxy HbS polymerizes into multi-stranded fibers, accompanied by an almost simultaneous liquid-to-gel phase transition; and by rheological changes which, in the red blood cell (RBC), form the pathophysiological basis of sickle cell disease. The kinetics and mechanism of polymerization and the gel fiber structure have been described in detail; less is known about alignment and compression of the polymers in concentrated systems. Using the osmotic stress method of Parsegian-Rand, we measured the thermodynamic parameters of liquid-to-gel transitions above solubility in near physiological conditions, and of liquid-to-crystal transitions at higher ionic strengths. The nature of the transition may influence gelation inhibitor testing.

Sickle cell anemia is a genetic disease, but its molecular mechanism and effects are best described in the languages of physical chemistry and of polymer chemistry. Except for genetic therapy, all current therapeutic approaches are based on modifying the physico-chemical behavior of the deoxy sickle cell hemoglobin (HbS) protein molecule, or of its polymeric form; even one of the most promising of the genetic therapies seeks to take advantage of the high degree of non-ideality in concentrated polymer solutions (*see "Approaches to Therapy", below*). The extensive and sophisticated understanding of the role of the sickle cell hemoglobin molecule in the pathophysiology of sickle cell disease which has been attained in the last fifteen years, is based primarily on physico-chemical studies on the macromolecular level. The molecular basis of sickle cell disease is extensively reviewed by Schechter et al. (1). Hofrichter and Eaton have reviewed the physical chemistry of the sickle cell hemoglobin polymer, and its formation and structure (2). On the whole, the biophysicists (theoreticians and physical chemists) who have worked with biochemists and physicians in this field, have not been polymer chemists; nor has this research been published in journals generally followed by polymer scientists. The symposium on *Macromolecular Assembly* and this issue of ACS Symposium Proceedings present an opportunity to reaffirm the important commonalities among diverging specialties.

We review briefly the molecular basis of sickle cell disease (SCD) and its symptoms, the kinetics and mechanism of polymerization and gelation of the deoxy HbS molecule, the fiber structure and behavior of the gel formed on polymerization, the rheology of the red blood cell (RBC), and the physico-chemical basis of therapies, with emphasis on those aspects which most reflect polymer physical chemistry.

We describe the "osmotic stress" method (3) for studying the energetics of macromolecular assembly and structure and its use in lipid bilayers, DNA double helices and other systems. In protein solutions we have used this approach to study the thermodynamics of phase transitions, to measure the work of concentrating protein gels, and to bring about crystallization under controlled conditions. We present results of experiments in which we have measured, by osmotic stress, the work of polymerization and gelation of deoxy HbS, and of aligning and compressing the resultant fibers as solvent is removed (4). Recently, we have tested the effect of increasing ionic strength of the phosphate buffer by using 0.5M and 1.0M buffers, and the effect of different anions on the nature of the condensed phase formed when deoxy HbS polymerizes (5). These results have implications for the testing of potential "gelation inhibitors" for their usefulness in the development of drug therapies for sickle cell disease.

Sickle Cell Disease

The normal red blood cell is a biconcave disc with an extremely flexible membrane; the RBC can squeeze through the capillaries (microvasculature) by thinning out and elongating. The presence of polymerized deoxy HbS causes increased rigidity in the red cell, and changes in rheology which (even in the absence of the morphological changes which gave the disease its name) are responsible, more or less directly, for the wide variety and severity of the symptoms of sickle cell disease. A "crisis" occurs when blood circulation is blocked by accumulation at the capillaries of sickled cells which are more rigid because of the presence of gels of deoxy HbS fibers; these cells are thus not flexible enough to squeeze through the microvasculature, as do normal cells. Blood vessels become blocked, tissues become deprived of oxygen, causing extreme pain in the extremities and joints. Other symptoms may include susceptibility to infection, diminished growth rate in children, anemia (sickle cells have about half the life span of "normal" RBC), damage to the spleen and other organs, and shortened life expectancy. Damage to the cell membrane including that caused by oxidative degradation of HbS, imbalance in RBC electrolytes, and other effects (2) have been implicated as causes of SCD pathology.

Hemoglobin Structure and Polymerization

The hemoglobin molecule as visualized in Figure 1 (roughly spherical -- diameter about 64 A) is composed of four polypeptide chains (for "normal" adult hemoglobin (HbA), two α, two ß) packed in a tetrahedral array. Each chain is coiled compactly in a specific structure and contains a heme group located in a crevice near the exterior; each heme contains an iron atom (Fe^{2+} ion) which can bind an oxygen molecule. The difference between HbA and sickle cell hemoglobin (HbS) involves a change in a single base in the gene for the hemoglobin ß-chain, which, in turn, causes a change in a single amino acid residue (ß-6 Lys-->Val, with a charge change of +1-->0). In deoxy HbS, the uncharged valine side chain acts as a "sticky spot" on the surface. In the process of giving up oxygen in the tissues, the Hb molecule undergoes a conformational change which brings potential complementary contact sites to the surface. Under physiological conditions, the result may be polymerization of the deoxy HbS protein into strands and fibers.

The fiber is known to be polymorphic, but the most common structure, determined by electron microscopy, is composed of seven double strands of monomers (Hb molecules) combined into a twisted helical rod as shown in Figure 2 (6). Contact sites in the double strand are accepted to be those found by X-ray diffraction to exist in the deoxy HbS crystal (7). Less is known about inter-strand contacts; the untwisted double strand may conform to the crystal structure; this is the subject of extensive research in X-ray diffraction (8) and electron microscopy (9). The gel formed by deoxy HbS fibers in the red blood cell may be composed of few or many domains, be more or less oriented, and have different effects on cell rheology and morphology, depending on polymerization conditions and kinetics. There have been extensive studies in vitro and in the red blood cell, of the effect of hemoglobin concentration and composition, rapidity and uniformity of deoxygenation, temperature, and other solution conditions on the kinetics and mechanism of polymerization. As polymer chemists would expect, these affect domain formation and size, appearance of spherulitic structures, etc; they are discussed extensively by Hofrichter and Eaton (1).

Polymerization Mechanism: Factors in Gelation All hemoglobin solutions are highly crowded at red blood cell concentrations (about 340g/L or 0.005 M) (2). Excluded volume effects result in a large degree of non-ideality (activity coefficients of about 50 at physiological Hb concentrations have been calculated (10)). This non-ideality is reflected in the measured osmotic pressures of hemoglobin solutions (see below). Although hemoglobin A remains monomeric and is very soluble both in the oxy and deoxy states under physiological conditions, deoxy HbS polymerizes above its solubility -- also called saturating concentration (c_{sat}),-- in the red cell, and in phosphate buffer solutions at about physiological ionic strength and pH. The polymerized deoxy HbS solution undergoes a liquid to gel phase transition. There is a delay time (t_d) which is very highly dependent on total Hb concentration, after which almost complete polymerization to fibers, and gelation occur almost simultaneously. (11,12) (Evidence of the transitory existence of either oligomers or of the liquid solution

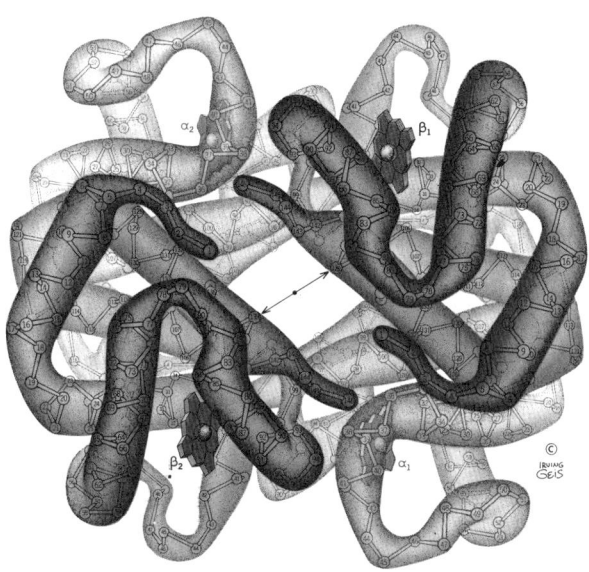

Figure 1. Arrangement of the four polypeptide chains, two α-(identical) and two ß-(identical), and the four hemes, in the hemoglobin molecule. The α- and ß- chains are similar but distinguishable in sequence and folding. (Copyright Irving Geis, with permission.)

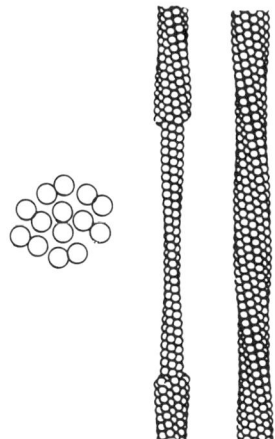

Figure 2. Basic molecular structures in the most common form of polymerized deoxy sickle cell hemoglobin. The twisted 14-strand model of the deoxy HbS fiber (schematic), and the arrangement of 7 pairs of anti-parallel strands in a space-filling model. (Adapted from Ref.9)

of deoxy HbS polymers, depends on interpretation of extremely exacting measurements.) The delay time is the result of a nucleation mechanism in HbS polymerization; in this sense, the process resembles classical nucleation and crystallization of salts and many small molecules, and that for nucleation and growth in binary mixtures of linear homopolymers (13). Depolymerization occurs spontaneously and without a delay time, when oxygen is introduced into the system or when the temperature is lowered. Figure 3 shows both the homogeneous nucleation mechanism first described by Hofrichter and Eaton (11,12) and the heterogeneous nucleation mechanism of Ferrone (14), which agrees more closely with the observed dependence of the delay time on protein concentration. It should be noted, however, that the discussion of nucleation mechanism arises from kinetic studies, and that therapeutic mechanisms based on increasing the delay time are also kinetically based.

Ferrone and his coworkers (15-17) have proposed a theory of the spatial dependence of the polymerization of sickle hemoglobin which should be of interest to polymer physical chemists. Without reference to studies of domain formation in the polymer field, it attempts to reconcile the double (heterogeneous) nucleation mechanism, the kinetics of polymerization, apparent rates of diffusion of monomer into growing polymer domains, and the birefringent domains and spherulitic forms observed in concentrated HbS gels (18). Whether a gel or crystalline solid results from the phase transition depends more on the thermodynamic state of the system than on kinetics: in the osmotic stress studies of sickle hemoglobin polymerization described below (4,5), the form and the concentration of the "equilibrium solid" depend finally on the chemical potential of the hemoglobin in the solution.

Gelation is an endothermic entropic process: both solubility and delay time decrease as temperature increases. Herzfeld and Briehl (19,20) have used statistical mechanical models based on excluded volume and aggregation tendency, to predict phase transitions in reversibly polymerizing systems, and have calculated that repulsive inter-fiber contacts are overcome by the anisotropic crowding in the gel at high hemoglobin concentrations. Gels made from deoxy HbS solutions have been observed to be birefringent (1,21). Although deoxy HbS gels made under quasi-physiological conditions have been observed to crystallize over long times, they are thermodynamically stable, on the physiological time scale of respiration and metabolism, on the crystallization time scale (months to years), and on that of most experiments.

The presence, in sufficient proportion, of non-polymerizing forms of hemoglobin (including oxygenated hemoglobin, normal hemoglobin (HbA), fetal hemoglobin (HbF) (22) or other proteins, may increase solubility -- if <u>total protein concentration</u> (and thus non-ideality) does not increase. It is for this reason that persons who have "sickle trait" -- e.g. whose blood contains both HbA (which does not polymerize) and HbS -- are non-symptomatic. On the other hand, increased total protein concentration, or changes which lead to co-polymerization, hybrid formation, or further crowding in the highly concentrated red cell milieu may cause both deoxy HbS solubility and delay time to decrease.

Therapeutic Strategies.

Most therapeutic approaches seek to modify the physical or chemical factors outlined above, which influence deoxy HbS solubility and/or kinetics. On the whole, these factors depend on the macromolecular nature of hemoglobin itself, and on the "supra-molecular assemblies" represented by its fiber and gel forms. Some therapeutic designs are outlined in Table I, which shows also the particular property of the sickle hemoglobin system which each attempts to alter. "HPFH" (high persistence of fetal hemoglobin) occurs normally in some individuals, and has been observed to have an ameliorating effect in SCD patients who exhibit it. Currently efforts are under way to use the drug hydroxyurea in sickle patients to modify the regulatory mechanism which "turns off" expression of the HbF ß-chain gene at birth, so that the non-polymerizing HbF will continue to be synthesized in the bone marrow -- using genetics to alter polymer chemistry (23). Although HbS production is not lowered, the cell appears to compensate for the increased total amount of hemoglobin by increasing cell volume (24). The shape of the sickle RBC becomes somewhat more spherical, but any adverse effects on cell rheology appear to be outweighed by the beneficial effect on HbS solubility and polymerization delay time which is gained by decreasing the proportion of HbS in the total hemoglobin content of the cell.

The problem addressed in this research project relates to the testing of non-specific "gelation inhibitors"-- compounds which can be expected to prevent polymer formation and sickling by increasing deoxy HbS solubility. The search for peptides which mimic complementary sites and thus compete specifically for contact sites but do not polymerize, has not been successful; however, nonpolar organic compounds, including aromatic amino acids and peptides (25), have been found to act non-specifically to increase the solubility ratio $[(c_{sat})_{add}/(c_{sat})_o]$ of HbS. Such an increase has been generally used as a first test of the effectiveness of a potential therapeutic agent (26). Many promising compounds are rare, or relatively insoluble, and HbS itself is not easy to obtain or purify. Therefore, tests have often been carried out in high ionic strength phosphate buffers in which the solubilities of both deoxy HbS and the additive are markedly decreased -- even though it has been reported that the form of the condensed phase obtained above c_{sat} was "aggregate" or "crystal" not "gel" (21). The aim of this research has been to study the effect of high ionic strength buffers on the structure and mechanism of formation of the condensed phase, as well as on HbS solubility, and to assess the importance of the nature of the solvent-determined phase transition on the ability of the added agent to prevent or delay polymerization.

The Osmotic Stress Method Applied to Proteins

We have adapted the "osmotic stress" method, developed by Parsegian and Rand for the study of intermolecular forces in lipids (3), to the study of the thermodynamics of phase transitions in protein solutions. In systems such as lipid bilayers, DNA double helices, and other macromolecular assemblies whose structural parameters may be readily determined by such techniques as x-ray

Table I. Approaches to Therapy

PHYSICOCHEMICAL PROPERTY	THERAPEUTIC STRATEGY
I. Modification of HbS	-Increase deoxyHbS solubility
A. Contact Sites	-Sterospecific inhibitors
B. Non-specific (hydrophobic) solvent effects	-Non-specific non-polar gelation inhibitors
C. Deoxy HbS conformation	-Covalent reagents bind to amino terminal (increase O_2 affinity)
D. Non-polymerizing hemoglobins (HbA for HbS)	-Replacement transfusion -decrease HbS conc., same total Hb conc.
II. Modification of Red Cell	-Decrease HbS concentration to increase delay time
A. Increase cell volume	-Decrease salt concentration (hyponatremia)
B. Modify Cell Membrane	-Modify ion transport
III. Genetic Modifications	-Alter genes to modify pattern of hemoglobin synthesis
Decrease relative conc. of HbS in total Hb conc.	-Hydroxyurea treatment to maintain expression of fetal hemoglobin

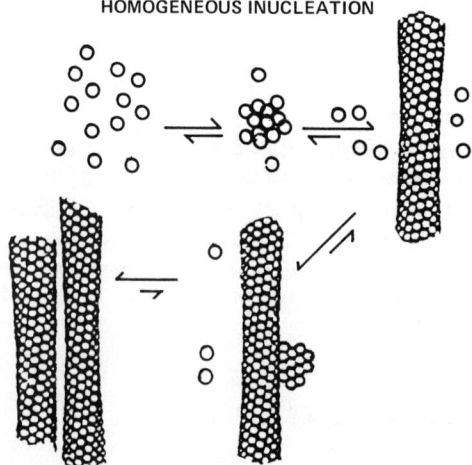

Figure 3. Double nucleation mechanism for deoxy HbS polymerization. (Adapted from Ref.2, based on data in Ref.9 and 14).

diffraction (27), osmotic stress has been used to measure inter- and intra-lamellar forces and to test model structures. This "secondary osmometry" technique is useful for determining the osmotic pressure (and thus macromolecular chemical potentials) in concentrated solutions and in gels and solids in which such measurements have not been practical previously. As applied to protein studies, the method involves equilibration of a number of solutions of protein in dialysis sacs against inert polymer solutions of known, and increasing, osmotic pressure, forming a series of "secondary osmometers". For liquid solution conditions, these yield protein solution osmotic pressures in good agreement with directly measured values. We have shown that protein solutions, brought to concentrations greater than their solubility by equilibration against inert polymers in this way, undergo the same phase transitions as when a temperature-jump (or other unstabilizing change) is used to induce phase transitions in liquid solutions. We are able to measure the osmotic pressure of gels and solids (which we refer to as condensed phases in this context), and thus to calculate the thermodynamic parameters of the phase transition.

On a practical level, if it is necessary to prepare condensed phases with large domains, osmotic stress can be used to increase polymer solution concentrations to values much greater than their solubility slowly and regularly, in such a manner that the expected phase transition can be brought about under controlled conditions. This is preferable to such measures as temperature jumps, which often lead to a large number of very small domains, and which may not succeed for the most concentrated solutions.

We may predict that the response of protein solutions to being concentrated by increasing osmotic stress will depend on the intermolecular forces present (protein-protein, protein-solvent, and solvent-solvent), and on excluded volume considerations. Where repulsive forces or hard sphere conditions exist, high non-ideality leads to sharply increasing osmotic pressures (or, stated another way, a limiting solution concentration with increasing osmotic pressure) as shown by the liquid solution curves in Figures 4 and 6. We plot concentration versus osmotic pressure as the independent variable. Above saturation, we have seen two kinds of phase transitions: in the first the product is a single non-compressible condensed phase -- this is formed by proteins (such as lysozyme) which in a test tube, yield a solid crystalline phase in equilibrium with a saturated liquid solution. The product in osmotic stress cells prepared at stresses greater than π_{sat} (the osmotic pressure of a saturated solution), is always a solid phase only, with no liquid solution present. The saturated liquid solution can *never* be in equilibrium with the stressing solution: solvent is always being withdrawn from the saturated stressed solution, which undergoes a continuing phase transition to the condensed phase until there is no saturated solution left. Figure 4 shows actual data for lysozyme (4) for which the product found in all tubes above π_{sat} was a white powder. Alternatively, we find (4) a compressible (gel) phase as in deoxy HbS (Figure 6), in which the final protein concentration is determined by its equilibrium osmotic pressure (imposed by the stressing inert polymer solution). Where attractive forces dominate (eventually causing phase separation), clustering of proteins may occur in the liquid solution, and may decrease osmotic pressures observed below that expected for the hard sphere excluded volume case. Tardieu et al (28) have reported this effect in γ-crystallin solutions.

Experimental Methods

Sickle cell hemoglobin is purified from whole blood using standard methods, as described in Ref.4. The inert polymer used for stressing is Dextran (T500 and T70) from Pharmacia. Osmotic stress cells are prepared from dialysis tubing and test tubes (Figure 5). (Generally about 0.5 mL of hemoglobin solution is immersed in 7-8 mL of Dextran solution in the same buffer.) Osmotic cells are assembled in the cold in a glove box under nitrogen. As an oxygen scavenger, sodium dithionite (final concentration = 0.02M) is added to both Hb and degassed Dextran solutions. Dextran solutions are made up by weight in the desired buffer; their osmotic pressures are determined before and after equilibration. Equilibration is carried out in a thermostatted shaker bath, and has been shown to be complete in four days. (Deoxy hemoglobin is a stable protein below about 45°C and does not show deterioration under osmotic stress, using standard tests.) Gelation of the hemoglobin is tested by cessation of flow. After equilibration the cells are opened, the gelled or crystallized hemoglobin is "melted" in air at 3°C, and hemoglobin concentrations are measured spectrophotometrically as cyanomet hemoglobin (29).

In our previous work, as well as in most other osmotic stress studies, final equilibrium osmotic pressures of the inert polymer solutions have been routinely calculated from concentrations obtained by refractive index or viscosity measurements, using calibration curves from carefully carried out standardization experiments. It has been found that, because of the high degree of non-ideality of the dextran or polyethylene glycol solutions used, neither the molecular weight of the polymer used, the temperature or the nature of the aqueous solvent has a significant effect on the polymer solution osmotic pressure. For dextran in 0.15M phosphate buffers, we used data from measurements in distilled water with confidence (4). This assumption has been shown to be valid also in the work of Parsegian, Rau and Rand and others on lipid bilayers (3), muscle and TMV (30), and DNA's (27), where pressures of several atmospheres or more are common. However, in the present research, we are primarily concerned with solvent and salt effects, using phosphate buffers up to 2.0 M, and also achieving high ionic strengths with other anions. In these media, the solubilities of both hemoglobin and dextran are reduced, as is the osmotic pressure at which the hemoglobin phase transition occurs, so that most of our critical measurements are at osmotic pressures below one-fourth of an atmosphere. This requires use of dextran solutions at quite low concentrations. We now use a UIC Osmomat 050 electronic membrane osmometer (with cellulose membranes from UIC, or Amicon membranes) to measure directly the osmotic pressure of the stressing dextran solutions. We find experimentally significant differences in dextran osmotic pressures in the high ionic strength media. We have not yet attempted to analyze these, but use the directly measured osmotic pressure in each osmotic stress equilibration experiment. One of the advantages of the osmotic stress method in thermodynamic studies is that non-ideality is built in: osmotic pressures

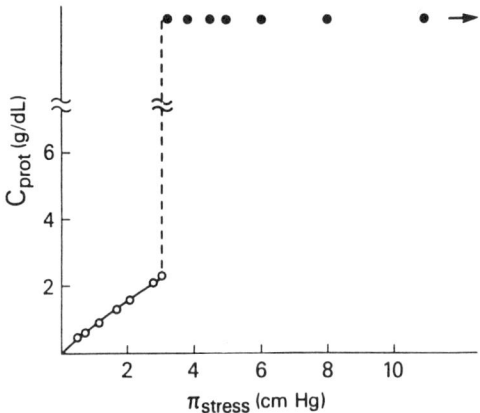

Figure 4. Osmotic stress induces formation of a crystalline solid phase where protein-protein interactions are strong. Below saturation, the solution is highly crowded and non-ideal. At any osmotic pressure greater than that of the saturated solution, no equilibrium is possible -- solvent is withdrawn until only solid remains in the osmotic cell. The protein is egg white lysozyme in 0.5M acetic acid buffer, ph 5.0 at 30°C. (Adapted from Ref.4)

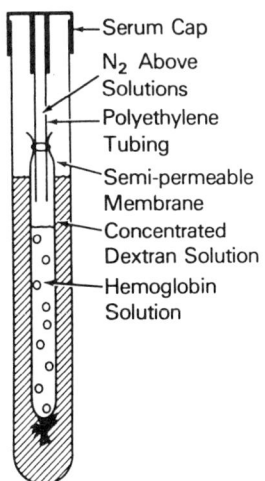

Figure 5. Test tube "secondary osmometer". (Adapted from Ref.4)

may be directly converted to chemical potentials without requiring activity corrections to concentrations. Initial studies were carried out in 0.15M phosphate buffer, pH 7.1, at 3, 20, 30 and 37 C (4). In recent work (5), osmotic stress runs were carried out in 0.5, 1.0, and 1.5M phosphate buffer at pH 7.1; and in 0.15M phosphate buffers, pH 7.1, with ionic strength made equivalent to that of 0.5 and 1.0M phosphate buffers by using sodium chloride or sodium sulfate salts.

Results and Discussion

In every osmotic stress cell the final concentration of hemoglobin and the state of the solution, liquid or gel, are determined by the chemical potential imposed by the osmotic pressure of the stressing solution. Extensive tests (4) have shown that the same equilibrium can be achieved from less and more concentrated solutions, from lower and higher temperatures and from liquid or gel initial states. Ferrone and co-workers (15-17) have used these data when equilibrium gel concentrations were required in testing theories of monomer diffusion in growing gels. The total hemoglobin concentration in the final or "equilibrium" gel is stated to be in agreement with that achieved by Prouty et al (4) in compressing dilute gels by osmotic stress.

Hemoglobin Solutions. Data for the osmotic pressures of deoxy HbS solutions, obtained by the osmotic stress method, in low phosphate buffer (0.15M) at several temperatures are represented by the curve in Figure 6. This curve fits the measured solution osmotic pressures from osmotic stress data at all temperatures, for HbA, HbS and an HbAS mixture (4). The results are also in excellent agreement with osmotic pressures obtained by Adair in a classic experiment in 1928, for sheep hemoglobin in distilled water at $0^{\circ}C$ (31). The curve is that calculated by Ross and Minton (10) by fitting a hard sphere excluded volume model to Adair's data. As long as the pH is near 7, most hemoglobins, oxy or deoxy, from many species, in different buffers at different temperatures fit the same curve.

Phase Transitions in 0.15M Phosphate Buffer. As shown in Ref.4, and seen in Figure 6, at all temperatures, just above a certain saturating concentration (c_{sat}) and osmotic pressure, a sharp increase in the concentration of deoxy HbS occurs; the material in the osmotic stress cell is observed to be a gel: it no longer flows. At all osmotic pressures up to 30cm Hg (the highest in this work), the gel looks clear and ruby red, and is birefringent. When the sacs are opened, the gels formed just above π_{sat} (the critical osmotic pressure corresponding to c_{cat}) have a gelatin consistency; those formed at much higher osmotic pressures are like tars. The gelation process at π_{sat} is accompanied by a large loss of solvent, corresponding at the minimum to the sharp rise in concentration of the c vs π plot at the saturating osmotic pressure. At each temperature, under osmotic stress, gelation occurs at the same c_{sat} as measured by the ultracentrifugation test.

Ultracentrifugation Test of deoxy HbS Solubility What occurs in osmotic stress cells is quite different from what happens when gels are formed in test tubes by warming a supersaturated solution ($c > c_{sat}$) of deoxy HbS from $4^{\circ}C$ to a temperature near or above room temperature. In this "stoichiometric" process there is no solvent loss on gelation (32,33). However, if the gel formed is then

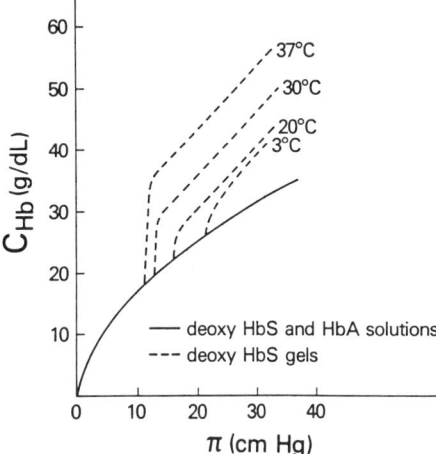

Figure 6. Protein-protein interactions cause formation of a compressible gel phase when interactions are moderate -- the phase transition is entropically induced. The osmotic stress diagram is for deoxy HbS polymerization and gelation in 0.15M phosphate buffer, pH 7.1. The solid curve shows hemoglobin solution osmotic pressure at all temperatures. At each temperature, a phase transition occurs at a particular osmotic pressure. The isotherms (dashed curves) for gel formation "peel off" the solution curve at concentrations corresponding to solubilities measured by other methods. Gel formation is accompanied by significant dehydration. Above the sharp transition region, hemoglobin concentration increases gradually, linearly with osmotic pressure. (Adapted from Ref 4.)

subjected to ultracentrifugation, it is separated into a packed gel and a supernatant solution of concentration c_{sat}. When more concentrated gels are centrifuged, the amount of packed gel increases, but not c_{sat}. This separation into two phases is the basis of the ultracentrifugation assay which is the generally accepted method of determining the solubility of deoxy HbS (11,21).

Polymer Packing and Alignment in Stressed Gels As shown in Figure 6, at each temperature, the gels formed at successively higher osmotic pressures than π_{sat}, are more concentrated in hemoglobin, as solvent continues to be removed from the gel. We have interpreted this linear region to be one of alignment and compression of the gel fiber structure, with little formation of new polymers, but some growth in the length of existing polymers. The slope $(dc/d\pi)$ of the c vs π line can be interpreted as a measure of the work of removing an increment of solvent from the gel; it is not clear why these slopes are identical: the implication is that the work (and thus the process) of removing an increment of buffer is the same at all concentrations and temperatures in the range measured. Hentschke and Herzfeld (34,35) have calculated by statistical mechanical models that, in addition to the compression and alignment of the polymer, some dehydration of the fibers themselves may be necessary to account for the linearity of the gel response to removal of solvent by osmotic stress. The large positive entropies and the great decrease in gel hydration with increasing temperature observed in Figure 6, may be evidence of long-range hydration forces similar to those observed in DNA (27).

Comparison of Condensation in Macromolecular and Simple Systems Physical chemists may find the data in Figure 6 more familiar when replotted as pressure (π) vs molar volume, V (1/c) in Figure 7. In this way we may compare this protein solution system with the phase diagram (PV isotherms) of the condensation of carbon dioxide gas as shown in a standard physical chemistry text (insert).

Phase Transitions in 0.5M Phosphate Buffer. The phase diagram (isotherm for 37°C) is shown in Figure 8. Both the form of the plot and the behavior of the condensed phase are the same as that for 0.15M phosphate buffer, although the phase transition occurs at a lower hemoglobin concentration, as expected, and in agreement with values measured by Asakura (17). It may be significant, in assessing gel structure, to note that the behavior of deoxy HbS gels, once formed, seems to be the same in different media. Here the gels formed in 0.50M phosphate buffer compress on the same line in Figure 8 as those formed in 0.15M phosphate at the same temperature. Previously (4), we found that HbS gels formed in a mixture of HbS and HbA compressed on the same line as Hbs gels in pure HbS solution (both in 0.15M phosphate at 30°C).

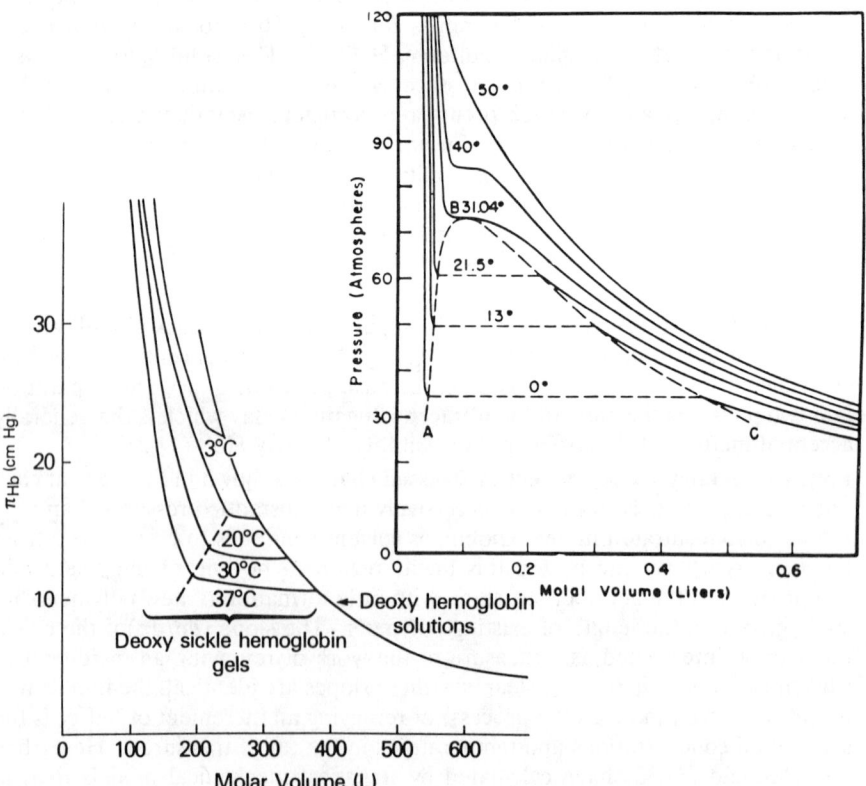

Figure 7. Isotherms for deoxy HbS in 0.15M Phosphate buffer, pH 7.1, plotted as osmotic pressure (π) versus molar volume (V). Insert shows PV isotherms for condensation of carbon dioxide gas, from a standard physical chemistry text. (Adapted from Ref.4)

Phase Transitions in 1.0M Phosphate Buffer. Deoxy HbS, equilibrated at 37°C, in this high ionic strength buffer behaves in the way crystal-forming proteins were predicted and found to behave (see Figure 5), both in the form of the condensed phase (Figure 9) and the phase diagram for 1.0M phosphate (Figure 10). Hemoglobin concentrations of the condensed phase removed from all the dialysis sacs where the osmotic pressure was greater than π_{sat} were close to 70 g/dL, which is the hemoglobin concentration measured in the crystal (8), and no residual liquid or gel can be seen in the osmotic stress cells shown in Figure 9. The aggregates have the regular appearance of macroscopic crystals, but we have not yet attempted X-ray diffraction studies. Asakura and coworkers (17) who studied gelation in hemoglobin solutions in high phosphate buffers, at Hb concentrations not far above c_{sat}, by a very different method, observed a transition between the formation of gels and "aggregates" at about 1.8M phosphate. Their measurements were carried out at 30°C where HbS solubility is greater, and represented essentially a "stoichiometric" experiment. We are planning further experiments to locate the ionic strength at which the form of the condensed phase changes, and to determine the phase diagram. Knowing critical ionic strength, critical osmotic pressure and solubility, and temperature it should be possible to quantify the effects which lead to a fibrous compact gel or to a crystal.

Phase Transitions in Mixed Salt Media When the ionic strength of the 0.15M phosphate buffers is increased to that of the 0.5 phosphate buffers by adding either NaCl or Na_2SO_4, crystalline rather than gel condensed phases are seen (Figure 11). Crystals appear to form more easily and are more distinct. We obtain only crystalline solids in dialysis sacs equilibrated against dextrans in buffers where NaCl or Na_2SO_4 is added to attain the ionic strength of 1.0M phosphate buffers (Unpublished results).

Interpretation of Gelation Inhibitor Tests These results urge caution in the use of high ionic strength phosphate buffers to test the effectiveness of gelation inhibitors on deoxy HbS solubility. The solubility ratio (c_{sat}/c_{sat}^o) of solutions with additive (c_{sat}) and without (c_{sat}^o) is only one of several factors to be considered: the effect of masking electrostatic interactions by increasing ionic strength is not limited to polymerization tendency (solubility) alone; inter-fiber interactions act to govern the nature of the condensed phase. One must consider the overall effects of generally non-polar additives on hydrophobic entropic processes taking place in aqueous media of varying ionic strengths. The results above indicate also that the particular ions present affect the nature of the phase transition: in 0.5 M phosphate buffer a gel phase is found at concentrations greater than c_{sat}, while in solutions with the same ionic strength, but made up of 0.15M phosphate and either NaCl or Na_2SO_4, the solid formed above the same solubility is crystalline. It is by no means clear that (as has been assumed) ranking a series of gelation inhibitors in order of their effect on deoxy sickle hemoglobin solubility 1.8 M phosphate buffer will give a reliable indication of their relative effectiveness in preventing gelation under physiological conditions.

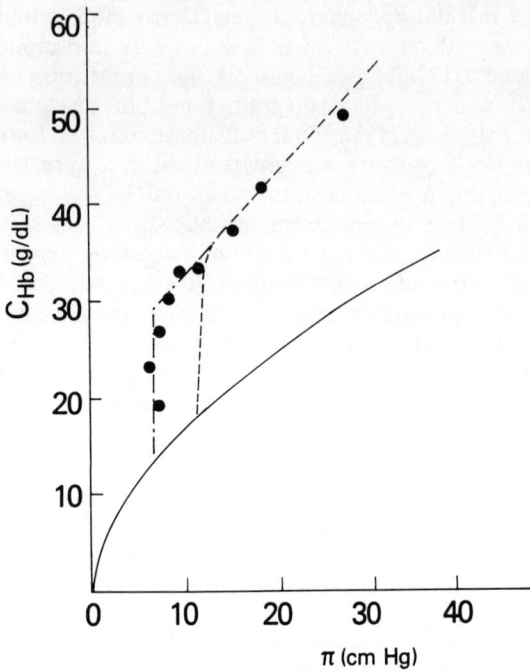

Figure 8. Osmotic Stress Phase Diagram for deoxy HbS in 0.5M Phosphate buffer, pH7.0, at 37°C. The solid curve represents liquid solution osmotic pressures for hemoglobin. (— —) is the gel phase curve for deoxy HbS in 0.15M phosphate buffer at 37°C shown in Figure 5. (—·—) and data points are for experiments in 0.5M phosphate buffer, pH 7.1

Figure 9. Osmotic Stress Cells containing deoxy HbS in 1.0M Phosphate buffer, pH 7.0, after equilibration with Dextran solutions of osmotic pressure 8.0 cmHg (1008) and 9.0 cmHg (1009).

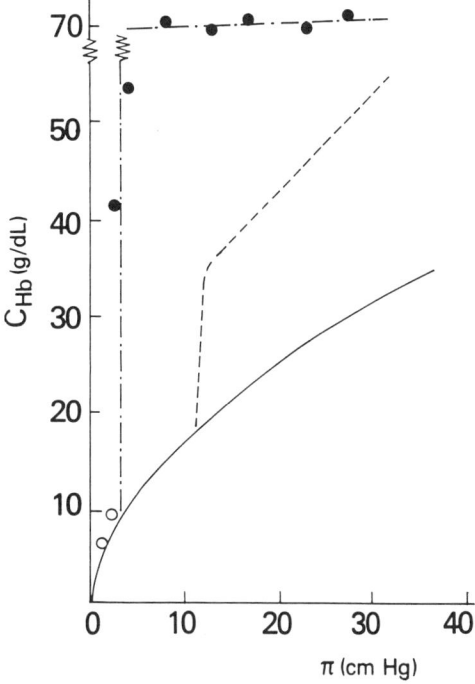

Figure 10. Osmotic Stress Phase Diagram for deoxy HbS in 1.0M Phosphate buffer, pH7.0, at 37°C. The solid and dashed curves are for liquid solution and gels, respectively, in 0.15M phosphate at 37°C; (— · —) and data points are for experiments in 1.0M phosphate buffer, pH 7.1.

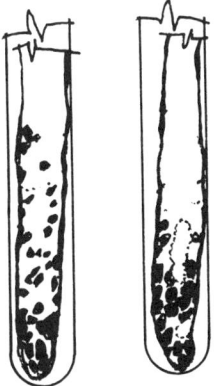

Figure 11. Osmotic Stress Cells containing deoxy HbS in 0.15M Phosphate buffer, pH 7.0, with ionic strength made equal to that of 0.5M Phosphate buffer, using sodium chloride (508C) or sodium sulfate (508S). Stressing solutions in these cells contained Dextran in the same buffers, with an osmotic pressure of 8.0 cmHg.

Because the pathophysiology of sickle cell disease results from the physicochemical properties of a system which is macromolecular and supra-molecular in nature, the principles of polymer physical chemistry must be used to analyze and explain the phenomena observed. Primarily, polymer chemistry should be taken into account in designing and testing therapeutic approaches at every stage. As is shown here, consideration of such principles is necessary in the verification of the effectiveness of agents for gelation inhibition -- and, by extension, in other therapies as well. Conversely, the polymerization and phase behavior of this protein *in vivo* and *in vitro* provide informative examples of the phenomenon of self-assembly in biopolymers.

Acknowledgments M.S.P. is supported by NIH Minority Biomedical Research Support Grant GM08005-20. MBRS students Erick Nana, Keisha Alexander and Marion Patterson helped carry out many of the experiments. Their assistance is gratefully acknowledged, as is that of A. Curtis and J. Martinelli.

LITERATURE CITED:

1. Eaton,W.A.;Hofrichter,J. In *Advances in Protein Chemistry*, U.S.A.,Anfinsen,C.B.; Edsal,J.T.;Richards,F.M.;Eisenberg,D.S., Eds;*Academic Press,Inc.,N.Y.,N.Y.*; 1990,*40*,pp.63-253.
2. Schechter,A.N.;Noguchi,C.T.;Rodgers,G.P. In *Molecular Basis of Blood Disease*; Stamatoyannopoulos,G.;Nienhuis,A.W.,Eds.; Saunders, Philadelphia,PA, **1987**, pp.179-218.
3. Parsegian,V.A.;Rand,R.P.;Rau,D.C. In *Methods in Enzymology*, Packer,L.,Ed.; *Academic Press,Inc.,N.Y.,N.Y.*; **1986**,*127*,400-416.
4. Prouty,M.S.;Schechter,A.N.;Parsegian,V.A.*J.Mol.Biol.* **1985**,*184*,517.
5. Patterson,M.;Nana,E.Y.;Alexander,K.;Prouty,M.S.*Biophys.J.* **1991**,*59*,285a.
6. Dykes,G.W.;Crepeau,R.H.;Edelstein,S.J.*J.Mol.Biol.* **1979**,*130*,451.
7. Wishner,B.C.;Ward,K.B.;Lattman,E.E.;Love,W.E.*J.Mol.Biol.* **1975**,*98*,179.
8. Padlan,E.A.;Love,W.E.*J.Biol.Chem.* **1985**,*260*,8272,2880.
9. Rodgers,D.;Crepeau,R.H.;Edelstein,S.J.*Proc.Nat.Acad.Sci.U.S.A.***1987**,*84*,6157.
10. Ross,P.D.;Minton,A.P.*J.Mol.Biol.* **1977**,*112*,437.
11. Hofrichter,J.;Ross,P.D.;Eaton,W.A.Proc.Nat.Acad.Sci.U.S.A.**1976**,*73*,3035.
12. Ross,P.D.;Hofrichter,T.A.;Eaton,W.A.*J.Mol.Biol.* **1977**,*115*,111.
13. Bates,F.S.*Science* **1991**,*251*,898.
14. Ferrone,F.A.;Hofrichter,J.;Sunshine,H.R.;Eaton,W.A.*J.Mol.Biol.* **1985**,*183*,611.
15. Basak,S.F.;Ferrone,F.A.;Wang,J.T.*Biophys.J.* **1988**,*54*,829.
16. Zhou,X.Z.;Ferrone,F.A.*Biophys.J.* **1990**,*58*,695.
17. Cho,M.R.;Ferrone,F.A.*Biophys.J.* **1990**,*58*,1067.
18. Cristoph,G.W.;Hofrichter,J.;Eaton,W.A.*Biophys.J.* **1991**,*59*,285a.
19. Briehl,R.W.;Herzfeld,J.*Proc.Nat.Acad.Sci.,U.S.A.* **1979**,*76*,2740.
20. Herzfeld,J.;Briehl,R.W.*Macromolecules* **1981**,*14*,379.
21. Adachi,K.;Asakura,T.*Biol.Chem.* **1979**,*254*,7765.
22. Sunshine,H.R.;Hofrichter,J.;Eaton,W.A.*J.Mol.Biol.* **1979**,*133*,435.
23. Noguchi,C.T.;Rodgers,G.P.;Serjeant,G.;Schechter,A.N.*N.Engl.J.Med.* **1988**,*318*,96.

24. Rodgers,G.P.;Dover,G.J.;Noguchi,C.T.;Schechter,A.N.;Neinhuis, A.W.*N.Engl.J.Med.* **1990**,*322*,1037.
25. Dean,J.;Schechter,A.N.*N.Engl.J.Med.* **1978**,*299*,863.
26. Poillon,W.N.;Bertles,J.F.*J.Biol.Chem.* **1979**,*254*,3462.
27. Rau,D.;Lee,B.K.;Parsegian,V.A.*Proc.Nat.Acad.Sci.U.S.A.* **1984**,*81*,2621
28. Veretout,F.;Delaye,M.;Tardieu,A.*J.Mol.Biol.* **1989**,*205*,205.
29. van Assendelft,O.W.;Spectrophotometry of Hemoglobin Derivatives; Royal Vanogram,Assen; **1970**; pp.110-112.
30. Millman,B.M.;Irving,T.C.;Nickel,B.G.;Loosely-Millman,M.E. *Biophys.J.* **1984**,*41*,551.
31. Adair,G.S.*Proc.Roy.Soc.London ser A.* **1928**,*129*,573.
32. Kowalczykowski,S.;Steinhardt,J.*J.Mol.Biol.* **1977**,*115*,201.
33. Kahn,P.C.;Briehl,R.W.*J.Biol.Chem.* **1979**,*257*,12209.
34. Hentschke,R.;Herzfeld,J.*Mater.Res.Soc.Sym.Proc.* **1990**,*177*,305.
35. Hentschke,R.;Herzfeld,J.*Phys.Rev.A* **1991**,*44*,1148.

RECEIVED October 14, 1991

Chapter 17

Reduction of Phospholipid Quinones in Bilayer Membranes
Kinetics and Mechanism

Charles R. Leidner, Dale H. Patterson, William M. Scheper, and Min D. Liu

Department of Chemistry, Purdue University,
West Lafayette, IN 47907-1393

Gel permeation chromatography, electron microscopy, ^{1}H NMR spectroscopy, UV-Vis spectroscopy, and stoppped-flow kinetics have been employed to determine the structural and redox properties of quinone-functionalized phosphatidylcholine liposomes. These unilamellar liposomes (ca. 25-30 nm diameter from sonication or 100 nm diameter from extrusion) typically contain 2 - 20 mol% phosphatidyl-choline anthraquinone (DPPC-AQ) which can be reduced and reoxidized by solution reagents. The transmembrane distribution of DPPC-AQ is controllable (58 - 98 %outer) via phospholipid compositions and liposome preparation methods. The rate law for $S_2O_4^{2-}$ reduction of DPPC-AQ/DOPC,

$$k_{obs} = k_1 k_2 [S_2O_4^{2-}] / (k_{-1} + k_2 [S_2O_4^{2-}])$$

indicates the presence of two kinetically-distinct forms of DPPC-AQ. Comparison with the corresponding homogeneous rate constant suggests the identities of the two pathways.

The recent (1) crystal structure of the bacterial photosynthetic reaction center provides a striking vision of how nature positions, orients, and assembles redox molecules in order to effect specific and efficient redox reactions (i.e., charge separation). Complimentary to photosynthetic energy transduction is respiratory energy transduction within the inner mitochondrial membrane wherein charge is transported across a phospholipid bilayer membrane, eventually leading to the reduction of oxygen and the formation of ATP (2). Figure 1 provides a representation of the Q-cycle, a key sequence of redox reactions occuring at specific protein sites within the membrane. The performance of these photosynthetic and respiratory "reaction centers" depends critically on the position, orientation, and assembly of the redox molecules. Both energy transduction processes involve the transport of electrons through a phospholipid bilayer membrane by membrane-bound

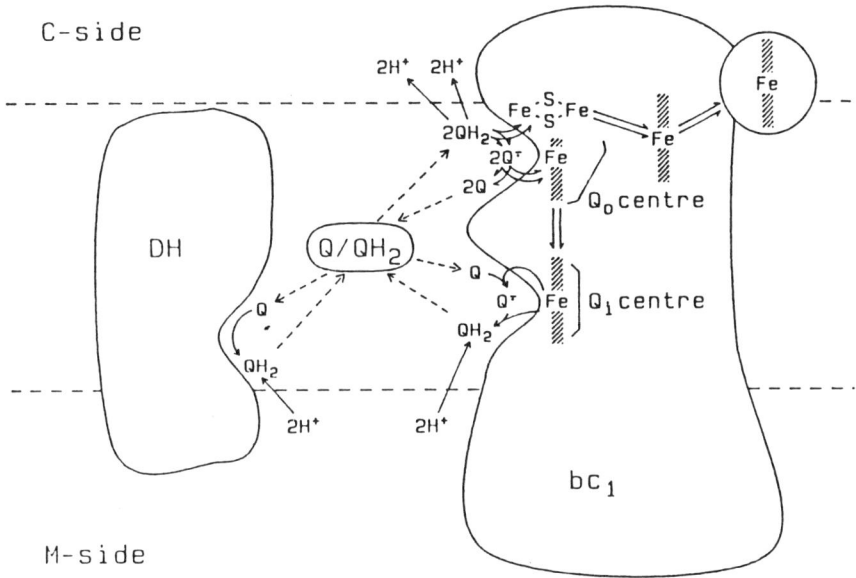

Figure 1. Representation of the Q-cycle in respiratory energy transduction. (Reproduced with permission from ref. 2. Copyright 1986 Plenum.)

quinones (2). The motivation to prepare structurally-defined redox assemblies, particularly containing quinones within phospholipid membranes, is obvious. To this end we initiated (3-8) a study of the stuctural, redox, and transport properties of quinone-functionalized monolayers and bilayers as simple, chemical models for quinone-mediated energy transduction.

DPPC-AQ is a phosphatidylcholine anthraquinone that closely resembles simple phospholipids like dipalmitoylphosphatidylcholine (DPPC) and its dioleoyl (DOPC) and ethanolamine (DPPE) analogs (Scheme I). Unilamellar, quinone-functionalized liposomes, prepared by the sonication (6) or extrusion (9) of DPPC-AQ and the simple phospholipids, provide chemical assemblies with which to study the redox and transport properties of membrane-bound quinones. We present herein the use of gel permeation chromatography, electron microscopy, ^{1}H NMR spectroscopy, UV-Vis spectroscopy, and stoppped-flow kinetics to provide a description of the structural and redox properties of unilamellar liposomes containing DPPC-AQ.

Experimental Section

The simple phospholipids were purchased from Avanti Lipids (DOPC, DPPC, DPPE, and MPPC (monopalmitoylphosphatidylcholine)); the anthraquinone (DPPC-AQ) and anthracene (DPPC-AN) analogs were prepared and purified as described previously (6). The purity (> 99%) of the phospholipids was verified by TLC or ^{1}H NMR. All lipids were stored in a dessicator at -10°C. All other chemicals were reagent-grade or better and were used without further purification.

Liposome preparation. Liposomes were prepared from a mixture of simple phospholipid (DPPC, DPPE, MPPC, or DOPC) and functionalized phospholipid (DPPC-AQ or DPPC-AN). Phospholipid mixtures were (20 mg) suspended in 1 mL of 50 mM tricine (pH = 8.0), 0.2 M KCl, and 1 mM EDTA solution under N_2. The resulting suspension was either sonicated (20-40 minutes) or extruded (9) through 100 nm Nuclepore membranes under nitrogen. After sonication the liposomes were fractionated on a Sephadex G50 or Sepharose 4B columns. All manipulations were performed at 52°C for DPPC (T_c = 42°C) or room temperature for DOPC (T_c = -22°C). ^{1}H NMR spectra were obtained on a Varian VXR500S spectrometer at 52°C using the standard Varian S2 pulse sequence. Electron micrographs were obtained with frozen, phosphotungstate-stained suspensions of liposomes.

Spectrophotometry and Stopped-Flow Experiments. Details of the spectrophotometric and kinetics experiments have been presented previously (6). In short, the liposome eluate from the Sephadex column was diluted in a cuvette in a thermostatted cuvette holder. Reagents were syringe-injected into the liposome solution. The titration experiments were performed on a Hewlett Packard 7450 diode array spectrophotometer. The lipid concentratation was typically 0.1 to 1 mM; the DPPC-AQ concentration was typically 2 to 200 μM. The DPPC-AQ concentration was calculated from the extinction coefficient of the [Me_3NCH_2AQ](Br) analogue (ϵ = 4750 $M^{-1}cm^{-1}$ at 322 nm). Stoppped-flow spectrophotometry was performed on a High-Tech Stopped-Flow Spectrophotometer interfaced to a Zenith 151 computer by a MetraByte Dash 16 A/D card. All solutions were thermostatted

SCHEME I. Phospholipids employed to prepare liposomes.

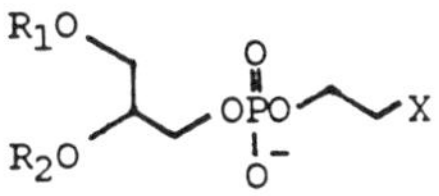

R	X	Phospholipid
$R_1 = R_2 = CH_3(CH_2)_{14}C(O)$	$(CH_3)_3N^+$	DPPC
$R_1 = R_2 = CH_3(CH_2)_{14}C(O)$	H_3N^+	DPPE
$R_1 = R_2 = CH_3(CH_2)_{14}C(O)$	$(CH_3)_2N^+$–CH$_2$–anthraquinonyl 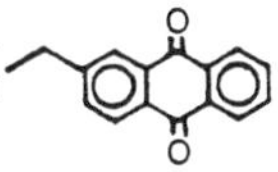	DPPC-AQ
$R_1 = R_2 = CH_3(CH_2)_{14}C(O)$	$(CH_3)_2N^+$–CH$_2$–anthracenyl 	DPPC-AN
$R_1 = CH_3(CH_2)_{14}C(O)$, $R_2 = H$	$(CH_3)_3N^+$	MPPC
$R_1 = R_2 = CH_3(CH_2)_7CH=CH(CH_2)_7C(O)$	$(CH_3)_3N^+$	DOPC

at 25 ± 0.1 °C. Excellent fit to a single exponential was observed for most of the absorbance (of the product) vs. time traces. Some traces (especially at high [$S_2O_4^{2-}$]) deviated at short times and were fit beyond the first half-life. Runge-Kutta simulations of [H_2Q] vs. time were performed on a Zenith 286 microcomputer using the Fortran program GEAR.

Results and Discussion

The structural similarity between DPPC-AQ and simple phospholipids (Scheme I) is evident. However, the presence of a bulky, anthraquinone head group affects amphiphile assembly; neat suspensions of the cone-shaped DPPC-AQ do not form liposomes. Mixtures with less than 25 mol% DPPC-AQ yield clear suspensions of liposomes. Liposomes containing 2-20 mol% DPPC-AQ are routinely prepared and are optically identical to those from DPPC, except for the quinone peak at 322 nm. As demonstrated by Figure 2, the elution volume and peak width of these functionalized liposomes from the size exclusion column (Sepharose 4B) matches that of unilamellar DPPC liposomes. Additionally, electron micrographs of frozen samples of DPPC and DPPC-AQ/DPPC liposomes are identical -- a majority of the structures are 25 - 35 nm; a few larger objects are observed, likely due to fusion of the smaller liposomes during cooling. Thus, sonicated DPPC-AQ/DPPC liposomes are unilamellar with an average diameter of ca. 25-30 nm (9). Sonicated DPPC-AQ/DPPC/DPPE, DPPC-AQ/DPPC/MPPC, and DPPC-AQ/DOPC liposomes likewise possess optical and size-exclusion chromatographic characteristics identical to those of the corresponding non-functionalized liposomes. The mol% DPPC-AQ in the DPPC-AQ/DPPC liposomes, calculated from the dry mass of the lipid mixture used to prepare the liposome, was verified using spectrophotometry and ^{1}H NMR spectroscopy (6). The dry mass mol% values are accurate to ± 15%.

We have performed little characterization of the extruded liposomes at this time, but we envision no reason for the extruded DPPC-AQ/DPPC or DPPC-AQ/DOPC liposomes to differ from the DPPC or DOPC analogs. Thus, we have unilamellar, 100 nm diameter liposomes containing 5-10 mol% DPPC-AQ.

Spectrophotometric experiments. The liposome-bound quinones can be reduced by external, aqueous $S_2O_4^{2-}$ or BH_4^- and reoxidized by exposure to oxygen or addition of $Fe(CN)_6^{3-}$. Figure 3 illustrates the UV-Vis spectra of a 8.7 mol% DPPC-AQ/DPPC liposome solution with sequential additions of external, aqueous $S_2O_4^{2-}$. Note that the quinone peak (λ_{max} = 322 nm) decreases from 100% to 0% as the hydroquinone peak (λ_{max} = 384 nm) increases from 0% to 100%. Exposure of the hydroquinone solution to oxygen or addition of $Fe(CN)_6^{3-}$ causes a fading of the yellow color and a concomitant regeneration of the quinone peak. Clearly, all of the DPPC-AQ amphiphiles within the liposome are redox-active. (Similar redox behavior is observed with all of the liposome systems.) $S_2O_4^{2-}$ penetrates the bilayer and therefore can reduce all quinones regardless of their location within the bilayer (10). In contrast, BH_4^- does not penetrate into the bilayer (10). By measuring the fraction of DPPC-AQ remaining upon addition of excess BH_4^-, we have

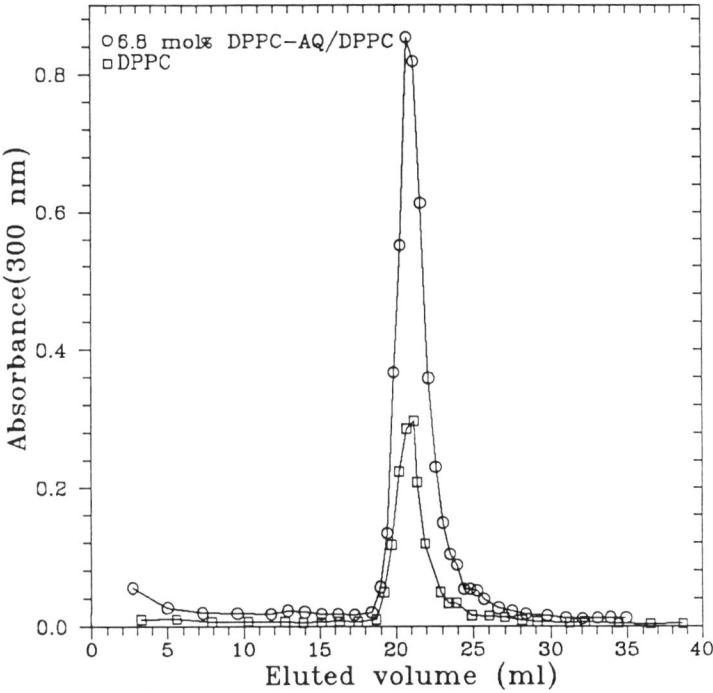

Figure 2. Size exclusion chromatograms for sonicated DPPC (□) and 6.6 mol% DPPC-AQ/DPPC (O) liposome solutions.

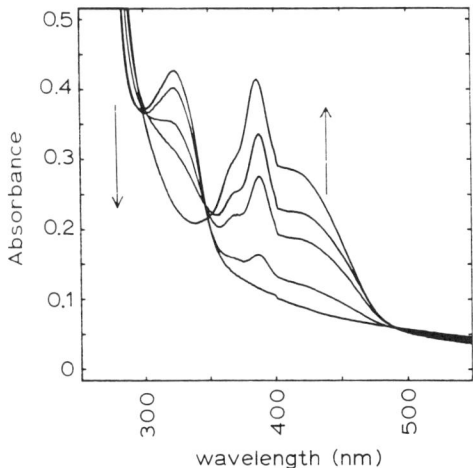

Figure 3. Spectrophotometric response of a 8.7 mol% DPPC-AQ / DPPC liposome solution to the sequential addition of external $S_2O_4^{2-}$.

a means of measuring the fraction of the quinone residing within the outer monolayer of the liposome. Table I lists the percent quinone reduced by BH_4^- in both DPPC and DOPC liposomes. The BH_4^- typically reduces 90 ± 4% of DPPC-AQ amphiphiles in DPPC liposomes containing 5-12% DPPC-AQ, the typical mol% DPPC-AQ used in these studies. DPPC-AQ/DOPC liposomes possess a constant %outer (86 ± 4) throughout the studied mol% range (3.7 to 24).

Statistical considerations (9) alone considering a symmetric phospholipid distribution predict that 70% of the quinone amphiphiles should reside within the outer monolayer for small, unilamellar vesicles (SUVs) of 25 nm diameter. Our data show that the cone-shaped (large head group) DPPC-AQ is incorporated preferentially (90% and 86% vs. 70%) into the outer, less hindered monolayer of the liposomes. This asymmetry can be manipulated through varying the phospholipid proportions and preparation methods. DPPC-AQ/DPPC liposomes with less than 5 mol% DPPC-AQ posssess a greater fraction of outer quinones (98% for 4 mol%) than those with 5-12 mol%. Apparently at low mol% the DPPC-AQ amphiphiles can be accommodated within the more favorable outer layer, but upon increasing mol% the inner monolayer is populated. Incorporation of the inverted cone-shaped (small head group) DPPE into DPPC liposomes pronounces the transmembrane asymmetry of DPPC-AQ (Table I). This system is an example of a completely asymmetric, functionalized liposome -- one in which all of the functionalized phospholipids reside on one side of the liposome. Incorporation of the cone-shaped MPPC into DPPC-AQ/DPPC liposomes has the opposite effect, only 72% (Table I) of the quinones are reduced by BH_4^-. Using the 100 nm extruded liposomes relieves the geometric strain of the SUVs and leads to more statistical distributions. With extruded DPPC-AQ/DPPC liposomes, 58% (vs. 53% statistical) resides on the outer monolayer. This ability to manipulate the DPPC-AQ distribution and thereby engineer liposomes with variable and controllable structure should prove useful in our attmepts to use quinone-functionalized liposomes as simple, chemical models for biological processes.

NMR Studies. The 500 MHz ^1H NMR spectra of (sonicated) DPPC-AQ/DPPC liposome solutions (Figure 4A) reveal the ^1H resonances (11) of the majority DPPC amphiphiles and the diminutive peaks from the anthraquinone portion of DPPC-AQ at 7.8-9.0 ppm (1:4:2 ratio). Several of the DPPC resonances possess the double-peaked shape indicating different magnetic environments for amphiphiles residing on the outer and inner monolayers of the liposome (11). These resonances are partitioned 67:33 on the average, indicating that our quinone-functionalized liposomes are structurally similar to the well-characterized DPPC liposomes -- they are unilamellar and possess an average diameter of 25-30 nm. The observation of only one set of quinone resonances indicates that the quinone amphiphiles exist predominently in one enviroment in the liposomes. This is in agreement with the spectrophotometric results that ca. 90% of the incorporated quinones reside in the outer layer. Observation of separate resonances for the remaining inner quinones is impractical due to their low concentration and/or the substantial linewidths of the quinone resonances.

TABLE I. The % quinone reduced upon addition of excess BH_4^- to liposomes containing DPPC-AQ

liposome[a]	%$Q_{reduced}$
DPPC-AQ/DPPC	
4-5 mol%	97 ± 3
6-16 mol%	90 ± 4
DPPC-AQ/DPPC (extruded, 100 nm)	
5-10 mol%	58 ± 5
DPPC-AQ/DPPC/DPPE	
9.5 mol% (9.5:37:53)	98 ± 2
DPPC-AQ/DPPC/MPPC	
7.6 mol% (7.6:83:9)	72 ± 7
DPPC-AQ/DOPC	
4-24 mol%	86 ± 4

[a]Sonicated, 25 nm diameter liposomes, except as indicated.

Further insight into the structure of DPPC-AQ/DPPC liposomes is provided by nuclear Overhauser experiments. Figure 4B illustrates that irradiation of the 2-glycerol H at 5.6 ppm results in substantial diminution of the 1-AQ H resonance at 8.9 ppm via a through-space coupling (nuclear Overhauser effect). Note the minimal effect on the other resonances in the spectrum. A similar, although less dramatic, effect is observed upon irradiating the NMe_3^+ resonance; irradiating the resonances within the alkyl chains has little effect. These simple experiments demonstrate that the 1-AQ H is located near the glycerol portion of the majority DPPC amphiphiles within the liposome. The quinone "head" group is located at a position near the hydrophilic - hydrophobic interface, not extended out into solution. This description is shown in Figure 5.

A subtlety of the NMR results in Figure 4 is the intensities of the quinone resonances with respect to those of DPPC. At low (< 4) mol% DPPC-AQ, the integrated intensities convert to the expected mol% DPPC-AQ; at higher mol% the integrations are too small (e.g., 4.3 mol% calculated vs. 6.0 mol% actual). This could indicate that a portion of the DPPC-AQ amphiphiles aggregate into a gel-like region and thus exhibits such severely-broadened resonances that they are effectively absent from the spectrum. The integrated intensities would reflect only the "fluid" DPPC-AQ amphiphiles. Increasing temperature could cause some of these immobile DPPC-AQ amphiphiles to "melt" or become more mobile; the integrated intensities of the quinone resonances would increase. Such an effect, although slight, is observed in our experiments. Endogenous quinones are known (12) to aggregate within liposomes, so our chemical models may be mimicking even this aspect of the membrane-bound quinones. Despite this fortuitous similarity, a gel-fluid equilibrium would complicate any detailed analysis of the redox, transport, and structural properties of our system. We are investigating more closely the possibility of aggregation of DPPC-AQ amphiphiles.

Kinetics. Reduction of (sonicated) DPPC-AQ/DOPC liposomes at room temperature proceeds at a rate readily measured using stopped-flow techniques. Figure 6 illustrates the reduction of DPPC-AQ/DOPC liposome solutions ([DPPC-AQ] = 2.3 μM) with $S_2O_4^{2-}$. Satisfactory fit to an exponential growth of H_2Q (———) is observed (see Experimental). The psuedo-first order rate constants (k_{obs}) for the reduction of DPPC-AQ/DOPC by $S_2O_4^{2-}$ were measured at various [DPPC-AQ] and [$S_2O_4^{2-}$]. These data are presented in Figure 7. Within the scatter in these data, no discernible trend of k_{obs} with [DPPC-AQ] is noted. The leveling effect (saturation kinetics) exhibited in Figure 7 is accounted for by the rate law:

$$k_{obs} = \frac{k_1 k_2 [S_2O_4^{2-}]}{k_{-1} + k_2 [S_2O_4^{2-}]} \tag{1}$$

with the best fit parameters:

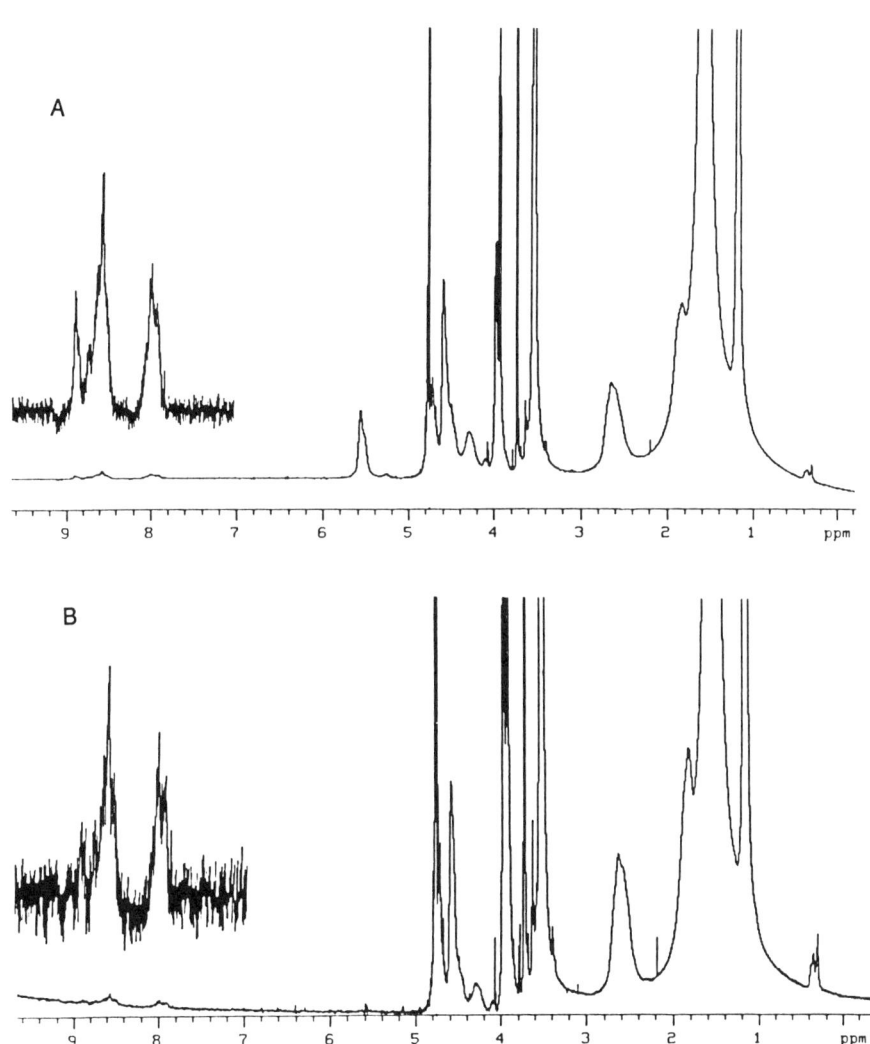

Figure 4. 500 MHz ^{1}H NMR spectra of 6.7 mol% DPPC-AQ/DPPC liposome solutions at 52°C (A); same with irradiation at 5.6 ppm (B). (Reproduced with permission from ref. 6. Copyright 1991 American Chemical Society.)

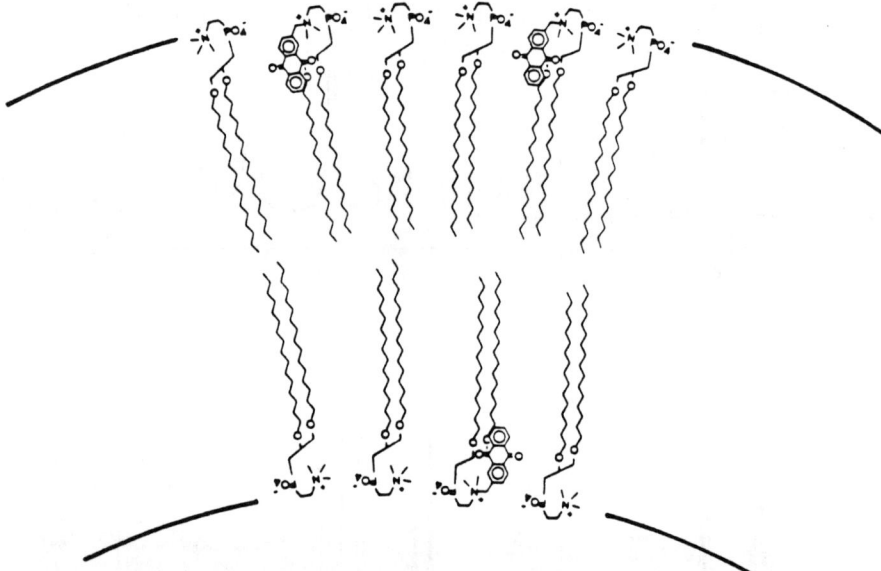

Figure 5. Representation of liposomes containing DPPC-AQ. (Reproduced from reference 6. Copyright 1991 American Chemical Society.)

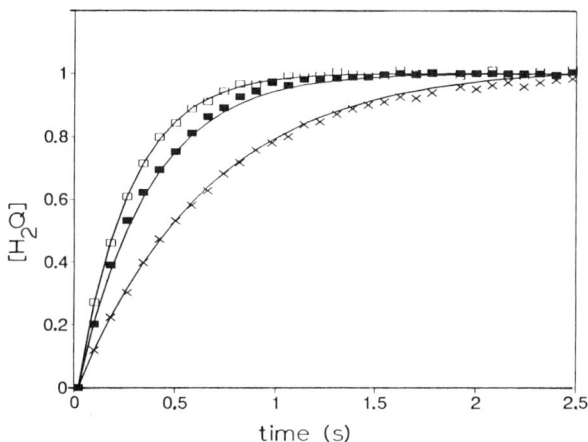

Figure 6. Time dependence of [H_2Q] following addition of $S_2O_4^{2-}$ to DPPC-AQ/DOPC liposome solutions ([DPPC-AQ] = 2.3 μM); [$S_2O_4^{2-}$] = 4140 μM (□), 513 μM (■), and 131 μM (X). Relative hydroquinone concentrations were obtained from absorbance changes at 385 nm after correction for baseline changes. Solid lines are exponential fits for k_{obs} = 3.9, 2.9, and 1.5 s^{-1}. (Reproduced from reference 6. Copyright 1991 American Chemical Society.)

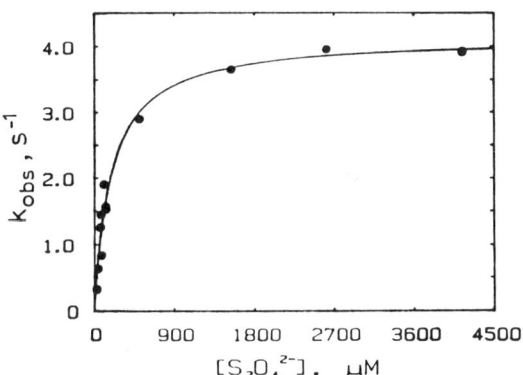

Figure 7. [$S_2O_4^{2-}$]-dependence of k_{obs} for reduction of DPPC-AQ/DOPC. [DPPC-AQ] = 1.78 - 2.28 μM. Solid line is fit to rate law (1). (Reproduced from reference 6. Copyright 1991 American Chemical Society.)

$$k_1 = 4.1 \ (\pm 0.2) \ s^{-1}$$

$$k_2/k_{-1} = 5.2 \ (\pm 1.0) \times 10^3 \ M^{-1}$$

(The significance of these rate constants is given below.) A plot of $1/k_{obs}$ vs. $1/[S_2O_4^{2-}]$ is linear for these data, while a $1/k_{obs}$ vs. $1/[S_2O_4^{2-}]^{1/2}$ plot exhibits pronounced curvature. Thus $S_2O_4^{2-}$, not $SO_2^{-\cdot}$, is the reducing species (13).

This kinetic behavior contrasts that of the solution analog $Me_3NCH_2AQ^+$ with $S_2O_4^{2-}$. The $Me_3NCH_2AQ^+$ data fit the simpler rate law:

$$k_{obs} = k[S_2O_4^{2-}] \qquad (2)$$

where $k = 1.1 \ (\pm 0.05) \times 10^5 \ M^{-1}s^{-1}$. Again $S_2O_4^{2-}$ is the operative reductant.

Mechanism of $S_2O_4^{2-}$ + DPPC-AQ reaction. The $[S_2O_4^{2-}]$-dependence exhibited in Figure 7, described by rate law (1), for the DPPC-AQ + $S_2O_4^{2-}$ reaction could arise from two plausible mechanisms:

quinone equilibrium:

$$Q_A \underset{k_{-1}}{\overset{k_1}{<--->}} Q_B \qquad (3)$$

$$Q_B + S_2O_4^{2-} \xrightarrow{k_2} H_2Q + 2 \ SO_2 \qquad (4)$$

$S_2O_4^{2-}$ binding:

$$S_2O_4^{2-}{}_{AQ} \underset{k_{-1}}{\overset{k_1}{<--->}} S_2O_4^{2-}{}_{LIP} \qquad (5)$$

$$S_2O_4^{2-}{}_{LIP} + Q \xrightarrow{k_2} H_2Q + 2 \ SO_2 \qquad (6)$$

where Q_A and Q_B represent two different forms of DPPC-AQ (DPPC-AQ in two different environments), $_{LIP}$ represents liposome-bound, and $_{AQ}$ represents aqueous. The mechanistic implications are drastically different, but the functional form of the rate law is identical in both cases.

The crucial aspect of the experimental data that permits dismissal of the $S_2O_4^{2-}$ binding is the observation of biphasic growth of H_2Q at high $[S_2O_4^{2-}]$, as illustrated in Figure 8. Note that a single exponential fits beyond the first half-life (Figure 8B), but not the entire region (Figure 8A). The $S_2O_4^{2-}$ binding mechanism is inconsistent with an initial fast phase, while the quinone equilibrium is entirely consistent when it is recognized that the quinone equilibrium is established at the outset of the reaction (addition of $S_2O_4^{2-}$). At high $[S_2O_4^{2-}]$ reaction (4) is extremely rapid ($k_2[S_2O_4^{2-}] \gg k_1, k_{-1}$) yielding the first phase, while the remainder of the reaction proceeds from Q_A to give the rate law (1). At low $[S_2O_4^{2-}]$ the disparity in the two phases is minimal and a single exponential fits the data. Fitting the $[H_2Q]$ vs. t data to the quinone equilibrium mechanism using Runge-Kutta

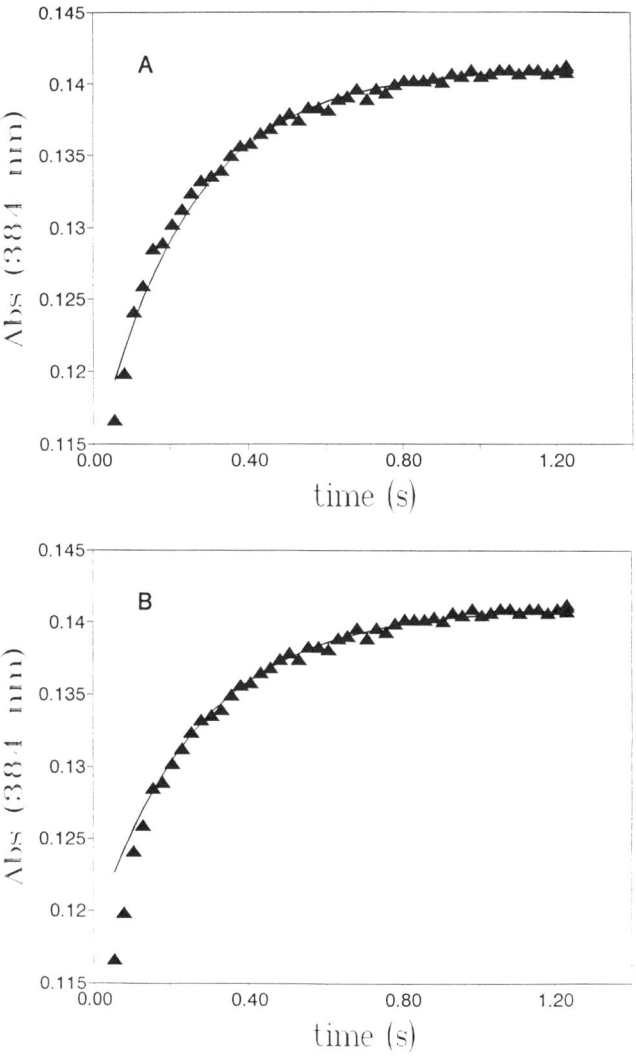

Figure 8. Experimental (▲) and calculated (———) [H$_2$Q] vs. time plots; solid lines were calulated with single exponentials for best fit over entire time range (A) and over long time (B).

simulations was reasonably successful (cf. Figure 9). The biphasic behavior is reproduced for the high [S$_2$O$_4^{2-}$] and the single exponential at low [S$_2$O$_4^{2-}$]. The best fit to the data is obtained for $k_1 \sim 4 - 5$ s^{-1}, $k_{-1} \sim 5 - 6$ s^{-1}, and $k_2 \sim 0.4 - 2.5 \times 10^4$ M^{-1}s^{-1}.

Estimating the values of the various rate constants is straightforward, but identifying the nature of the quinone equilibrium is more difficult. Fortunately, the NMR results

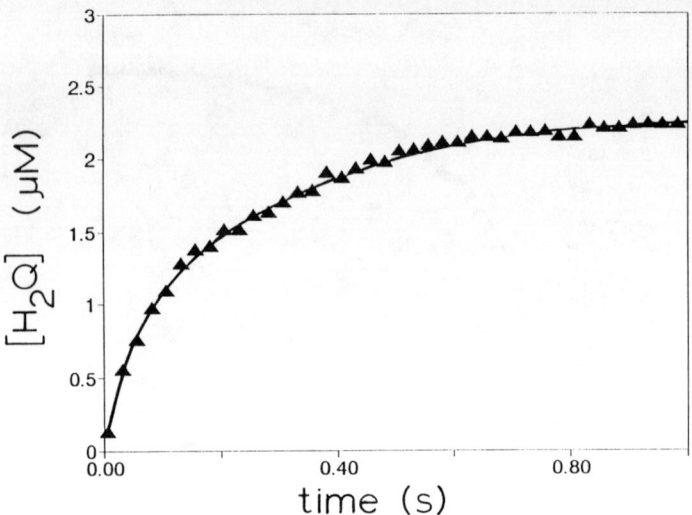

Figure 9. Experimental (▲) and calculated (———) $[H_2Q]$ vs. time plots; solid line was calulated with $k_1 = 4.5$ s^{-1}, $k_{-1} = 5.6$ s^{-1}, and $k_2 = 4000$ M^{-1}s^{-1}.

described above provide two possible scenarios identifying Q_A and Q_B: gel + fluid and membrane-embedded + solution-extended. In both cases, one form of the quinone (Q_B) would react much more rapidly than the other form (Q_A). Within the membrane-embedded + solution-extended scenario the Q_B form should resemble a solution species, so use of k for $Me_3NCH_2AQ^+$ as an initial estimate of k_2 is reasonable. k_{-1} thus calculated is 21.2 ($\pm$ 0.2) s^{-1}, providing an estimate of $k_1/k_{-1} = 0.19 \pm 0.01$. This estimate is reasonable for the membrane-embedded + solution-extended scenario, but is inconsistent with the curve-fitting described above ($k_1/k_{-1} \sim 0.8$). The other scenario is consistent with $k_1/k_{-1} \sim 0.8$, since this leads to $[Q_A]/[Q_B] \sim 0.45$. k_2 thus estimated ($\sim 2 \times 10^4$ M^{-1}s^{-1}) is significantly less than for the corresponding aqueous reaction (1.1×10^5 M^{-1}s^{-1}). Such a difference in rate constants could be due to the distance between the aqueous $S_2O_4^{2-}$ and the lipid-embedded AQ, restricted access to the AQ (a steric component), or due to the change in dielectric of the medium at the hydrophobic / hydrophilic interface. Although a more detailed data analysis will be necessary to differentiate between the two scenarios, the gel-fluid scenario presently provides the best explanation of the kinetic data.

Conclusions

DPPC-AQ is the first example (3) of a quinone-functionalized phospholipid. Quinone-functionalized liposomes with varying phospholipid composition and transmembrane distribution can be prepared with DPPC-AQ. Incorporation of DPPC-AQ into DPPC and DOPC liposomes has no effect on liposome size, although it does increase liposome permeability. NMR spectroscopy reveals that the AQ "head group" of DPPC-AQ resides near the hydrophobic / hydrophilic interface of DPPC liposomes. The quinone-functionalized liposomes undergo facile redox reactions with solution reagents. The mechanism of $S_2O_4^{2-}$ reduction involves electron transfer between solution $S_2O_4^{2-}$ and liposome-bound DPPC-AQ.

Our primary interest in quinone-functionalized amphiphilic assemblies stems from the desire to provide simple, chemical models for quinone-mediated energy transduction. The present systems are a successful beginning; however, the biological quinones (2) reside deep within the bilayer membrane, possess considerable mobility, and interact strongly (if not necessarily) with membrane-bound proteins. We must incorporate these concepts into our quinone-functionalized liposomes so that we can prepare well-defined, chemical models of the endogenous systems.

Acknowledgements. The authors thank the Purdue Research Foundation for financial support and Karie M. Horvath for preparing samples of DPPC-AQ. Prof. Dale Margerum provided access to the High-Tech stopped flow spectrophotometer and useful comments on the kinetic data analysis. NMR experiments were performed on instruments funded by NIH Grant RR01077 and NSF/BBS-8714258.

Literature Cited

1. Feher, G.; Allen, J. P.; Okamura, M. Y.; Rees, D. C. Nature 1989, 339, 111.
2. von Jagow, G.; Link, T. A.; Ohnishi J. Bioenerg. Biomemb. 1986, 18, 157.
3. Leidner, C. R.; Liu, M. D. J. Am. Chem. Soc. 1989, 111, 6859.
4. Leidner, C. R.; Simpson, H. O'N.; Liu, M. D.; Horvath, K. M.; Howell, B. E.; Dolina, S. J. Tetrahedron Lett., 1990, 31, 189.
5. Liu, M. D.; Leidner, C. R. J. Chem. Soc., Chem. Comm. 1990, 383.
6. Liu, M. D.; Patterson. D. H.; Jones, C. R.; Leidner, C. R. J. Phys. Chem., 1991, 95, 1858.
7. Liu, M. D.; Leidner, C. R.; Facci, J. S. J. Am. Chem. Soc., submitted.
8. Liu, M. D.; Duevel, R. V.; Corn, R. M.; Leidner, C. R. J. Phys. Chem., submitted.
9. Thomas, P.D.; Poznansky, M.J. Biochim. Biophys. Acta 1989, 978, 85.
10. Ulrich, E. L.; Gervin, M. E.; Cramer, W. A.; Markley, J. L. Biochemistry 1985, 24, 2501.
11. Michaelis, L.; Moore, M. J. Biochim. Biophys. Acta. 1985, 821, 121.
12. Hinz, H.-J.; Korner, O.; Nicalou, C. Biochim. Biophys. Acta 1981, 339, 111.
13. Lambeth, D. O.; Palmer, G. J. Biol. Chem. 1973, 248, 6095

RECEIVED December 10, 1991

Chapter 18

Proteinaceous Microspheres

Mark W. Grinstaff and Kenneth S. Suslick

School of Chemical Sciences, University of Illinois at Urbana–Champaign, Urbana, IL 61801

Using high-intensity ultrasound, we have synthesized aqueous suspensions of proteinaceous microspheres. Microspheres filled with nonaqueous liquids (*i.e.*, microcapsules) or air-filled (*i.e.*, microbubbles) were examined by optical microscopy, scanning electron microscopy, and particle counting. High concentrations of microspheres were observed with Gaussian size distributions. Microsphere formation is strongly inhibited by the absence of O_2, by free radical traps, by superoxide dismutase (but not by catalase), and by the lack of free cysteine residues in the protein. It is proposed that superoxide, which is produced by acoustic cavitation, cross-links the microcapsule protein by oxidizing cysteine residues to form disulfide bonds.

The organization of macromolecules into even larger structures determines the physical properties of much of the macroscopic world. The cellular structure of life itself is an obvious example of the general case of the formation of micrometer-sized structures from nanometer-sized macromolecules. As a class of macromolecular assemblies, such microencapsulation has proved to be a valuable technique in modern science and has found numerous technological applications. Some important uses include encapsulation of active metals, deodorants, dyes, perfumes, cosmetic ointments, and pesticides. The pharmaceutical industry has found this technology especially valuable with applications ranging from encapsulated drugs and vitamins *(1-6)* to contrast agents for sonography *(7-10)* and magnetic resonance imaging *(11)*.

The most common microspheres are composed of liposomes (*i.e.*, lipid bilayer microspheres) or synthetic polymers. Specific compositions, however, are usually complex formulations with proteins often added to increase biocompatibility or to modify the microsphere's properties. For use *in vivo*, the ideal material would have a long shelf life, low toxicity and micron size.

A sonochemical technique has been developed for the synthesis of nonaqueous liquid-filled microcapsules and air-filled microbubbles entirely composed of proteins that meet these criteria *(12-15)*. These materials are an assembly of protein molecules linked together by disulfide bonds. The chemical crosslinking responsible for these microspheres is a direct result of the chemical effects of ultrasound on aqueous media.

Morphology of Protein Microspheres

The morphology of these materials was determined using scanning electron microscopy and light microscopy. A scanning electron micrograph of dodecane-filled proteinaceous microcapsules shows their morphology and size (Figure 1). The air-filled and liquid-filled (dodecane, decane, toluene) microspheres both have Gaussian size distributions with an average diameter of approximately 5.5 μm and 2.3 μm, respectively (Figure 2). Proteinaceous microspheres have been synthesized with a high intensity ultrasonic probe (Heat Systems, W375, 20 kHz, 0.5 in. Ti horn) from various proteins including bovine serum albumin (BSA), human serum albumin (HSA) and hemoglobin (Hb) as describe in detail elsewhere *(12-15)*. A typical experiment to synthesize proteinaceous microcapsules involves irradiating a toluene and 5% w/v BSA solutions for three minutes at an acoustic power of ≈ 150 W/cm^2, with an initial reaction cell temperature of 23°C, and at pH 7.0. We find that more microcapsules and microbubbles are produced with increased acoustic power starting from the same initial temperature (Figure 3).

Mechanism of the Sonochemical Synthesis of Microspheres

How are the microspheres formed and what holds them together? Ultrasonic irradiation of liquids is well known to produce both emulsification *(16)* and cavitation *(17-19)*. In forming the microspheres, the nonaqueous liquid or air is dispersed into the aqueous protein solution. Ultrasonic emulsification does occur in this bi-phasic system. Emulsification alone, however, is insufficient: emulsions produced by vortex mixing, instead of ultrasonic irradiation, produced no long-lived microspheres (Figure 2). Furthermore, the vortex emulsions are not stable and phase separation occurs immediately. In contrast, the proteinaceous microspheres are stable for many hours at room temperature and for several months at 4°C (Figure 4). Preliminary experiments on leakage from interior liquids indicates that the stability of the microsphere content is on the order of hours at 38°C.

Hydrophobic or thermal denaturation of the protein after the initial ultrasonic emulsification might be responsible for microsphere formation. High concentrations of microspheres, however, are only observed when the mixture

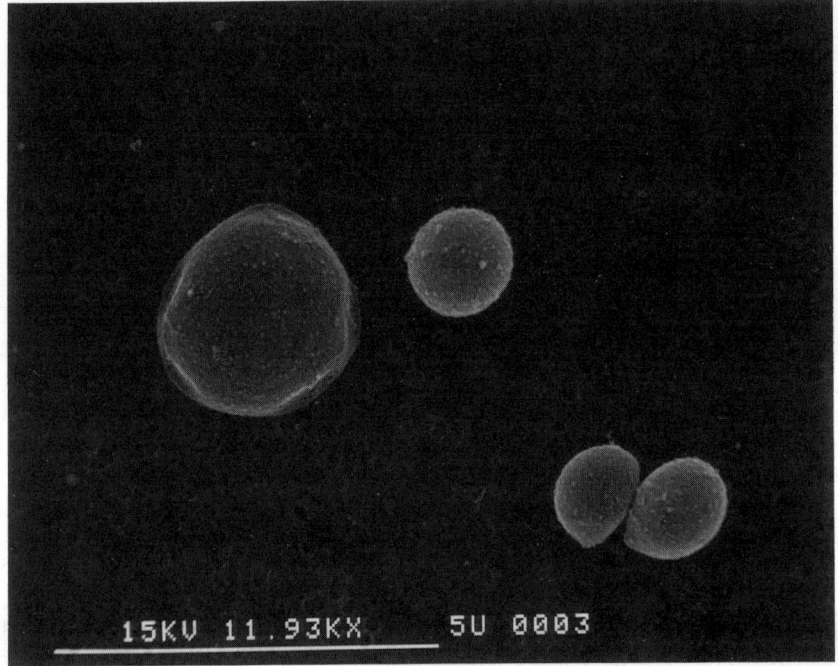

Fig. 1 Scanning electron micrograph of dodecane-filled proteinaceous microcapsules. The microcapsules were prepared for SEM by secondary cross-linking with glutaraldehyde and coating with Au/Pd. Volatile nonaqueous liquids gave non-spherical microcapsules due to evaporation during sample preparation.

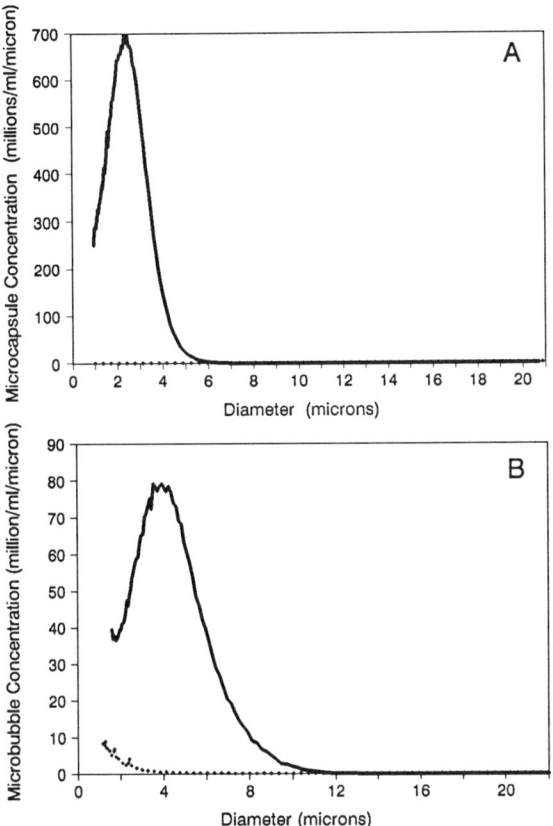

Fig. 2 Particle distribution of an aqueous suspension of proteinaceous microspheres, determined with an Elzone particle counter (Model 180XY).

(A)

———— Toluene-filled microcapsules were synthesized from ultrasonic irradiation toluene and of 5% w/v BSA solutions for three minutes at an acoustic power of ≈150 W/cm^2, with an initial cell temperature of 23°C at pH 7.0.

..... Toluene and 5% w/v BSA solution vortexed for three minutes at pH 7.0.

(B)

———— Air-filled microbubbles were synthesized from ultrasonic irradiations of 5% w/v BSA solutions for three minutes at an acoustic power of ≈150 W/cm^2, with an initial cell temperature of 50°C at pH 7.0.

..... 5% w/v BSA solution vortexed for three minutes at pH 7.0.

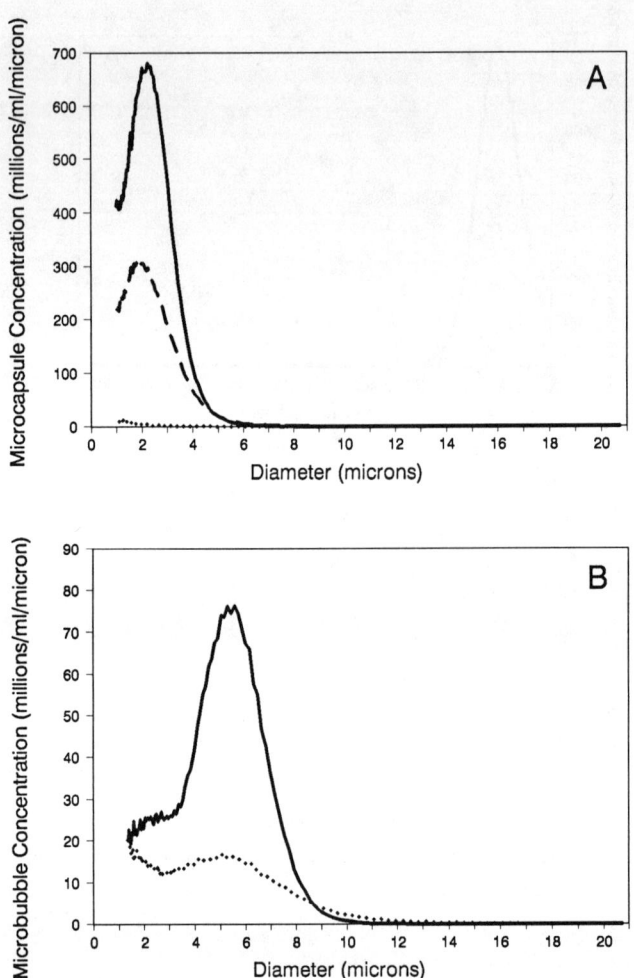

Fig. 3 The effect of acoustic power on microcapsule and microbubble formation. Acoustic powers were taken from the manufacturer supplied nomograph and are only approximate.

(A) Microcapsule formation.
　　　———　　≈ 150 W/cm^2
　　　- - - -　　≈ 70 W/cm^2
　　　........　　≈ 40 W/cm^2

(B) Microbubble formation.
　　　———　　≈ 150 W/cm^2
　　　........　　≈ 40 W/cm^2

is sparged with air or O_2. If the reaction is run under an inert atmosphere (He, Ar, or N_2) microcapsules are not formed. Thus, thermal or solvent denaturation (for which O_2, N_2, and Ar should give similar results) cannot explain the microsphere permanence.

Another, *chemical* process must be involved. As mentioned above, ultrasonic irradiation of liquids generates acoustic cavitation: the formation, growth and collapse of bubbles in a liquid *(17-19)*. The collapse of these bubbles produces high energy chemistry. Specifically, aqueous sonochemistry produces OH• and H• *(20-22)*. The radicals so produced by ultrasound form H_2, H_2O_2, and (in the presence of O_2) superoxide, HO_2 *(22-24)*. Hydroxyl radicals, superoxide, and peroxide are all potential protein cross-linking agents.

To identify the specific oxidant involved, the formation of microspheres was examined in the presence of radical traps (Figure 5). The addition of nonspecific traps (*e.g.*, 2,6-di-t-butyl-4-methylphenol or glutathione), catalase (which decomposes hydrogen peroxide to oxygen and water *(25)*) and superoxide dismutase (which decomposes superoxide to oxygen and hydrogen peroxide *(26)*) were tested. Both microcapsule and microbubble formation were inhibited by nonspecific traps and by superoxide dismutase, but not by catalase. Catalase activity was confirmed after ultrasonic irradiation; ultrasonic irradiation did not, therefore, destroy the functioning of this enzyme. We propose that the important oxidant involved in microsphere formation is superoxide.

To determine the specific effect of superoxide on the proteins, several experiments were performed. Cysteine is easily oxidized by superoxide *(27)* and is present in BSA, HSA, and Hb. In fact, ultrasonic irradiation of proteins has been reported to oxidize cysteine residues *(28)*. If the microspheres are held together by protein cross-linking through disulfide linkages from cysteine oxidation, Hb and myoglobin (Mb) provide an interesting test: they have very similar sequences and monomeric structures, but Mb has no cysteine amino acid residues. Ultrasonic irradiation of Mb solutions does not form microspheres; Hb does. In another set of tests, the addition of a disulfide cleavage reagent, dithioerythritol *(29)*, destroys Hb-toluene or BSA-toluene microcapsules. Finally, the oxidation of cysteine residues can be inhibited by alkylation with N-ethylmaleimide *(30)*, and microsphere formation from Hb solutions so treated is greatly reduced. These results confirm the importance of disulfide bond formation in microsphere formation.

Summary

In summary, ultrasound can produce proteinaceous microcapsules and microbubbles of a few microns diameter at high concentrations with narrow size distributions. The process involves two separate acoustic phenomena: emulsification and cavitation. The dispersion of gas or nonaqueous liquid into the protein solution, coupled with the chemical cross-linking of the protein molecules, produces stable protein microspheres.

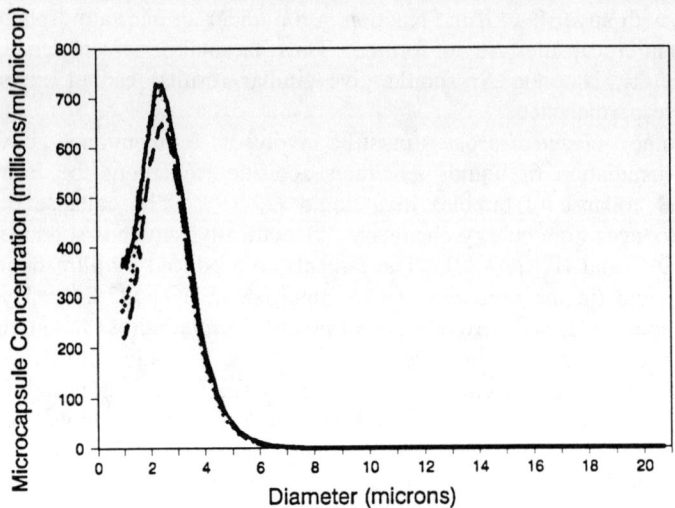

Fig. 4 The stability of toluene-filled proteinaceous microcapsules at 4°C.

————	20 Minutes
········	42 Days
- - - - -	83 Days

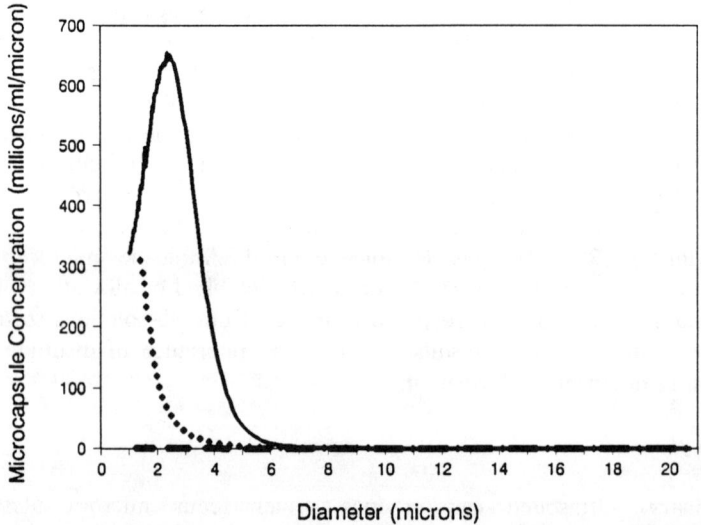

Fig. 5 The effect of radical traps on microcapsule formation. 5% w/v aqueous solution of BSA and toluene were irradiated in the presence of catalase, glutathione or superoxide dismutase. Inhibition of microcapsule formation also occurred with 2,6-di-t-butyl-4-methylphenol.

————	0.09%	Catalase
· · · · · ·	0.1 M	Glutathione
- - - - -	0.1%	Superoxide Dismutase

Acknowledgments

This work was supported by N.S.F. and N.I.H. MWG gratefully acknowledges receipt of the Procter and Gamble Graduate Fellowship of the Colloid and Surface Chemistry Division of the American Chemical Society.

Literature Cited

1. Langer R. *Science* **1990**, *249*, 1527.
2. Lee, T. K.; Sokoloski, T. D.; Royer, G. P. *Science* **1981**, *213*, 233.
3. Heller, J.; Bake, R. W. *Controlled Release of Biomaterials* Academic Press, New York, 1980.
4. Burgess, D. J., Davis, S. S.; Tomlinson, E. *Int. J. Pharm.* **1987**, *39*, 129.
5. Morimoto, Y.; Sugibayashi, K.; Kato, Y. *Chem. Pharm. Bull.* **1981**, *29*, 1433.
6. Jalsenjak, V.; Stolink, S.; Jalsenjak, I. *Acta Pharm. Jugosl.* **1988**, *38*, 297.
7. Kerber, R. E., ed. *Echocardiography in Coronary Artery Diseases* Future, New York, 1988.
8. Pieroni, D. R.; Varghese, J.; Freedom, R. M. et. al. *Cathet. Cardiovasc. Diagn.* **1979** 5, 5.
9. Dick, C. D.; Feinstein, S. B. *Pract. Cardiology* **1988**, *14*, 71.
10. Feinstein, S. B.; Keller, M. W.; Dick, C. D. *J. Am. Coll. Cardiol.* **1987**, *9*, 111A.
11. Thaku M. L. et. al. *Magnetic Resonance Imaging,* **1990**, *8*, 625.
12. Keller, M. W.; Feinstein, S. B.; Briller R. A. *J. Ultrasound Med.,* **1986**, 5 493.
13. Feinstein, S. B. *U.S Patent 805975*, Jan. 5, 1985.
14. Suslick K. S.; Grinstaff, M. W. *J. Am. Chem. Soc.,* **1990**, *112*, 7807.
15. Grinstaff, M. W.; Suslick K. S. *Proc. Natl. Acad. Sci. USA,* **1990**, *88*, 7708.
16. Rooney, J. A. in *Ultrasound: Its Chemical, Physical, and Biological Effects*; Suslick, K. S., ed.; VCH, New York, 1988: pp 74-96.
17. Suslick, K. S. *Science* **1990**, *247*, 1439.
18. Suslick, K. S. *Sci. Am.* **1989**, *260*, 80.
19. Suslick, K. S., ed. *Ultrasound: Its Chemical, Physical, and Biological Effects*; VCH, New York, 1988.
20. Makino, K.; Mossobo, M. M.; Reiz, P. *J. Phys. Chem.* **1983**, *87*, 1369.
21. Riesz, P.; Berdahl, D.; Christman, C.L. *Environ. Health Perspect.* **1985**, *64*, 233.

22. Weissler, A. *J. Am. Chem. Soc.* **1959,** *81,* 1077.
23. Del Duca, M.; Jeager, E.; Davis, M. O.; Hovarka, F. *J. Acoust. Soc. Am.* **1958,** *30,* 301.
24. Lippitt, B.; McCord, J. M.; Fridovich, I. *J. Biol. Chem.* **1972,** *247,* 4688.
25. Felton, R. H. in *The Porphyrins*; Dolphin, D., ed.; Academic Press, New York, 1978, vol. 5, pp 111-115.
26. Fee, J. A. in *Metal Ion Activation of Dioxygen;* Spiro, T. G., ed.; Wiley, New York, 1980, pp 209-239.
27. Asada, K.; Kanematsu, S. *Agr. Biol. Chem.* **1976,** *40,* 1891.
28. Bronskaya, L. M.; El'piner, I. Ye. *Biophys.* **1963,** *8,* 344.
29. Jocelyn, P. C. *Biochemistry of the SH Group*; Academic Press, New York, 1976.
30. Haugaard, N.; Cutler, J.; Ruggieri, M. R. *Anal. Biochem.* **1981,** *116,* 341

RECEIVED October 1, 1991

Chapter 19

Molecular Recognition in Gels, Monolayers, and Solids

Kevin L. Prime, Yen-Ho Chu, Walther Schmid, Christopher T. Seto, James K. Chen, Andreas Spaltenstein[1], Jonathan A. Zerkowski, and George M. Whitesides[2]

Department of Chemistry, Harvard University, Cambridge, MA 02138

> This paper describes work in four areas: affinity electrophoresis of carbonic anhydrase in cross-linked polyacrylamide-derived gels containing immobilized derivatives of aryl sulfonamides; inhibition of the hemagglutination of erythrocytes induced by influenza virus using water-soluble polyacrylamides bearing sialic acid groups; the application of self-assembled monolayers (SAMs) of alkyl thiolates on gold to the study of protein adsorption on organic surfaces; and the use of networks of hydrogen bonds to generate new classes of non-covalently assembled organic materials, both in solution and in crystals.

This paper summarizes research in two areas of molecular recognition: affinity polymers and molecular self-assembly. We illustrate these areas by examples drawn from affinity gel electrophoresis, soluble synthetic macromolecular inhibitors of binding of influenza virus to erythrocytes (1), protein adsorption on self-assembled monolayers (2, 3), and self-assembling hydrogen-bonded molecular aggregates (4–6).

Affinity Polymers: Molecular Recognition in Gels

Affinity gel electrophoresis (AGE) uses the biospecific equilibrium binding of a protein to an immobilized ligand to reduce the electrophoretic mobility of that protein selectively. AGE is a useful technique for studying receptor–ligand interactions (7). It combines the selectivity of affinity chromatography with the high sensitivity of gel electrophoresis. AGE allows both the qualitative examination of the specificity of binding between protein and covalently immobilized ligand and the quantitative determination of the dissociation constant of the protein–

---

[1]Current address: Burroughs Wellcome Company, Research Triangle Park, NC 27709
[2]Corresponding author

ligand complex. By observing how the dissociation constant changes with the structure of the ligand, it is possible to probe the chemical characteristics and topology of the ligand-binding site.

We chose carbonic anhydrase B (CAB, E.C.4.2.1.1) as a model protein for our initial studies of AGE. CAB is a well-characterized protein *(8–11)*. It is inhibited by a number of aryl sulfonamides, with dissociation constants ranging from 10^{-6} to 10^{-8} M *(11)*. The active site of the enzyme is known from X-ray crystallography and can be described qualitatively as being located at the bottom of a conical pocket approximately 15 Å deep and 15 Å wide.

In order to test the sensitivity of AGE to the topology of a binding pocket, we prepared the series of glycyl-linked monomers **1** and formed gels by copolymerizing them in different

1a $n = 0$
1b $n = 1$
1c $n = 2$
1d $n = 3$
1e $n = 4$
1f $n = 5$
1g $n = 6$
1h $n = 7$

concentrations with acrylamide and cross-linking agent. These gels were used as the stationary phase in electrophoresis experiments. The retention ratio, R_f, of CAB on electrophoresis in these gels was a function of the concentration of immobilized sulfonamide in the gel, [L], as illustrated in Figure 1 for gels based on poly(**1f**-*co*-acrylamide). The retention ratio is related to the dissociation constant, K_d, by equation 1 *(7)*. When [L]R_f is plotted as a function

$$[L]R_f = K_d - K_d R_f \quad (1)$$

of R_f, both the slope and the [L]R_f-intercept give the value of K_d. Figure 2 shows the measured values of K_d for the affinity ligands **1**, as a function of the number of glycine residues in the spacer. From these data, we conclude that the binding pocket of CAB is insensitive to linking chains longer than three glycine residues. Use of a linker connecting the sulfonamide to the

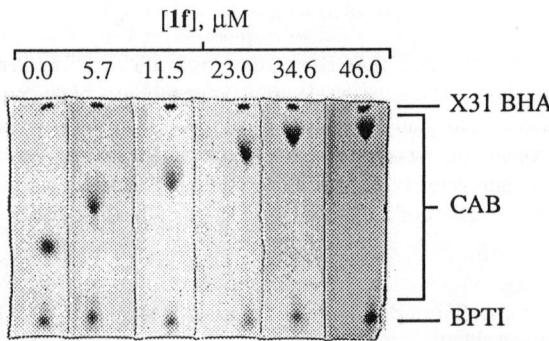

Figure 1. Affinity electrophoresis of bovine carbonic anhydrase B (CAB) on polyacrylamide slab gels containing various concentrations of affinity ligand **1f**. Bovine pancreatic trypsin inhibitor (BPTI) and the bromelain-released hemagglutinin of influenza virus X-31 (X31 BHA) were used as internal standards.

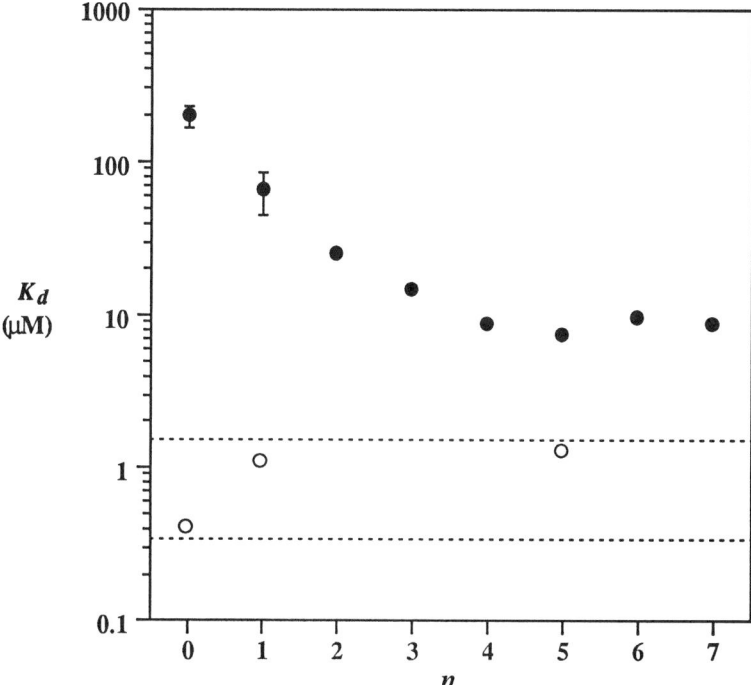

Figure 2. Dependence of the dissociation constants, K_d, of complexes of bovine carbonic anhydrase B (CAB) and immobilized (filled circles) or soluble (hollow circles) affinity ligands 1 on the number of glycine residues, n, in the ligand.

polymer backbone shorter than (gly)$_4$ gives an apparent K_d that is larger than the solution value, reflecting (we presume) unfavorable steric interactions between the protein and the backbone. This value is in agreement with our estimate from the crystallographic dimensions of the binding pocket.

Affinity-Polymer Inhibition of Influenza-Induced Agglutination of Erythrocytes. We have also begun to design soluble, polymeric affinity ligands to interact with proteins on biological surfaces. We have explored the inhibition of the agglutination of erythrocytes induced by influenza virus in greatest detail. Hemagglutinin (HA) present on the viral surface binds to sialic acid (SA) residues on glycoproteins and glycolipids located at the surface of the cell (12–14). Unlike the tight-binding ($K_d = 10^{-6}$–10^{-8} M) CAB–sulfonamide system, the HA–SA complex is weakly bound ($K_d = 2$ mM) (15). Although there is no corresponding value for the binding of virus to erythrocyte, the binding of genetically altered fibroblasts expressing HA on their surface to erythrocytes has a substantially lower dissociation constant ($K_d \cong 7 \times 10^{-10}$ M) (16). We and others believe that the difference in strength between the interactions of HA with sialic acid and of influenza virus with erythrocyte can be traced to the polyvalency of the latter (17–20). We wished to test the hypothesis that an appropriate polyvalent molecule presenting many sialic acid residues to the virus would be an effective inhibitor of the binding of influenza to erythrocytes.

The naturally occurring hemagglutination inhibitors are structurally complex glycoproteins (21, 22), and rather than attempting to mimic these proteins (23), we chose to include sialic acid residues in acrylamide-derived polymers. We hoped that the flexibility of the acrylamide backbone would allow multiple sialic acid residues per polymer chain to bind to the surface of the virus particle, and that this multipoint attachment would result in strong inhibition of the binding of virus to erythrocytes. Acrylamide-derived polymers are well suited for this purpose, since they can be prepared easily, their structures can be varied readily, and they are water-soluble.

We synthesized monomer 2 and copolymerized it with a number of acrylamide monomers (3a-g). Figure 3 shows the inhibition constant of the soluble polymer, K_i, determined by a hemagglutination assay, as a function of the mole fraction, χ_{SA}, of 2 in the mixture of 2 and 3a used to form the polymer (1). The values of K_i were calculated on the basis of sialic acid groups in solution. Polymers having values of $K_i > 0.625$ mM (the horizontal line in Figure 3) were not examined quantitatively; the hollow points represent upper limits. The values of K_i for proteins and analogs of sialic acid were obtained from the literature and are shown on the right. The strongest inhibition occurred over a broad range of χ_{SA} (0.2–0.6) and was within an order of magnitude of the best naturally occurring inhibitors. Copolymers derived from the other co-monomers (3b-g) showed little difference in inhibition but were less water-soluble than those derived from 3a. Copolymers derived from analogs of 2 containing shorter spacers, however, showed significantly lower inhibition constants. We do not yet understand why sialic acid residues bound to short spacers are less efficient inhibitors than those linked to long spacers, but the X-ray structure of hemagglutinin shows that the binding site is *not* in a pocket. This observation suggests that the change in inhibition with the length of the spacer may arise from changes in the structure of the affinity polymer itself, or from interaction between the HA and the polymer backbone. Similar results were found by other groups (24, 25). We are currently working to optimize the performance of these polymers and to determine the relationship between their structures and their inhibition constants.

Self-Assembly: Molecular Recognition in Monolayers and Solids

The term "self-assembly" is used to describe a variety of processes, all of which involve the spontaneous organization of dispersed molecules into an ensemble with a defined structure. Self-assembled structures are ubiquitous in nature: the double helix of DNA, many multi-unit

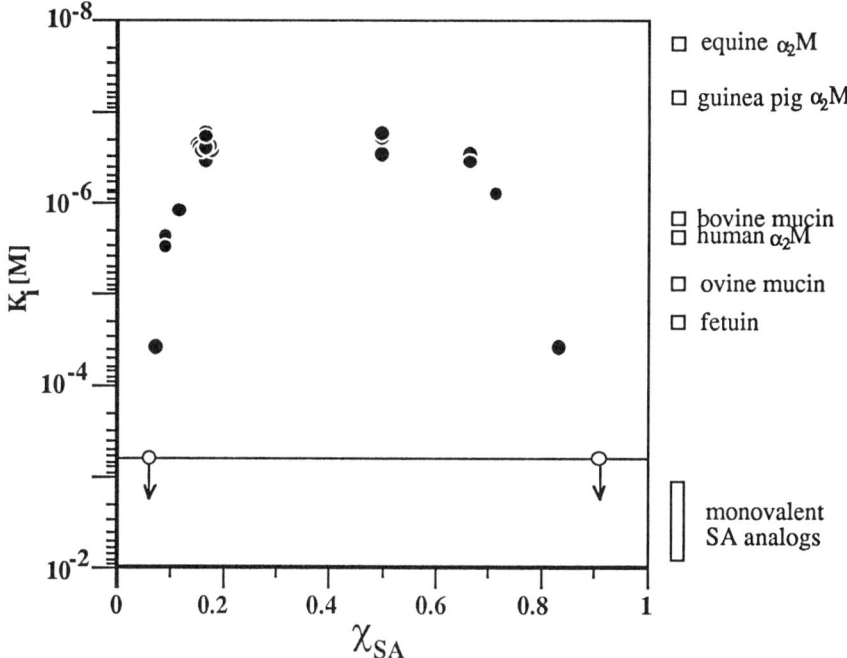

Figure 3. Inhibition of hemagglutination of erythrocytes by poly(2-co-acrylamide). (Reproduced from ref. 6. Copyright 1991 American Chemical Society.)

enzymes, structural proteins, ribosomes, and viruses assemble spontaneously into their native structures from solutions of their constituent parts (26).

Self-assembly is also a practical synthetic strategy in the laboratory (albeit at a simpler level than in nature!). For example, long-chain surfactants with terminal groups capable of bonding to solid surfaces (head groups) self-assemble into ordered, oriented monomolecular films when solutions of the surfactants contact the solid surfaces. Such self-assembled monolayers (SAMs) are known for alkanoic acids on a variety of metal oxides (27); trichlorosilanes on oxide surfaces (28), such as silica (29–32) and alumina (33); alkanethiols, dialkyl sulfides, and dialkyl disulfides on gold, silver, and copper (34); and alkyl isonitriles on platinum (35). The monolayer–air interface of a SAM comprises principally an ordered array of the tail group (the end of the molecule opposite from the Au–S interface). By synthetic variation in the tail groups, SAMs can be prepared that exhibit a wide variety of properties (36).

In this section, we describe first the use of SAMs as model systems for studying the adsorption of proteins on organic surfaces. We then turn to an example of a different strategy for self-assembly: the use of hydrogen-bonded networks to prepare large, self-assembling complexes.

Self-Assembled Monolayers as Substrates for Studying the Mechanisms of Adsorption of Proteins to Man-Made Surfaces. SAMs formed by the adsorption of alkanethiols onto gold have received considerable attention in our laboratories (34). Two attractive features of this form of SAM are the variety of polar functional groups that are compatible with Au–S binding and the ease of preparing SAMs containing mixtures of tail groups from solutions containing mixtures of different alkanethiols (*36*).

We have used mixed SAMs to model polymer surfaces that contain poly(ethylene glycol) and different amounts of hydrophobic material (*3*). Figure 4 suggests schematically the structure of the monolayer–water interface of one of the SAMs. The tail group is flexible and drawn roughly to scale. We immerse the SAMs in solutions of proteins and observe (by ellipsometry and X-ray photoelectron spectroscopy) the amount of protein that is retained on the SAM after rinsing it with water (*2*).

Figure 5 shows the amount of fibrinogen adsorbed to SAMs containing a mixture of two coadsorbed thiolates ("mixed SAMs") prepared from various mixtures of **4** and **5**. The

$$CH_3(CH_2)_{10}SH \qquad RO(CH_2CH_2O)_n(CH_2)_{11}SH$$

$$\mathbf{4} \qquad \qquad \mathbf{5}$$

5a	$n = 0$	
5b	$n = 1$	
5c	$n = 2$	$R = H$
5d	$n = 4$	
5e	$n = 6$	
5f	$n_{ave} = 17$	$R = CH_3$

thicknesses of the adsorbed films of protein, d, were determined by ellipsometry. The values of χ represent the surface mole fraction of **5** in the SAM, as determined by XPS. Each datum represents the average of three measurements taken on one SAM. The scatter in each average falls within the size of the symbol used to represent it. The results are reproducible: the curves for $n = 0, 2, 4,$ and 6 represent two independent sets of experiments each. The data have been offset in increments of 20 Å for clarity. The dashed lines on the right side of the graph represent the location of $d = 0$ Å for each set of experiments; the symbols to the right of the dashed lines indicate to which set of data each zero line applies.

Two observations can be made from these data: First, there is a qualitative difference between the adsorption of fibrinogen to SAMs containing hydroxyl- or mono(ethylene glycol)-terminated chains and to SAMs containing di(ethylene glycol)- or oligo(ethylene glycol)-terminated chains. The advancing contact angles of water upon SAMs formed from **5a**, **5b**, **5c**, and oligo(ethylene glycol)-terminated alkanethiols are 0°(*34*), 20°, 27°, and 33°(*3*), respectively (unpublished results unless otherwise noted). We do not yet know whether or how these two sets of observations are related. Second, adsorption of fibrinogen to mixed SAMs containing chains terminated by shorter oligomers of ethylene glycol ($n = 2$–6) reaches a

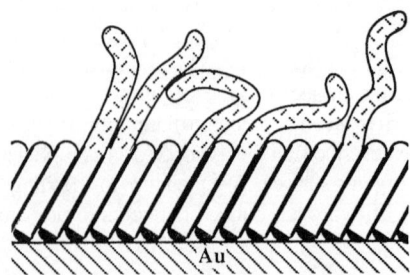

Figure 4. A schematic representation of mixed SAMs of **4** and **5e**. (Reproduced with permission from ref. 2. Copyright 1991 American Association for the Advancement of Science.)

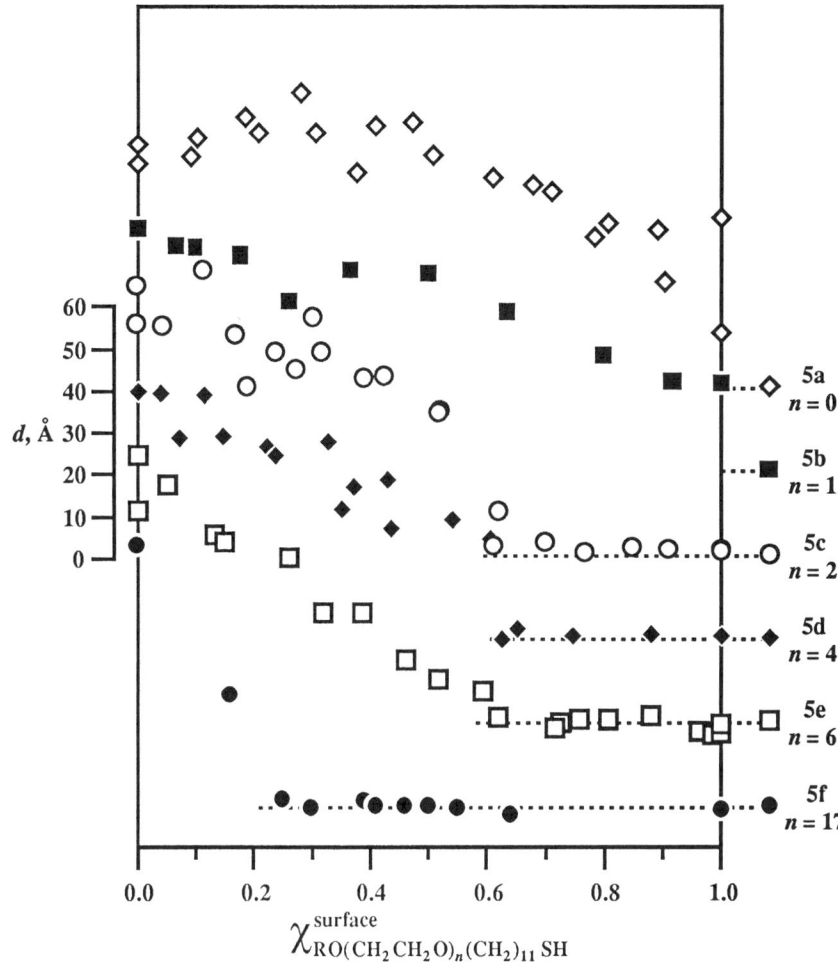

Figure 5. Adsorption of fibrinogen to SAMs formed from mixtures of 4 and 5.

minimum value within experimental error of the same concentration of oligo(ethylene glycol) chains in the SAM ($\chi \approx 0.65$). This observation suggests the possibility that a uniform, nonadsorbent interfacial structure is reached by all these SAMs around $\chi = 0.65$, and that the thickness of this structure is not an important parameter in determining the adsorption of protein to the SAM. These data suggest that the elimination of protein adsorption on polymer surfaces by end-grafted poly(ethylene glycol) chains depends more upon the grafting density than upon the length of the grafted chain (37, 38).

These SAMs are models for polymer surfaces. They should make it possible to analyze the relative importance of the effects at the solid–water interface influencing protein adsorption, and perhaps in time, to develop materials with selective adsorptivities for use *in vivo*.

Self-Assembly in Three Dimensions: New Classes of Non-Covalent Macromolecules.
Hydrogen bonding is a principal source of the enthalpic driving force used both for molecular recognition and for self-assembly in biological systems. Perhaps the best-known example is the pairing of the bases in polynucleotides.

In an analogous non-biological system, a remarkably stable, solid, 1:1 complex of melamine (M, **6**) and cyanuric acid (CA, **7**) forms spontaneously when aqueous solutions of the two components are mixed together (*39–41*). Figure 6 shows an idealized section of the polymeric network that is produced; the structure shown is not proven unequivocally. For comparison, the inset shows the adenine–uracil base pair.

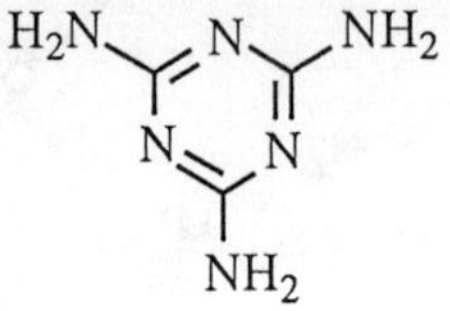

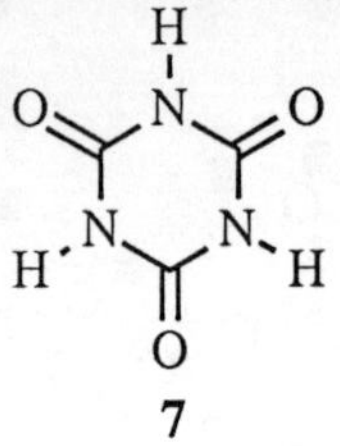

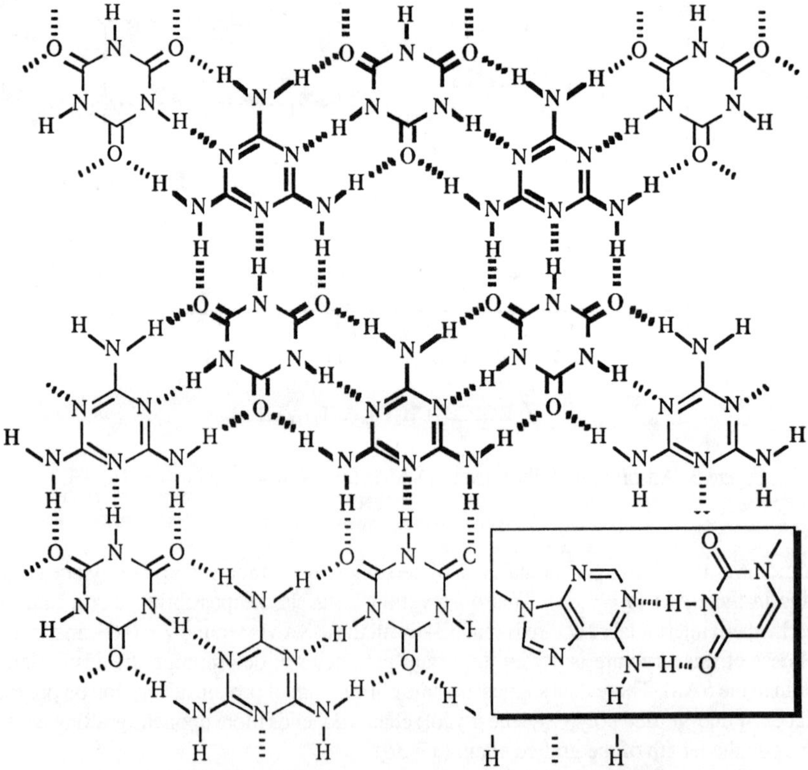

Figure 6. An idealized structure of the 1:1 complex between melamine and cyanuric acid (**6**–**7**). The inset shows the adenine–uracil base pair.

We have been developing this three-dimensional self-assembling system into two new classes of compounds. The first is a series of ribbon-like polymers prepared from derivatives of **6** and **7** where one or more of the N–H bonds is replaced with an N–alkyl bond (*4*). The alkyl substituents strongly affect the solid-state structure of these ribbons. Figure 7 shows the X-ray structure of diphenylmelamine–diethyl barbituric acid (**8**–**9**) complex. We

are currently attempting to obtain a predictive correlation between the structure of the alkyl substituent and the solid-state structure of the hydrogen-bonded polymer. For other approaches to the design of the solid state, see references *42–44*.

The second system involves the formation of smaller hydrogen-bonded complexes of a single structure (*5, 6*). (For other supramolecular complexes, see references *45–47*.) Using molecular models, we predicted that the compounds we call hubM$_3$ (**10**) and R(CA)$_2$ (**11**) would self-assemble into a discrete 2:3 complex (Figure 8). We synthesized hubM$_3$ and R(CA)$_2$ and found that the product that resulted from their mixture was, indeed, the 2:3 complex. Figure 9 shows an NMR titration of hubM$_3$ with R(CA)$_2$: as R(CA)$_2$ is added, the three broad peaks arising from hubM$_3$ disappear and are replaced by sharp peaks arising from the complex. Exchange between the complex and hubM$_3$ in solution is slow on the NMR time scale. We believe that the small peaks in the baseline of the upper spectrum correspond to conformational isomers of the 2:3 complex. These minor peaks are not impurities in either of the individual components. R(CA)$_2$ alone is too insoluble to give a detectable spectrum at saturation (< 0.1 mM) in CDCl$_3$ at the instrument gain used here. The NOE spectra, ultraviolet titrations, and molecular weights obtained from vapor-pressure osmometry are all consistent with the formation of a 2:3 complex.

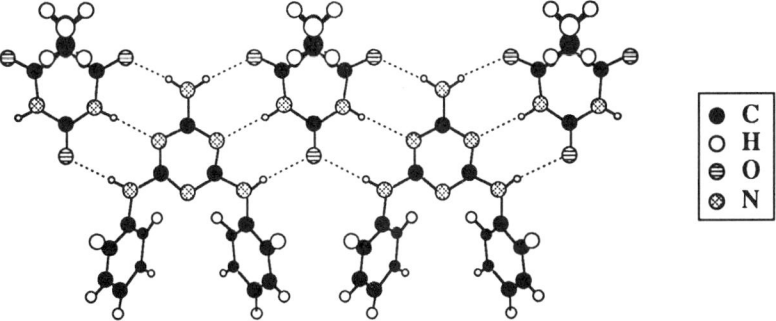

Figure 7. An X-ray crystal structure of the polymeric diphenyl melamine–diethyl barbituric acid (**8**–**9**) complex.

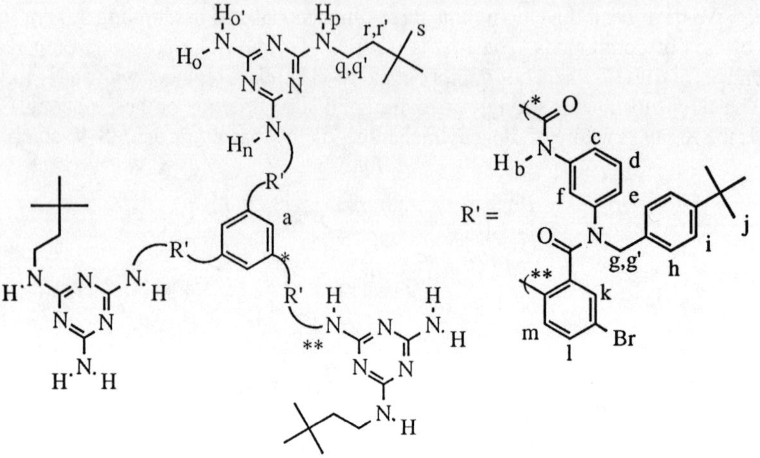

hubM₃, **10**

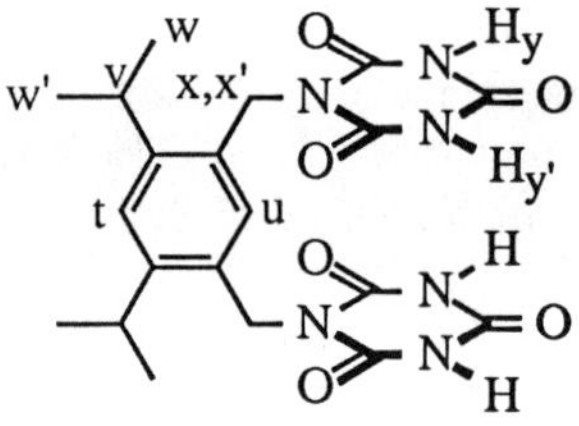

R(CA)₂, **11**

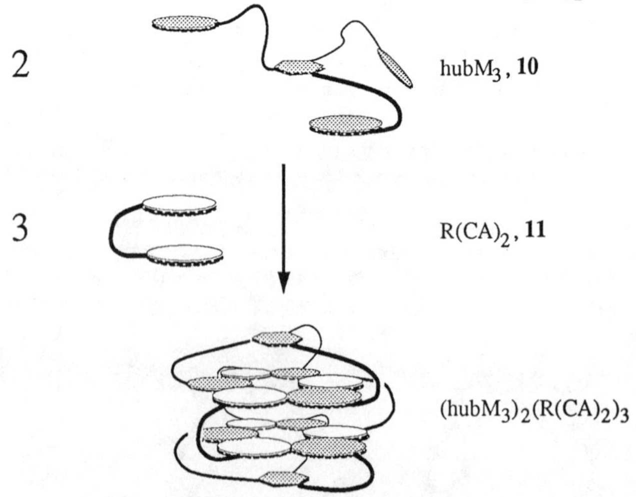

2   hubM₃, **10**

3   R(CA)₂, **11**

(hubM₃)₂(R(CA)₂)₃

Figure 8. A schematic representation of the self-assembly of hubM₃ (**10**) with R(CA)₂ (**11**) to give a supramolecular complex. (Adapted from ref. 6. Copyright 1991 American Chemical Society.)

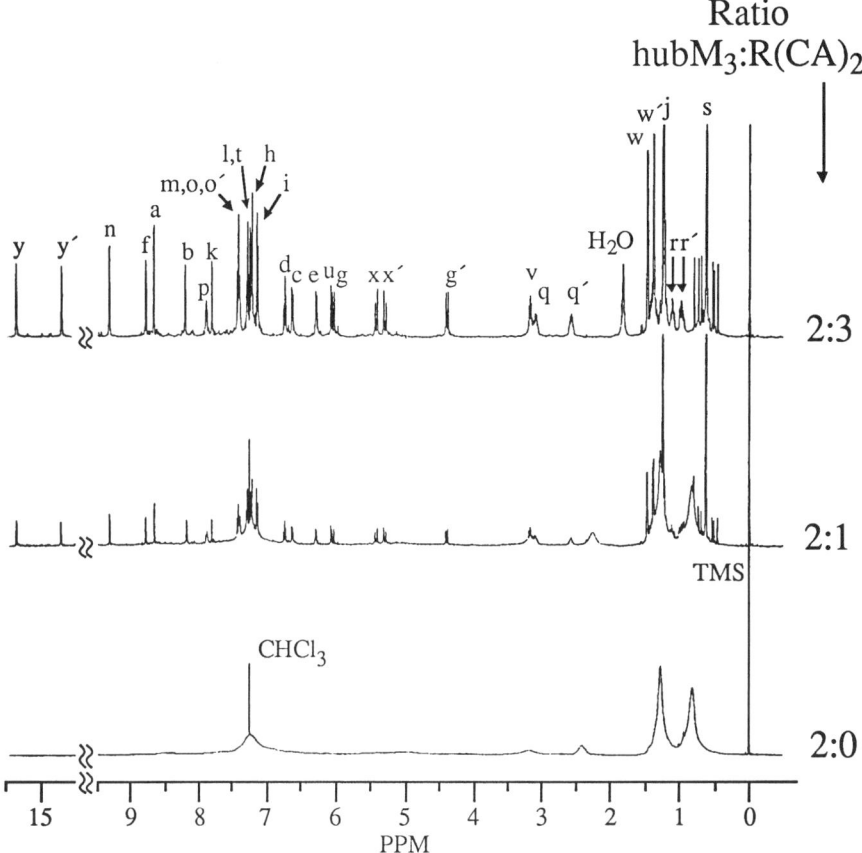

Figure 9. ^{1}H NMR titration of hubM$_3$ (**10**, 500 MHz, 10 mM in CDCl$_3$) with R(CA)$_2$ (**11**). The peak assignments are shown at the top of the figure and correspond to the labels on the structures of **10** and **11** shown in the text. (Reproduced from ref. 1. Copyright 1991.American Chemical Society.)

Conclusion

Studies of molecular recognition and of synthetic polymers combine in a number of useful approaches to new analytical tools, to drug design, and to new materials. Polymers permit the interactions provided by the structural elements used for molecular recognition to be localized, addressed, and amplified through cooperativity or polyvalency. The synthetic methodologies available from polymer science provide convenient routes to complex, multifunctional substances; molecular recognition and self-assembly provide new strategies for the synthesis of high molecular weight assemblies.

Acknowledgments

The work described here was taken from projects supported by a number of agencies: affinity electrophoresis [National Science Foundation (NSF) under the Engineering Research Center Initiative through the Massachusetts Institute of Technology Biotechnology Process Engineering Center (BPEC, Cooperative Agreement CDR-88-03014)]; polymers containing sialic acid [National Institutes of Health (GM39589)]; hydrogen-bonded networks [NSF (CHE-8812709); Harvard University Materials Research Laboratory (MRL), an NSF-funded facility]; and thiolate monolayers on gold [BPEC, the Defense Advanced Research Projects Agency (DARPA), and the Office of Naval Research]. The X-ray diffractometer and the NMR spectrometer were provided to Harvard University through grants from the NSF (CHE-8000670 and CHE-8410774). The XPS was provided to the MRL through a University Research Initiative grant from DARPA.

Literature Cited

1. Spaltenstein, A.; Whitesides, G. M. *J. Am. Chem. Soc.* **1991**, *113*, 686–687.
2. Prime, K. L.; Whitesides, G. M. *Science* (Washington, DC) **1991**, *252*, 1164–1167.
3. Simon, E. S.; Pale-Grosdemange, C.; Prime, K. L.; Whitesides, G. M. *J. Am. Chem. Soc.* **1991**, *113*, 12–20.
4. Zerkowski, J. A.; Seto, C. T.; Wierda, D. A.; Whitesides, G. M. *J. Am. Chem. Soc.* **1990**, *112*, 9025–9026.
5. Seto, C. T.; Whitesides, G. M. *J. Am. Chem. Soc.* **1990**, *112*, 6409–6411.
6. Seto, C. T.; Whitesides, G. M. *J. Am. Chem. Soc.* **1991**, *113*, 712–713.
7. Takeo, K. In *Advances in Electrophoresis*, Vol. 1; Chrambach, A.; Dunn, M. J.; Radola, B. J., Eds.; VCH: New York, **1987**; pp. 229–279.
8. Silverman, D. N.; Lindskog, S. *Acc. Chem. Res.* **1988**, *21*, 30–36.
9. Deutsch, H. F. *Int. J. Biochem.* **1987**, *19*, 101–113.
10. Pocker, Y.; Sarkanen, S. *Adv. Enzymol. Relat. Areas Mol. Biol.* **1987**, *47*, 149–276.
11. Coleman, J. E. *Ann. Rev. Pharmacol.* **1975**, *15*, 221–242.
12. Wiley, D. C.; Skehel, J. J. *Ann. Rev. Biochem.* **1987**, 365–394.
13. Paulson, J. C. In *The Receptors*; Conn, P. M., Ed.; Academic Press: New York, **1985**; Vol II, p. 131.
14. Weis, W.; Brown, J. H.; Cusack, S.; Paulson, J. C.; Skehel, J. J.; Wiley, D. C. *Nature* **1988**, *333*, 426–431.
15. Sauter, N. K.; Bednarski, M. D.; Wurzburg, B. A.; Hanson, J. E.; Whitesides, G. M.; Skehel, J. J.; Wiley, D. C. *Biochemistry* **1989**, *28*, 8388–8396.
16. Ellens, H.; Bentz, J.; Mason, D.; Zhang, F.; White, J. M. *Biochemistry* **1990**, *29*, 9697–9707.
17. Eilat, D.; Chaiken, I. M. *Biochemistry* **1979**, *18*, 790–795.
18. Chaiken, I. M. *J. Chromatogr.* **1986**, *376*, 11–32.
19. Lee, T. T.; Lin, P.; Lee, Y. C. *Biochemistry* **1984**, *23*, 4255–4261.
20. Gopalakrishnan, P. V.; Karush, F. J. *Immunology* **1974**, *113*, 769–778.
21. Hanaoka, K.; Pritchett, T. J.; Takasaki, S.; Kochibe, N.; Sabesan, S.; Paulson, J. C.; Kobata, A. *J. Biol. Chem.* **1989**, *264*, 9842–9849.
22. Pritchett, T. J.; Paulson, J. C. *J. Biol. Chem.* **1989**, *264*, 9850–9858.
23. Roy, R.; Laferrière, C. A. *Can. J. Chem.* **1990**, *68*, 2045–2054.

24. Mastrosovich, M. N.; Mochalova, L. U.; Marinina, V. P.; Byramova, N. E.; Bovin, N. V. *FEBS Lett.* **1990**, *272*, 209–212.
25. Roy, R.; Laferrière, C. A. *Carbohydr. Res.* **1988**, *177*, c1–c4.
26. Alberts, B.; Bray, D.; Lewis, J.; Raff, M.; Roberts, K.; Watson, J. D. *Molecular Biology of the Cell*; Garland: New York, **1983**; pp. 121–127.
27. Allara, D. L.; Nuzzo, R. G. *Langmuir* **1985**, *1*, 45–52.
28. Ulman, A. *J. Mater. Educ.* **1989**, *11*, 205–80.
29. Wasserman, S. R.; Tao, Y.-T.; Whitesides, G. M. *Langmuir* **1989**, *5*, 1074–1087.
30. Maoz, R.; Sagiv, J. *Langmuir* **1987**, *3*, 1034–1044.
31. Maoz, R.; Sagiv, J. *Langmuir* **1987**, *3*, 1045–1051.
32. Haller, I. *J. Am. Chem. Soc.* **1978**, *100*, 8050–8055.
33. Kessel, C. R.; Granick, S. *Langmuir* **1991**, *7*, 532–538.
34. Laibinis, P. E.; Whitesides, G. M. *Langmuir* **1990**, *6*, 87–96.
35. Hickman, J. J.; Zou, C.; Ofer, D.; Harvey, P. D.; Wrighton, M. S.; Laibinis, P. E.; Bain, C. D.; Whitesides, G. M. *J. Am. Chem. Soc.* **1989**, *111*, 7271–7272.
36. Bain, C. D.; Evall, J.; Whitesides, G. M. *J. Am. Chem. Soc.* **1989**, *111*, 7155–7164.
37. Jeon, S. I.; Andrade, J. D.; de Gennes, P. G. *J. Colloid Interface Sci.* **1991**, *142*, 149–158.
38. Jeon, S. I.; Andrade, J. D. *J. Colloid Interface Sci.* **1991**, *142*, 159–166.
39. Ostragorich, G.; Bacaloglou, R. *Timisoara, Studii Cercetari Stiint. Chim.* **1962**, *9*, 273–289.
40. Finkel'shtein, A. I.; Rukevich, O. S. *Zh. Prikl. Spectrosk.* **1983**, *38*, 327–330.
41. Wang, Y.; Wei, B.; Wang, Q. *J. Crystallogr. Spectrosc. Res.* **1990**, *20*, 79–84.
42. Desiraju, G. R. *Crystal Engineering: The Design of Organic Solids*; Elsevier: New York, **1989**.
43. Etter, M. C. *Acc. Chem. Res.* **1990**, *23*, 120–126.
44. Hagler, A. T.; Dauber, P. *Acc. Chem. Res.* **1980**, *13*, 105–112.
45. Rebek, J., Jr. *Angew. Chem., Int. Ed. Engl.* **1990**, *29*, 245–255 and references therein.
46. Bryant, J.; Ericson, J.; Cram, D. *J. Am. Chem. Soc.* **1990**, *112*, 1255–1256.
47. Ashton, P.; Goodnow, T.; Kaifer, A.; Reddington, M.; Slawin, A.; Spencer, N.; Stoddart, J.; Vincent, C.; Williams, D. *Angew. Chem., Int. Ed. Engl.* **1989**, *28*, 1396–1399.

RECEIVED November 1, 1991

SCANNING PROBE MICROSCOPY OF MACROMOLECULAR ASSEMBLIES

The use of scanning probe microscopy, such as scanning tunneling and atomic force microscopy, has become an important analytical methodology to obtain images and molecular parameters of macromolecular assemblies. The resolution in scanning probe microscopy is now of such quality that individual molecules can be imaged routinely within a macromolecular structure. For example, the two-dimensional structure of monolayers and multilayers can be quantified in terms of the spacing between neighboring molecules and the orientation of the subgroups or atoms on molecules. The adage "seeing is believing" is extremely important in science, particularly in research on macromolecular assemblies.

Chapter 20

Scanning Probe Microscopy of Surfactant Bilayers and Monolayers

J. A. N. Zasadzinski, J. T. Woodward, M. L. Longo, and B. Dixon-Northern

Department of Chemical and Nuclear Engineering, University of California, Santa Barbara, CA 93106

The Scanning Tunneling Microscope (STM) and Atomic Force Microscope (AFM) offer exciting new ways of imaging surfactant monolayers and bilayers with resolution to the sub-molecular scale. We present images of biological and synthetic surfactant monolayers and bilayers obtained with these techniques that demonstrate the possibilities and limitations of each. STM is used to image freeze-fracture replicas of the ripple phases of saturated phosphatidylcholines to show a previously unknown secondary ripple recently confirmed by X-ray diffraction. AFM is used to study the structure and defects of Langmuir-Blodgett films to a resolution never before attained, both under water and in air. The molecular spacing and lattice symmetry of the films are visible in phosphatidylethanolamine bilayers, and the images are suggestive of hexatic organization. In monolayers of cadmium arachidate, we have visualized point defects less than 10 nm in extent - about an order of magnitude smaller than any previous technique.

In the 300 years of optical and 50 years of electron microscopy development, nothing has prepared us for the incredible imaging power and simplicity of the scanning tunneling and atomic force microscopes (1). The STM has the greatest proven resolution of any imaging technique; for ideal samples the lateral resolution is about 1Å and the vertical resolution is less than 0.1Å. The operating principle of the STM is surprisingly simple. A metal tip is brought close enough to the surface to be imaged that, at a convenient operating voltage (2 mV - 2V), electrons begin to tunnel between the tip and the surface. The tip is scanned over the surface while the tunneling current is measured. A feedback network changes the height of the tip to keep the tunneling current constant in the so-called "constant current mode" or the current is monitored at constant tip height in the "constant height mode." In the more-commonly used constant current mode, if the current can be kept constant to 2%, then the gap between the surface and tip remains constant to within .01Å (2). An image consists of a map of the tip height vs lateral position. The result is a three-dimensional image of the scanned surface with atomic resolution.

The Atomic Force Microscope (AFM) was developed in 1986 to image non-conductors and should be the ideal instrument for visualizing biological and organic surfaces at high resolution (3). The AFM records interatomic forces between the apex of a cantilevered spring tip and the surface of the sample. In the "constant deflection mode" of imaging, the atomic force microscope traces out contours of constant force (as opposed to the scanning tunneling microscope which traces out contours of constant electron density). The AFM is compatible with a variety of environments, including liquids like water or saline, at varied temperatures and is capable of imaging with a tracking force of only about 10^{-9} -10^{-11} Newtons under water (4). However, it is not yet known over what area the force is operating during imaging with the AFM.

Conventional transmission electron microscopes (TEM) can only give two dimensional projections, and because the high energy electrons penetrate deep into matter, the TEM has limited applications to surface investigations. A three-dimensional image of a surface with atomic resolution is only possible with the STM or the AFM. However, the surfactant bilayers and monolayers to be examined here are far from ideal surfaces for the AFM or STM. The STM is best suited to image rigid, conductive surfaces such as metals, silicon, graphite, etc; the STM is virtually useless for imaging materials that are fluid, non-conductive, or both. Although the AFM can image non-conductors, the local pressures developed on the sample surfaces can lead to deformations and degradation during imaging. The AFM can also act as a molecular "broom," sweeping objects aside as it seeks a firm base on which to scan. Even with these drawbacks, the incredible resolution and three-dimensional imaging capability of scanning probe microscopes makes it imperative to develop a way of applying them to the investigation of the relationship between structure and function in surfactant monolayers and bilayers. Like samples for the electron microscope, surfactant samples for the STM and AFM must be modified to make them suitable for imaging at high resolution.

Principles of Operation of Scanning Probe Microscopes

Electron tunneling, the fundamental physical principle underlying the operation of the STM, is yet another manifestation of the dual wave-particle nature of electrons. The free electrons in a metal, usually one per atom, can be thought of as mobile particles moving in a container defined by the positive metal ion lattice. The free electrons have energies roughly up to the Fermi energy, E. The Fermi energy can be identified with the chemical potential of the electrons in the metal. The vacuum gap between the metals constitutes a potential energy barrier of energy V. The difference between the energy of the electrons in the metal and the energy of an electron in vacuum, V-E, is the work function of the particular metal, that is, the energy required to remove an electron from the metal (The work function of the tungsten or platinum-iridium tips used in most STM's are of the order of a few electron volts.) Classically, at room temperature, the electrons in the metal do not have sufficient energy to leave the metal and enter the vacuum. However, there is a small, but finite probability, that an electron wave can tunnel through the vacuum barrier into a second metal, if the separation between the two metal surfaces is small enough. Solutions of the Schrödinger equation show that the tunneling current, T, through the vacuum gap is proportional to (5):

$$T \alpha \exp -\{[\frac{2m\pi^2}{h^2}(V - E)]^{1/2} S\}$$

S is the gap separation, or 2a, m is the mass of the electron, and h is Plank's constant. Clearly, there is a non-zero probability that the electron can tunnel from the tip through the vacuum gap and appear in the other metal. This exponential dependence of tunneling current on separation is what makes the STM work so well.

For typical values of the work function, the tunneling current changes by an order of magnitude for a 1Å change in separation. This sensitive dependence of tunneling current on separation is used to control the motion of the STM tip. As the tip moves away from the sample, the tunneling current decreases, and the feedback mechanism of the STM moves the tip closer to the sample to maintain the tunneling current at the original value. Conversely, if the tip moves closer to the sample, the tunneling current increases, and the feedback mechanism of the STM moves the tip away from the surface. The motions of the tip are magnified and recorded to form the STM images. If the current can be kept constant to within 2%, then the gap, and hence the vertical resolution of the STM, remains constant to within 0.01Å. The tunneling image should properly be interpreted as a map of the surface of constant electron density at the particular energy determined by the applied bias voltage. However, the simplified equation above allows for the interpretation of most images of chemically homogeneous surfaces as simple topographies. In practice, a bias voltage of a few millivolts to a few volts depending on the materials in question, is applied between the tip and the sample, and the tunneling current is usually a few tenths of a nanoamp.

To make best use of the unique properties of the tunneling current, it is necessary to be able to hold and position a metal probe within a nanometer or less of the sample surface while at the same time being able to raster this tip across the surface. In the original STM's this was accomplished with three independently controlled piezoelectric transducers of lead titanate - lead zirconate ceramic fixed in a tripod arrangement (2) . Piezoelectrics expand a given amount depending on the voltage applied to them, usually a few nanometers per volt, to provide a combined x-y-z range of motion. The tunneling tip, which is usually a fine needle of tungsten, platinum - iridium alloy, or gold, is then firmly mounted to the tripod. The distance between the tip and the surface is controlled by a voltage applied to the z piezoelectric transducer. This voltage is determined by the feedback circuit that measures and controls the tunneling current between the tip and the sample. As the x piezoelectric transducer moves the tip laterally across the surface, the feedback circuit adjusts the voltage to the z piezoelectric transducer, which raises and lowers the tip to keep the tunneling current constant. As the x scan is finished, it is displaced by the y piezoelectic transducer a given distance from the previous scan. In this way, a raster image is built up, very similar to the operation of the scanning electron microscope. In modern, and in most commercial instruments, the tripod arrangement has been replaced by a single, cylindrical piezoelectric tube to which the tunneling tip is attached (6). The electrodes on the outside of the tube are divided into four segments on opposite sides of the tube that control ±x and ± y. A single electrode on the inside of the tube controls the ±z. The tube scanner is significantly faster and more resistant to vibrations than the tripod arrangement. Single tube piezoelectric transducers with ranges less than a nanometer to more than 100 microns have been designed. Hence, the modern STM can now determine structural information from the atomic scale well into the range that is visible with the optical microscope. In addition, a typical STM image takes only a few to tens of seconds to acquire.

The sharpness of the tip, or more exactly, its radius of curvature, actually determines the lateral resolution of the STM, given state of the art piezoelectric transducers. Fortunately, it is possible to make a tip with atomic lateral resolution very simply. In fact, it is generally observed that a metal tip that appears sharp under a high quality optical microscope will give atomic resolution almost every time. This degree of sharpness means that the end of the tip has a radius of 0.1- 1 microns. If tunneling occurred with equal probability from all points of this tip, the lateral resolution would be on order of the tip radius, and the STM would not be much of a microscope. What actually happens is that on virtually every tip, naturally occurring protuberances project farther from the tip than other atoms. In an ideal case, the protuberance might end in a single atom. Because of the exponential decay of tunneling current with separation, the 2 to 3 Å less separation for this atom or group of atoms means that most, if not all of the tunneling current will originate with these atoms. Hence, the actual radius for tunneling is of atomic dimensions. In practice, tips are made either by mechanical grinding of metal wires or by chemical etching. Both techniques routinely produce tips with atomic resolution.

The AFM traces out contours of constant force (as opposed to the STM which traces out contours of constant electron density) by rastering a sharp tip across the sample surface. The AFM tip is pushed very gently against the surface to be imaged. As the tip is scanned across the surface, the cantilever is deflected by the variations in the surface contours. The deflection of the cantilever can be measured with electron tunneling, an interferometry, or by an optical lever (4). All that is required is an electrical signal that varies rapidly with the deflection. This signal is then sent to the same type of electronics used for a STM. A feedback loop controls the voltage to the vertical (z) piezo element on which the cantilever is mounted so that the force is held constant as the tip is scanned across the surface of the sample with the horizontal (x) piezo element. As in the STM, one scan is a plot of the voltage applied to the z piezo as a function of the x position. An image is built of of many scans, each offset in the y direction from the previous one.

The resolution of the instrument is determined by the reliability and sensitivity of detecting the cantilever motion and the resonant frequency of the cantilever. The optical lever AFM (4) amplifies the motion of the cantilever to produce a motion of the reflected beam at the detector that is greater by a factor of $2L/l$ where L = 4 cm is the distance for the cantilever to the photodiode and $l = 100$ µm is the length of the cantilever. The factor of 800 is sufficient that the instrument is not limited by the photodiode lateral sensitivity, but rather by sound and building vibrations. The microfabricated cantilevers are made from silicon oxide or silicon nitride and have resonant frequencies of order 100 kHz, resulting in a noise level small enough to permit routine atomic resolution imaging of non-conducting solids and crystals (4). The lateral resolution of the instrument is better than 2.5Å, and the vertical resolution is better than 1Å.

The AFM is very compatible with a variety of environments, including liquids like water or saline. Operating the AFM with the sample and cantilever under water allows not only for more realistic environments, but for better control of the applied forces. Surfaces in air are typically covered with an adsorbed layer of water and other unknown contaminants. When the AFM tip (which is also covered with a layer of contaminants) approaches a surface that is also covered, capillary forces drive the tip toward the sample (7,8). This adhesion produces a force of order 10^{-7} N. When both cantilever and surface are completely covered by water, the capillary forces are eliminated and the applied force is much less. The AFM is capable of imaging with a tracking force of about 2×10^{-9} Newtons under water, which allows much softer surfaces to be imaged (7). However, even these small forces appear to be too large for nondestructive imaging of many biomaterials. It is not yet known over what area the force is operating during imaging with the AFM. If the minimum force is applied over an area of 100 $Å^2$, the resulting pressure is about 10,000 atmospheres. (Although not usually appreciated, capillary and other forces of 10^{-7} N or more can occur between the tip and sample in a STM, often modifying the structure of even relatively hard materials). Calculations suggest that the applied force should not exceed 10^{-11} N (9) for biological surfaces and 10^{-9} N (10) for "hard" surfaces. These estimates are very conservative as many biological and hard surfaces have been imaged in air with the STM and AFM with good results. Still biological samples for AFM imaging must be firmly supported and immobilized, made as rigid as possible, and their orientation controlled for best results.

Sample Modification for Scanning Probe Microscopy

One of the greatest advantages of the STM and AFM is that they can operate in air at atmospheric pressure, and even under liquids including water (1). Air and most liquids are insulators for electrons, and electrons tunnel through the thin layer of air or liquid just as they would tunnel through a vacuum. Most organic materials are also extremely poor conductors of electricity. Hence the STM treats biological materials as if they were part of the potential barrier, if they are thin enough to tunnel through. The biological layers modify the tunneling barrier; as such, in certain circumstances, they can be imaged. Thicker layers of biological or other non-conductive materials cause the STM to cease operating. The tunneling electrons cannot be carried off at a sufficient rate because of the insulating nature of the material. The surface of the sample rapidly charges up to the tip potential, and the tunneling stops. For the STM to work properly, about 10^{10} electrons/second (one nanoamp) must flow in a 0.1 nm^2 area. This sort of conductivity is not found in most organic materials. An additional complication, for both STM and AFM, is that at room temperature, membranes, proteins, and macromolecules are dynamic structures. Brownian and other thermal fluctuations can be large in comparison to the molecular features of interest. These limitations must be overcome by appropriate sample modifications that make minimal changes in the specimen microstructure while providing sufficient conductivity for the STM and rigidity for the STM and the AFM (11).

Freeze-Fracture Replication STM. For surfactant bilayers, membranes and more fluid materials that cannot be simply adsorbed or freeze-dried onto a substrate, a more elaborate "fixation" procedure for STM is necessary. The best general fixation technique for fluid materials is freeze-fracture replication. Freeze-fracture replication involves 1) quick-freezing the sample to kinetically trap fluid structures; 2) fracturing the sample under vacuum at cryogenic temperatures to expose internal structure; 3) replication of the fracture surface with an evaporated layer of platinum and carbon. In this process, the fluid and non-conductive material is replaced by a rigid, conductive surface replica that is an almost ideal specimen for STM imaging (12).

For freeze-fracture STM, the initial quick-freezing step is the most important to achieve minimal disruption of the structure. If the cooling rate is too slow, the material can reorganize into some new structure, not always the one we wish to study. To obtain the best results, the specimen must be frozen at such a rate that the molecules do not have sufficient time to rearrange themselves. In practice, this means a cooling rate of about 10,000° C/sec. To achieve these rates, we sandwich a 10 - 50 µm thick layer of sample liquid between two 100 µm thick copper plates in an enclosed, temperature-controlled oven (±0.1° C), then rapidly cool the sandwich by placing it between two opposed high velocity jets of liquid propane, cooled by liquid nitrogen to < 100 K . This cooling rate is sufficient to freeze water in a vitreous state and preclude rearrangement of dispersed macromolecules or biomembranes . After freezing, the sample sandwich is fractured at cryogenic temperatures (100 K) in a high vacuum ($\approx 10^{-9}$ torr) freeze-etch apparatus. The complementary fracture surfaces are then immediately coated with a 1.5 nm thick shadowing film of platinum-carbon, followed by a 15 nm thick layer of carbon at normal incidence to strengthen the shadowing film. The samples and replicas are brought to ambient temperature and pressure, and any remaining sample is cleaned from the replica films.

The metal surface films form near-exact replicas of the fracture surface and are ideal specimens for both transmission electron and scanning tunneling microscopy as they are rigid, highly resistant to radiation or electric field damage, have virtually zero vapor pressure at room temperature, and are highly conductive. Figure 1a is a STM image of a freeze-fracture replica of the P$_\beta$' or "ripple" phase of a naturally occurring phospholipid liquid crystal, dimyristoylphosphatidylcholine dispersed in water (13) along with the corresponding transmission electron microscope image of a similar freeze-fracture replica (Figure 1b). The replicas were oriented so that the side directly contacting the sample surface was imaged by the STM. The improvement in resolution and three-dimensional imaging capability of the STM is clearly evident, especially as we go to higher magnifications, as in Figures 2a and 2b. The characteristic three-dimensional ripples of the P$_\beta$' phase are readily apparent as regular mounds spaced about 130 Å apart. What is surprising is the fine scale features such as the smaller periodicity perpendicular to the ripples. These undulations are only 5 to 10 Å in height and 30 Å in lateral separation. These are the highest resolution features ever seen on an unsupported bilayer surface; they may show how the molecules that make up the biomembrane are organized. In fact, recent X-ray diffraction work by Hentschel and Rustichelli (14) have shown that these secondary ripples may be the result of the hydrocarbon chain tilt direction.

The resolution of of a freeze-fracture replica is far superior when imaged by the STM than by the TEM. One explanation for this is that the STM only images the side of the replica film in direct contact with the original sample surface. Presumably, this side of the replica matches the original fracture surface most exactly. There is less influence of replicating metal migration, recrystallization, etc. on the images because the STM is only influenced by the topography of the replica surface. In the TEM, the image is formed by scattering variations caused by differences in the replica thickness. The electron beam samples the entire replica and any variations or imperfections in the replica film caused by local crystallization, preferential attachment, surface migration, etc., are imaged, thereby degrading the resolution.

Atomic Force Microscopy of Langmuir-Blodgett Films. With the AFM it should, in principal, be possible to visualize the organization of both monolayers and bilayers of membrane lipids, proteins incorporated into lipid layers, antibodies, etc., at the molecular level. However, the AFM puts its own set of constraints on the sample to be imaged. The force between the tip and sample is of order 10^{-9} N; however, this force is applied over an area as small at 100 Å^2 resulting in pressures of thousands of atmospheres. Biological materials such as cell walls, lipid bilayers, or proteins are incapable of withstanding such localized pressures without modification. The AFM also can act as a molecular "broom," sweeping objects aside as it seeks a firm base on which to scan. If the AFM is to be useful in investigations of biological structures at the molecular level, samples must be prepared so that the molecules of interest are <u>rigidly mounted</u> and <u>immobilized</u> with <u>well defined orientations</u>. Such sample requirements can be met by Langmuir-Blodgett deposition of lipid or lipid-protein monolayers onto mica or glass substrates.

The Langmuir trough has long been a standard tool in surface and colloid science for depositing chemically and physically well-defined monolayers (15). It allows for controlled deposition of lipids on a variety of surfaces, and in preliminary experiments, we have found it to be particularly suitable for depositing monolayers and bilayers of lipids on mica for use in the atomic force microscope (See Figs. 3,4). Langmuir-Blodgett troughs allow for the precise control of surface pressure, Π, average area per molecule, A, temperature, and the lateral distribution of the molecular components at the air-water interface. These monolayers can then be quantitatively transferred to rigid substrates such as mica or glass. Supported L-B films are remarkably robust and resistant to physical damage and wear. They are being used and investigated extensively as boundary lubricants and have been shown to be undamaged at pressures greater than 5000 atmospheres (16).

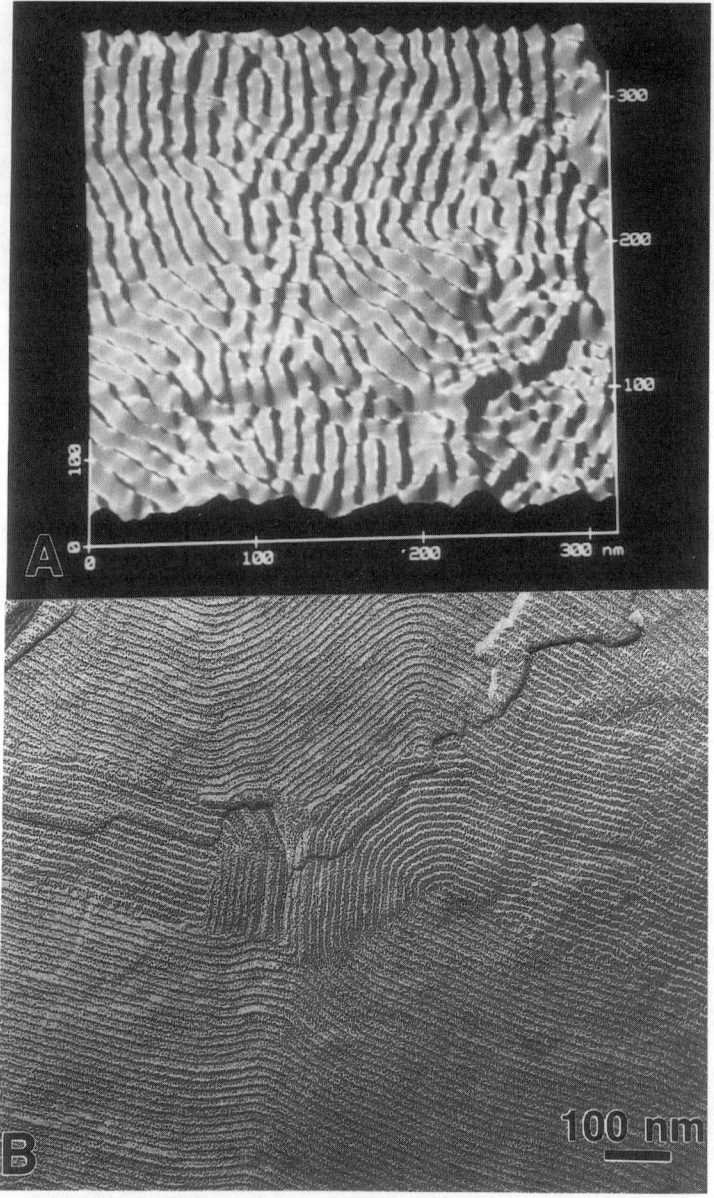

Figure 1. A) Three-dimensional scanning tunneling micrographs of a freeze-fracture replica of the ripple phase of dimyristoylphosphatidylcholine (DMPC).
B) Transmission electron micrograph of a similar freeze-fracture replica of the ripple phase. The resolution is clearly inferior to the STM image in A. In both images, the regular wavelength of the ripples is easily seen, although the ripple heights are much better defined in the STM image.

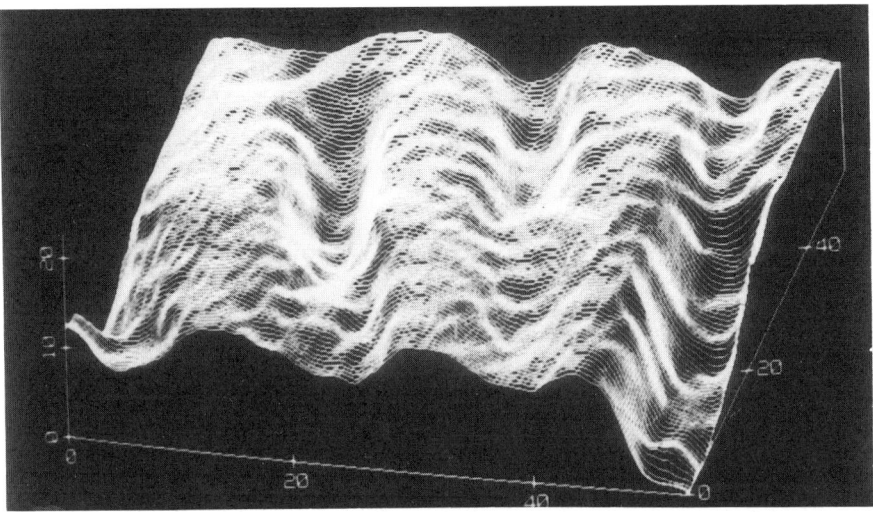

A

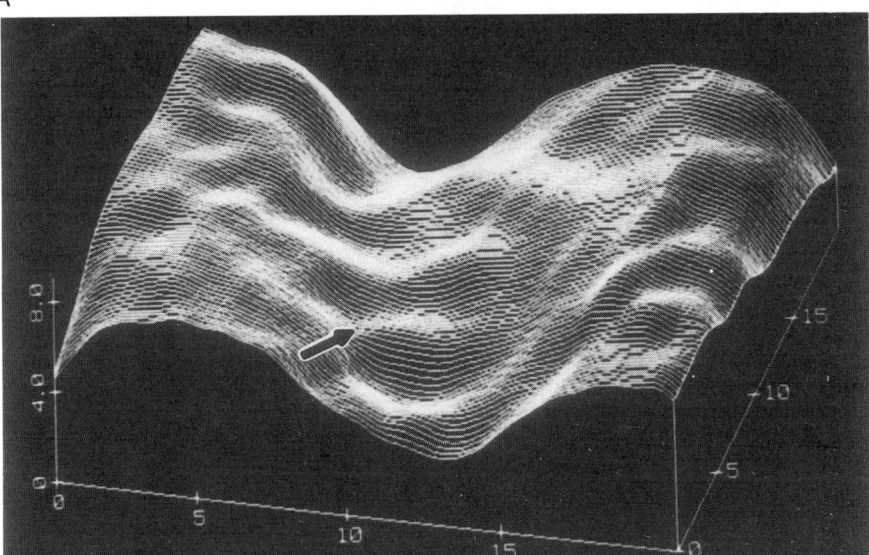

B

Figure 2. A) Higher magnification view of a freeze-fracture replica of the ripple phase of DMPC in water. The three-dimensional view of the ripple structure is impossible to obtain by any other technique.

B) One "ripple" magnified approximately 4,000,000 times. The wavelength of the ripple is about 13 nm, and the amplitude or height of the ripple is about 4.5 nm. At the arrow is a secondary "ripple" roughly orthogonal to the larger ripple. This secondary structure is believed to be an indication that the molecular lattice is oriented roughly perpendicularly to the main ripple direction.

(Adapted from reference 13.)

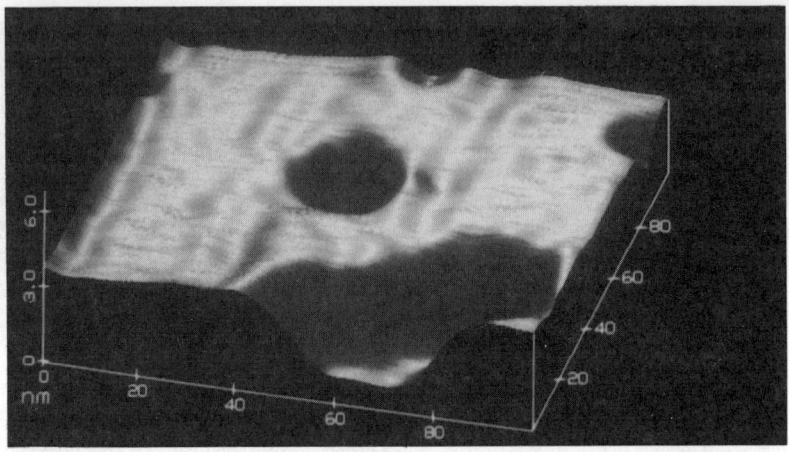

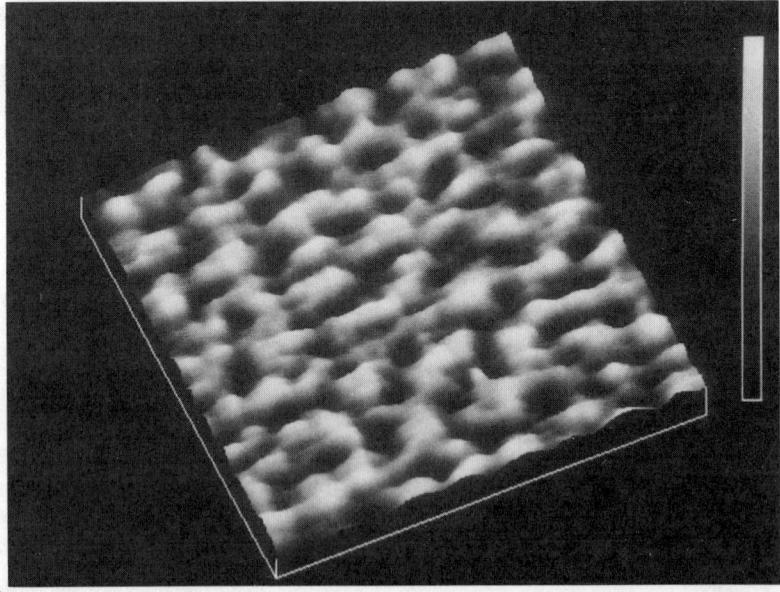

Figure 3. a) AFM image of cadmium-arachidate monolayer on mica. This is a side view of pinhole defects in the monolayer, the smaller hole is less than 20 nm in diameter, which is about an order of magnitude smaller than any previous observation. The larger hole at the bottom right of the image can be seen to be roughly 3 nm deep, which corresponds to the thickness of one monolayer.
b) Higher magnification view of the area in a. Image at low force shows less than 0.3 nm roughness on surface of the monolayer, but only short-range ordering which is consistent with X-ray diffraction measures. Total magnification is approximately 10,000,000 times. (Adapted from reference 22.)

Unfortunately for many of their potential applications, LB films are not always free from defects and are quite fragile and subject to damage (17). One requirement for making uniform, defect-free films is to have a sensitive, non-destructive technique for evaluating the quality of the films. Although the structure of LB films have been extensively studied by various methods, these techniques are not sensitive to microscopic defects such as pinholes, tears and folds within layers. Film defects have been examined primarily with electron microscopy, although investigations are limited to special substrates or result in destruction of the films (18-21). The AFM can be used to image defects as small as a few nanometers in cadmium arachidate monolayers on mica substrates without damaging the sample (22). The AFM provides a straightforward, non-destructive technique for determining film quality with superior resolution on any substrate. The AFM can also image films in air or under water or other fluids, which is of particular importance for biological and other applications of molecular films.

Arachidic acid was dissolved at 1 mg/ml in chloroform and was deposited behind the moveable barrier of a simple Langmuir-Blodgett trough. The subphase water was treated in a Millipore Milli-Q system; 0.5 mM cadmium chloride was added to the subphase water. Monolayers were deposited on freshly cleaved mica substrates by raising the mica through the interface with one hand while moving the barrier of the home-built trough with the other hand to keep a "constant" transfer pressure of 35±3 mN/m. A Nanoscope II-FM AFM was used in the constant deflection mode in air. Figure 3a shows a typical AFM image of a side view of a cadmium arachidate monolayer in air which contains several holes, 30 to 140 nm in size, in a 100 x 100 nm field. Subsequent images of the same field taken over the next few minutes showed the same pattern of holes. The surface of the films could be imaged repeatedly without damaging the sample. The smaller holes are more than twice as small as those detected using electron microscopy techniques (18-21). Dark areas correspond to the lower areas in the image, with about 3 nm difference between the bottom of the holes and the top of the monolayer. This is consistent with the expected length of the cadmium arachidate molecule, about 2.8 nm. The monolayer can be imaged repeatedly at low force (approx. 10^{-8} N) without damaging the surface or altering the size or location of the holes.

At higher magnification, the LB films showed a surface roughness of less than 0.3 nm (Figure 3a). No clear periodic structure could be seen from a Fourier transform of the high resolution images, suggesting that the surface is somewhat disordered at the molecular scale. This is consistent with X-ray diffraction data of Skita *et al.* who noted that a cadmium arachidate monolayer in contact with air was disordered, with an average length less than that for an all *trans* chain, as compared to a cadmium arachidate monolayer within a multilayer (23). They also noted that the terminal methyl groups of the monolayer interface with air had more degrees of freedom than the terminal methyl groups of monolayers in a multilayer. However, periodic molecular structure has been seen in STM images of the terminal methyl groups of LB films of Cd-arachidate and dimyristoylphosphatidic acid in air (24,25). The mechanism of imaging with STM and the interpretation of the images has been questioned (26); it is not yet clear why crystalline structure was seen with the STM but not with the AFM for LB films in air. Regardless of the details of the ordering of the methyl ends of the monolayer, the AFM is clearly the instrument of choice for non-destructive testing of Langmuir-Blodgett film quality.

AFM of Phospholipid Bilayers One of the reasons we are interested in high quality monolayers is to use such monolayers as substrates for additional layers (27). As an example, we present images of the polar or headgroup regions of bilayers of dimyristoyl-phosphatidylethanolamine (DMPE), deposited by Langmuir-Blodgett deposition onto mica substrates at high surface pressures and imaged under water at room temperature with the AFM. The lattice structure of DMPE is visualized with sufficient resolution that the location of individual headgroups can be determined (Figure 4a and 4b). The DMPE molecules in the bilayer appear to have relatively good long-range orientational order along the rows, but rather short-range and poor positional order within the rows. These results are in agreement with X-ray measurements of unsupported lipid monolayers on the water surface (28,29), and with electron diffraction of adsorbed monolayers (30, 31) which suggest hexatic order in LB bilayers.

Chromatographically pure, synthetic DMPE was purchased from Sigma Chemical Company (St. Louis, Mo.) and used without further purification. DMPE was dissolved at a concentration of 7 mg/ml in chloroform/methanol (3:1) and was deposited quantitatively behind the moveable barrier of a Lauda mini-trough to give directly the area per molecule. The solvent was allowed to evaporate over the course of several minutes before any measurements were taken. The surface pressure, π, was measured with a Wilhelmy balance detecting the differential force on a piece of filter paper partially submerged in the subphase. The subphase water was distilled, then treated in a Millipore Milli-Q system. A pressure/area isotherm showed that the fluid-gel coexistence pressure, π_c, occurred at 12.7 mN/m at a specific molecular area of .6 nm^2 and the gel phase occurred at a surface pressure, π_s, of 29.7 mN/m at a specific molecular area of .41 nm^2. Both of these values are consistent with previous studies (28-31).

Circular, 1 cm diameter mica substrates were repeatedly cleaned in absolute ethanol and Millipore-treated water, then cleaved immediately prior to deposition. The mica disc was then held with tweezers and lowered into the trough at low π, high A. The surface area was decreased to approximately .40 nm^2 and a surface pressure of 40 mN/m so that the monolayer was well within the gel phase. Deposition occurred at a rate of 1 cm/min. During this process the monolayer pressure was held constant by electronic feedback-control of the film area. To deposit the second monolayer, the procedure of Tamm and McConnell (32) was used. A monolayer covered mica disc was gently placed onto a fresh monolayer deposited on the trough at a constant surface pressure of 40 mN/m. Then the mica disc was rapidly pushed through the interface into the water subphase. The second monolayer adsorbs with the hydrophilic heads pointed outwards. This second monolayer is only stable under water and will desorb on being retracted from water. The bilayer-coated substrate was then placed (below water) into a glass petri dish with a shallow well, with the bilayer coated side facing down. The water subphase was then aspirated out of the L-B trough. The assembly can then be removed from the trough and transferred to the AFM without exposing the bilayer to air. A water-filled cell was created in the Nanoscope II FM (Digital Instruments, Goleta, California) AFM by placing an O-ring on top of the bilayer coated mica disc, which then is mounted on the sample stage (4).

Figure 4A and B show topographic and grey scale AFM images of the polar region of a bilayer of DMPE deposited on mica. These images were taken under water at ambient temperature and pressure. The dominant features of the images are long, uniformly spaced rows roughly .7 - .9 nm in spacing. A modulation also can be seen along the rows, with rounded bright spots roughly every .5 nm. We believe that the individual bright spots along the rows correspond to the individual headgroups of the DMPE molecule. The area per molecule in the AFM image, about .4 nm^2, is also in agreement with the .4 nm^2 area per molecule measured before deposition. The lattice spacing and symmetry are substantially different than that of the mica substrate (4) and it was possible to push through the DMPE bilayer to image the mica lattice underneath by increasing the applied force on the cantilever (data not shown). The white (or bright) areas in Fig. 4A are the highest regions, about .5 nm higher that the lowest regions, which are black. The topographic images make

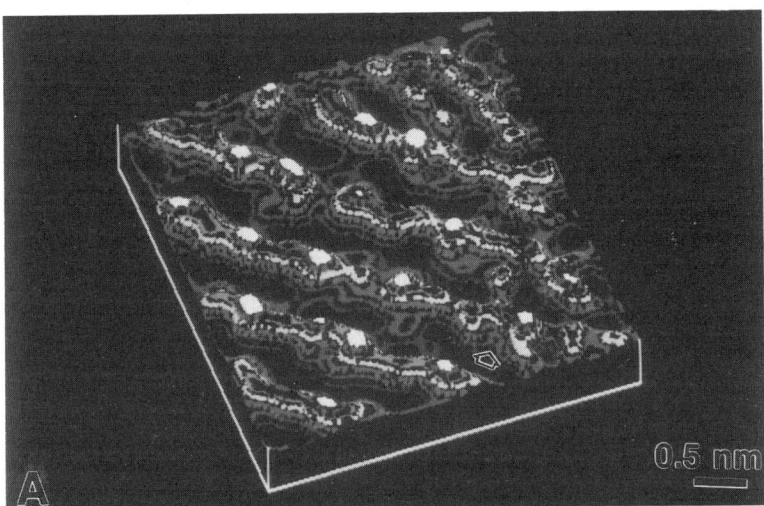

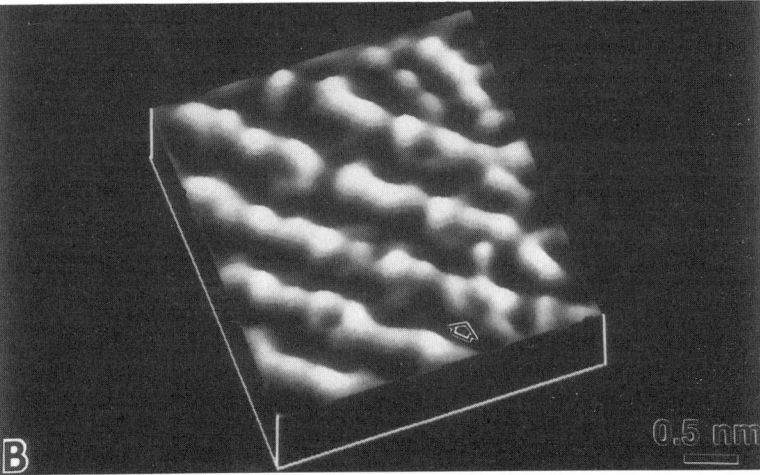

Figure 4. A) Topographic AFM images of symmetric dimyristoylphosphatidylethanolamine bilayer corresponding to the grey scale image in B. Along the rows are visible individual headgroups of DMPE separated by about .5 nm. Bright (white) corresponds to high points, about .4 nm above the lowest points which are black. The organization along the rows is imperfect, with many defects and vacancies, although the rows extend for several nm. This type of organization is suggestive of the hexatic phase in agreement with X-ray and electron diffraction.
B) Grey scale AFM images of the polar region of a bilayer of DMPE deposited by the Langmuir-Blodgett technique at a specific molecular area of .41 nm^2 and a surface pressure of 40 mN/m on a freshly cleaved mica substrate. The long, uniformly spaced rows are roughly .7 - .9 nm in spacing. The modulation along the rows, with rounded bright spots roughly every .5 nm, correspond to the individual headgroups of the DMPE molecule. The area per molecule in the AFM image is about .4 nm^2. The arrow marks similar areas on both images for reference. The total magnification of these images is about 20,000,000 times. (Adapted from reference 27.)

it easier to see the molecules along the rows. These images also show that the DMPE packing is very imperfect, that there are a number of defects and vacancies in the lattice, although the rows of molecules extend for several nanometers. This is suggestive of hexatic ordering, although certainly not conclusive. The general features are in agreement with both X-ray and electron diffraction which suggest that LB films should exhibit hexatic ordering. It is also possible to see that the headgroups are not at a uniform height; the bilayer is rough at the nanometer scale. This molecular roughness may be important to recent theories of the so-called hydration force between bilayers in aqueous solution (33). These are the highest resolution images of a hydrated, room temperature surface ever taken.

Discussion and Suggestions for Further Work

Neither the STM or the AFM will provide routine images of untreated biomaterials without significant improvements in both instruments. However, both instruments are much more compatible with hydrated and dynamic surfaces than the electron microscope, so less sample alteration is necessary. Both the STM and the AFM can operate at room temperature and pressure, and even under fluids, including water. Both instruments are much less destructive of samples than the electron microscope because the energy input, a few millivolts vs hundreds of kilovolts, is much less. The energy in the electron beam of a conventional transmission electron microscope is sufficient to break most organic chemical bonds, and usually causes gross structural rearrangements on the time scales required for imaging. The higher the magnification of a transmission electron microscope, the higher the electron flux required for imaging. Also, electron micrographs are two-dimensional projections, as opposed to the three-dimensional topographies provided by the STM and AFM. Hence, even at their present level of development, the STM and AFM can provide a wealth of new information on biological structure. The commercial availability of the STM and AFM at relatively modest prices will also spur new workers to join in the hunt. Still, the scientist who hopes for molecular and eventually atomic resolution images of biological materials faces many challenges before this goal can be reached. However, it is important to remember that scanning probe microscopy is still in the early stages of development, and that many sample preparation techniques need to be optimized before these instruments can fully realize their potential.

Acknowledgments

I thank A. Weisenhorn, H. Hansma, and S. Gould for helpful comments and discussions. I also especially thank Paul Hansma for being willing to share his vast knowledge of tunneling microscopy with a relative newcomer like myself. Financial support was provided by the Office of Naval Research, Grant #N00014-90-J-1551, a Whitaker Foundation Biomedical Engineering Grant, and by a National Science Foundation Presidential Young Investigator Award # CBT 86-57444.

Literature Cited

1. Hansma, P. K., Elings,V. B.,Marti, O., and Bracker,C. E. *Science*, **1988**, *242*, 209.
2. Binnig, G., Rohrer, H., Gerber, Ch. and Weibel, E. *Phys. Rev. Lett.*, **1982**, *49*, 57.
3. Binnig, G., Quate,C. F. and Gerber, Ch. *Phys. Rev. Lett.* **1986**, *56*: 930.
4. Drake, B., Prater, C. B., Weisenhorn, A. L., Gould, S. A. C., Cannell, D. S., Hansma,H. G., Hansma, P. K., Albrecht, T. R. and Quate, C. F.*Science*, **1989**. *243*, 1586.

5. Hansma, P. K. and Tersoff, J. *J. Appl. Phys.*, **1987**, *61*, R1.
6. Binnig, G. and Smith, D. P. E. *Rev. Sci. Instrum.* **1986**, *57*, 1688.
7. Weisenhorn, A. L., Hansma, P. K. Albrecht, T. R., Quate, C. F. *J. Appl. Phys.*, **1989**, *54*, 2651.
8. Woodward, J. T., Zasadzinski, J. A. N., Hansma, P. K., *J. Vac. Sci. Technol. B,* **1991**, *9*, 1231.
9. Persson, B. N. J. *Chem. Phys. Lett.,* **1987**, *141*, 366.
10. Abraham, F. F., Batra, I. P. and Ciraci, S. *Phys. Rev. Lett.,* **1988**, *60*, 1314.
11. Zasadzinski, J. A. N. *Biotechniques,* **1989**, *7*, 174.
12. Zasadzinski, J. A. N., and Bailey, S. M., *J. Elect. Mic. Tech.*, **1989**, *13*, 309.
13. Zasadzinski, J.A.N., Schneir, J., Gurley, J., Elings, V. and Hansma, P. K. *Science,* **1988**, *239*, 1014.
14. Hentschel, M. P. and Rustichelli, F. *Phys. Rev. Lett.*, **1991**, *66*, 903.
15. Adamson, A. W. *Physical Chemistry of Surfaces*, 3rd Ed. Wiley: New York, 1976.
16. Merkel, R., Sackman, E., and Evans, E. *J. Phys. France*, **1989**, *50*, 1535.
17. Swalen, J. D., Allara, D. L., Andrade, J. D., Chandross, E. A., Garoff, S., Israelachvili, J., McCarthy, T. J., Murray, R., Pease, R.F., Rabolt, J. F., Wynne, K.J., and Yu, H., *Langmuir*, **1987**. *3*, 932.
18. Ries, H. E., Jr., Matsumoto, M., Uyeda, N., and Suito, E., *J. Colloid Interface Sci.*, **1976**, *57*, 396-398.
19. Fryer, J. R., Hann, R. A., and Eyres, B. L., *Nature*, **1985**, *313*, 382.
20. Uyeda, N., Takenaka, T., Aoyama, K., Matsumoto, M., and Fujiyoshi, Y., *Nature*, **1987**, *327*, 319.
21. Luk, S. Y., Wright, A. C., and Williams, J. O., *Thin Solid Films,* **1990**, *186*, 147.
22. Hansma, H. G., Gould, S. A. C., Hansma, P.K., Gaub, H. E., Longo, M. L., and Zasadzinski, J. A. N. *Langmuir*, **1991**, in the press.
23. Skita, V., Filipkowski, M., Garito, A. F., and Blasie, J. K., *Phys. Rev. B,* **1986**, *34*, 5826.
24. Smith, D. P. E., Bryant, A., Quate, C. F., Rabe, J. P., Gerber, Ch., and Swalen, J. D., *Proc.Natl. Acad.Sci. USA,* **1987**, *84*, 969.
25. Hörber, J. K. H., Lang, C. A., Hänsch, T. W., Heckl, W. M., and Möhwald, H., *Chem. Phys. Lett.,* **1988**, *145*, 151
26. Coombs. J. H., Pethica, J. B., and Welland, M. E., *Thin Solid Films*, **1988**, *159*, 293.
27. Zasadzinski, J. A. N., Helm, C. A., Longo, M. L., Weisenhorn, A. L., Gould, S. A. C., and Hansma, P. K. *Biophys. J.* **1991**, *59*, 755.
28. Helm, C. A., Möhwald,H.,Kjaer, K., and Als-Nielsen, J. *Biophys. J.* **1987**, *52*, 381.
29. Kjaer, K., Als-Nielsen, J., Helm, C. A., Laxhuber, L. A., and Möhwald, H. Ordering in lipid monolayers studied by synchotron x-ray diffraction and fluorescence microscopy. *Phys. Rev.Lett.* **1987**, *58*, 2224.
30. Fischer, A. and Sackmann, E. *J. Phys. (France)* 1984, 45, 517.
31. Garoff, S., Deckman, H. W., Dunsmuir, J. H., Alvarez, M. S., and Bloch, J. M.. *J. Phys.(France)* **1986**, *47*, 701.
32. Tamm, L. K. and McConnell, H. M. *Biophys. J.*, **1985**, *47*, 105.
33. Israelachvili, J. N., Wennerström, H. *Langmuir,* **1990**, *6*, 873.

RECEIVED September 24, 1991

Chapter 21

Elucidation of Macromolecular Assemblies
Use of Scanning Tunneling Microscopy for Molecular Bioengineering of Cellular Self-Assemblies in Molecular Device Design

Vincent B. Pizziconi and Darren L. Page

Department of Chemical, Bio-, and Materials Engineering, Arizona State University, Tempe, AZ 85287–6006

The molecular structure and cell adhesion properties of the extracellular matrix (ECM) proteins, laminin and fibronectin, were investigated using optical and scanning tunneling microscopy (STM) to assess their potential to construct cellular self-assembly systems useful in molecular device design. Optical micrographs show that both laminin and fibronectin were able to bind cells effectively but only fibronectin was able to effectively interface cells with synthetic microstructures used in the fabrication of a whole cell biosensor. STM was used to gain insight into the molecular structure of these two key ECM proteins as a means to further assess their role in biological cellular assembly. High resolution STM images of these structures were obtained and correlated well with reported molecular structure and dimensions based upon transmission electron microscopy studies. Further, their molecular features were much more evident using STM. Thus, STM can provide the highest resolution yet possible of biological structures in a totally hydrated molecular state. STM offers the potential to study macromolecular self-assembly processes, and possibly their associated universal molecular recognition binding sequences, responsible for cellular self-assembly.

There is a growing interest by life scientists, medical researchers and bioengineers in understanding the molecular recognition processes associated with biological self-assembly. A deeper understanding of cellular self-assembly processes will provide insight into fundamental aspects of molecular biology processes. Additionally, this new knowledge will serve as a basis for the development of molecular heuristics that will significantly contribute to the newly developing field of molecular bioengineering. Such developments in molecular bioengineering will undoubtedly pave the way for molecular device/system design in bioengineering and biotechnology (1). Examples of applications would include the design of novel biomaterials (2-5), hybrid artificial organs (6), biosensors (7), cell culture systems (8), molecular biochips (9), as well as, providing a more fundamental understanding of biocompatibility (10), cancer research (11-14) and others.

This study represents a two-pronged approach to the characterization of both the molecular and macroscopic properties of cellular adhesive proteins. These are a unique class of proteins that are primarily responsible for cell attachment, migration, and growth on biological support matrices known as extracellular matrices (ECM). The ECM proteins and/or their sequences may be of great utility in providing novel materials and/or methods for molecular device design. For example, these structures or their analogues may provide highly specific mechanical linkages between synthetic microstructures and biological systems. One practical application is the design of whole cell biosensors which require relatively large (high density) cellular assemblies in direct contact with a transducer.

The ability of ECM proteins to promote cellular adhesion has been attributed to their very specific, yet widely generic, cell-surface-receptor binding sequences. However, the molecular recognition mechanism(s) associated with these systems are not well understood at the present time. One approach used in the first phase of this study to further characterize their molecular features is scanning tunneling microscopy (STM). STM is a relatively new imaging tool that has great potential to study and characterize biological structures at unprecedented resolution and in near native aqueous environments. A second, more macroscopic, phase of this work addresses the ability of ECM proteins to anchor whole cells onto biomaterial synthetic surfaces currently being evaluated in our laboratories as components of an immunobiosensor. An introductory discussion of STM and ECM protein systems precedes the results of these studies.

Scanning Probe Microscopy

One objective of this study is to assess the potential utility of scanning tunneling microscopy (STM) which is one of the two types of scanning probe microscopy (SPM) now commercially available. SPM presently includes both STM and atomic force microscopy (AFM) and, as a class, represent a novel tool to help elucidate molecular recognition mechanisms responsible for adhesion processes associated with cellular self-assembly. Both STM and AFM are capable of imaging biological structures (15) and offer several inherent advantages over other conventional microscopies, e.g. optical and electron microscopy. First, STM and AFM offer the highest resolution of any surface microscopy technique attaining 1.0 Å lateral resolution and .01 Å vertical resolution (16). Second, they can operate with the sample completely immersed in water (17). Thus, STM and AFM are particularly attractive tools for imaging and characterizing biological structures because of their ability to image these typically labile, viable structures at ambient pressure directly in an aqueous environment in three-dimensional molecular (and possibly atomic) resolution. This capability is particularly relevant as it allows imaging of biological structures in their near native biological environments. In addition, it may also afford the unique opportunity to conduct dynamic studies directly while scanning in real time by changing the appropriate experimental condition(s) in situ. Although this study focused on STM as the SPM tool of choice, AFM would also be amenable for biological imaging (18), the main advantage of the latter is its ability to image non-conductive structures. Together, STM and AFM may provide independent yet complementary topographical information on complex biological structures.

The operating principle of STM is based on the electron 'tunneling' phenomenon. It is one of several types of conduction mechanisms now known to be a common phenomenon occurring on the atomic scale in insulating materials and not predicted by classical physics (19). Yet, tunneling had been verified experimentally over a quarter of a century ago whereby electrons were shown to 'tunnel' across relatively thin insulating materials or regions bounded by two conductors by applying an appropriate bias voltage (20, 21). This discrepancy was eventually resolved by quantum mechanics which attributes tunneling as a mechanism in which electrons travel through the overlapping electron clouds emanating from conductive material surfaces.

Thus, when two conductive materials are placed in very close proximity separated only by a very small gap, electrons can be made to flow from one conductive material to the other by applying an appropriate bias voltage as shown schematically in Figure 1. As the tip is scanned over the surface, the tunneling current is tracked in either one of two scanning modes and from which images representing surface morphology can be produced. It is important to note, however, that STM images actually represent a material's surface electron charge density and not actual surface topography.

Although the first crude tunneling experiments of surface structure were demonstrated over two decades ago (22,23), the utility of tunneling as a surface microscopy tool on the atomic scale was first demonstrated less than a decade ago by Binnig et. al. (24,25). Part of their success was due to their careful attention to isolation of vibrations, the preparation of atomically clean surfaces, the development of a gentle tip approach and the control of very small gap spacings. Many advances since then have focused on computerizing tip control and image processing.

In practice, a tunneling current is induced by positioning a sharp conductive (metal) probe in close proximity (typically 1-10 Å) to a conductive surface across which a small bias voltage (typically in the range of a millivolt to 1 or 2 volts) is applied. A computer controlled servo drives x, y, and z piezo positioners that sweep the tip across the surface while maintaining either a constant tunneling current or a constant gap distance between the tip and the sample surface. For atomically flat surfaces, relatively rapid scanning can be achieved in the constant height mode. However, scanning in the constant current mode is preferred when imaging highly irregular surfaces such as that presented by large macromolecular assemblies, such as, proteins and even cells that are immobilized in some fashion (e.g., adsorbed) to the conductive substrate surface as depicted in Figure 2.

Atomic resolution with STM is possible because of the exponential dependence of the tunneling current with the tip distance (gap) from the surface as described in equation 1;

$$J = \exp(-A \Theta^{1/2} S) \qquad [1]$$

where J is the tunneling current, Θ is the work function of the material, S is the distance across the gap separating the two material surfaces and A is ~1 if Θ is measured in electron volts (eV) and S in angstroms. It is this extreme sensitivity of J with S (e.g., the tunneling current changes approximately an order of magnitude for every angstrom change in S for a local work function of 4 eV) that is exploited whereby the vertical tip movement is translated to a high vertical resolution image of the surface. More complete reviews on the basics of SPM, STM (26-29) and AFM (30), are available.

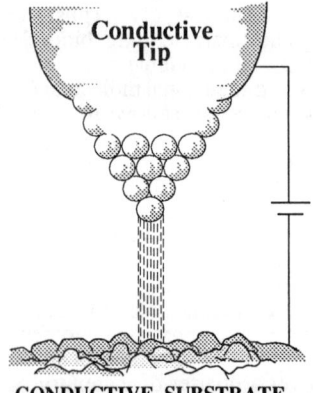

Figure 1. Conceptual representation of the operating principle of STM (adapted from ref. 24)

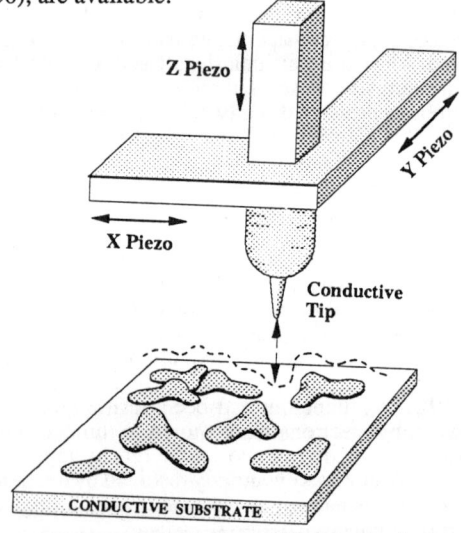

Figure 2. Schematic of STM tip hardware (adapted from ref. 26)

STM of Biological Structures

STM made its initial impact on characterizing nonbiological, conducting and semiconducting material surfaces at atomic resolution (24-26). This early success was followed shortly thereafter with the pioneering work in successfully imaging biological material, i.e., DNA using STM (31). There are now many reported studies on imaging a wide variety of biological structures. These studies are summarized elsewhere (32).

However, in spite of the growing number of reports on STM imaging of biological structures, imaging biological materials with STM poses additional and somewhat unique problems (33) including the reported creation of ambiguous, biological-like structures resulting from substrates alone i.e., artifacts can be created by the STM tip as it rasters over graphite line defects (34). This and other circumstances have led to some skepticism of STM as a generic tool for imaging biological structures. The major criticisms relate to the poorly understood constitutive properties of biological materials in regards to their conduction mechanisms, surface interactions and mechanical properties, all which hamper high resolution and image reproducibility, making them less amenable to STM (35,36). Additionally, proven preparatory techniques for the immobilization of biomolecules to the substrate are lacking. These factors have resulted in (a) the displacement of the deposited material from or along the surface; this outcome is due to inadequate surface forces associated with adsorption, direct chemical linkage and even electrochemical deposition relative to the force induced by the tip during scanning (estimated to be on the order of 10^{-8} to 10^{-6} Newtons); (b) the distortion of the structure itself as biological structures are typically very 'soft' and flexible and are easily deformed by the tip; subsequently, bending, twisting, or stretching of these 'soft' macromolecules results in a streaky or cloudy image; (c) the removal and even the subsequent readsorption of the deposited material to the tip itself.

In addition, biological structures can form multiple structures (individual molecules and aggregates) in solution and at solid-liquid interfaces, as well as undergo conformational changes. Thus, various forms of these complex structures may be present which probably do not reflect their true native, 3-dimensional conformations or physiological states. Further, there are presently only a few conductive substrates that have smooth surface features relative to the size of the biological structures to be imaged. The substrates most typically used are highly oriented pyrolytic graphite (HOPG) and gold coated mica.

Some approaches to circumventing these problems include metal coating of the sample, drying samples and freeze fracture replication (36) which, with the exception of the latter, will result in loss of physiological conformation and/or loss of resolution. More recently, attention is being paid to preparatory techniques, such as, tailored protein/cell immobilization techniques, exploitation of inherent interfacial forces e.g., hydrophobic effects, etc. which depend upon the specific biological structure, its local chemical environment and the chemistry of the substrate and others.

Despite the problems of imaging biological materials, STM has been used successfully to image a constantly expanding list of widely diverse biological structures including DNA (37-55), DNA binding mutagens and drugs (56), enzymes (36,57), amino acids (36), polypeptides (56,58), collagen (59,60), fibronectin (61), albumin (36), concanavalin A (62), fibrinogen (36), colloidal gold labeled immune complexes (63), microtubules (64,65), biomembranes (66,67,68), and viral phages T4 (69), T7 (44) and f29 (70). Additionally organic molecules have been imaged as isolated molecules (71) and as liquid crystals (72,73). Note that these images were produced with various sample preparations including metal coating of the material (38,61,66,69), material transformed to its dry state (36,40,42,43,57,63,64,70), material deposited in solution via adsorption and electrochemical deposition (37,39,41,67), by freeze fracture replication (68), as well as, material assembled into Langmuir-Blodgett films (62).

The proliferation of studies on STM of biological materials has now provided an impetus to propose theoretical arguments to explain why one can obtain images of relatively thick and poorly conductive biological structures. Several theories have been recently proposed based upon (a) energy relaxations coupled with biopolymer chain disorder (74) and (b) quantum interference (75) that support the validity of STM for imaging uncoated biological structures.

Biological Cellular Assembly

The attachment of cells to their surroundings is prerequisite to their maintaining proper cell function and shape, as well as, tissue integrity (76). The most noticeable change associated with this process is the dramatic shape transformation from rounded to a flattened cell morphology. Cell adhesion is known to occur in a series of steps which has been described as a multistep paradigm of adsorption, contact, attachment and spreading (77) as depicted schematically in Figure 3. In general, the physicochemical properties of interfaces govern the nature of adsorption forces. These include both long-range (e.g., van der Waals and electrostatic) and short range (dipole interactions, hydrogen bonding, hydrophobic effects, etc.) intermolecular forces. The long-range forces have been associated with non-specific cell adsorption processes, whereas the short-range forces are thought to govern 'biologically-directed' cell adhesion allowing for cellular attachment and assembly (78,79). Non-specific mechanisms include interactions between the

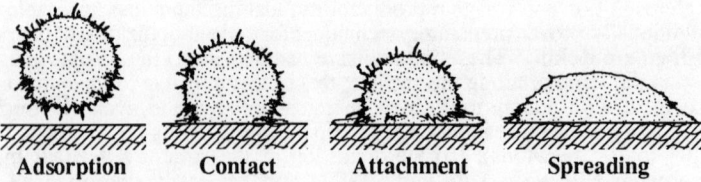

Figure 3. Multistep paradigm of cell adhesion (adapted from ref. 77)

negatively charged cell membrane and a positively charged substrate or similarly through hydrophobic interactions (80,81). Thus, the molecular structure dominates at the short range and the forces involved are correspondingly large and specific. It follows then that the chemistry and structure of the substratum dictate to a great degree whether or not cell adhesion will occur. For example, the presence or absence of proteins, either as intrinsic components of the substrata surface or adsorbed to the surface from solution, dictate whether the cell undergoes nonspecific or specific adsorption to the substrata and whether cell spreading occurs. Thus, the degree of cell anchoring to synthetic surfaces relies heavily on the presence of an adsorbed protein provided from media containing sera, from cell microexudate, or from other complex biological fluids (e.g., blood). A more comprehensive review of cell adhesion processes can be found elsewhere (77).

The Extracellular Matrix

The biological substrata to which cells adhere to in natural cellular environments (other than neighboring cells) is the extracellular matrix (ECM). The ECM is a biological composite structure comprised of an organized meshwork of structural and adhesive fibrous proteins embedded in a polysaccharide gel. Specific components are (a) cell attachment proteins including laminin/merosin, nidogen (entactin), fibronectin, collagen IV, thrombospondin, osteopontin, tenascin, and possibly others (b) other structural proteins including other collagens and elastin and (c) polysaccharides including hyaluronic acid, chondroitin sulfate, heparan sulfate, dermatan sulfate, keratan sulfate, and heparin. Most of these are ubiquitous components of the ECM that are secreted locally by cells. Others only appear at specific times such as wound healing, hemostasis, morphogenesis, angiogenesis, and/or metastasis. Examples include thrombin, von Willebrand factor, fibrinogen, vitronectin (76) and SPARC i.e., secreted protein acidic and rich in cysteine (86). Each protein type must be considered a family since isoforms and structural variants occur in different species, tissues, and periods of maturation (82,83). The construction of the ECM is accomplished by molecular self-assembly and cellular deposition of materials (84,85).

Cellular adhesion occurs via receptor-ligand interactions between cell receptor proteins (integrins) and ECM adhesive proteins. Examples of ECM ligand structures include laminin and fibronectin. Laminin promotes the attachment of epithelial cells to the basal lamina while fibronectin attaches fibroblast and other cells to connective tissues. A model of cell adhesion shown schematically in Figure 4 depicts the molecular arrangement of ECM proteins to integrin proteins.

Integrins are a family of heterodimer transmembrane proteins that act as transmembrane linkers. They mediate interactions between cytoskeletal protein structures residing within the cell interior and the adhesive protein residing in the ECM. The point of interface with the cytoskeletal filaments is known as the focal point and the point of contact between an integrin and an ECM located exterior to the cell membrane is sometimes called a fibronexus. The integrin subfamilies are categorized by the types of 'a' and 'b' subunits that they contain. Members of the integrin family include the very late antigen (VLA) proteins (87,88,89), the leukocyte integrins (LFA-1,MAC-1, and p150,95), the platelet adhesions gpIIb/IIIa, and the chicken integrin complex, among others. Each integrin binds to distinct sequences whether on ECM proteins or cell membranes.

As a cell makes contact with the ECM surface, the linkage between the integrin receptor and the ECM adhesive protein induces a triggering of cytoskeletal organization.

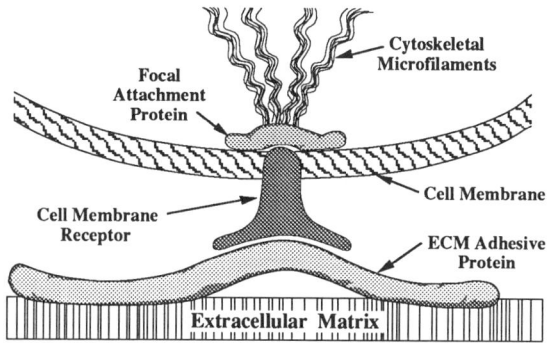

Figure 4. Illustration of spatial arrangement and order of macromolecular structures responsible for cell adhesion (adapted from ref. 76)

It is apparent from optical fluorescent studies on cell culture systems that the points of contact or adhesion between either cytoskeletal microfilaments or local regions of a culture plate surface correspond to precisely these focal attachment points after cytoskeletal microfilament reorganization has occurred (90).

Universal Molecular Recognition Cell Binding Sequences

It is now apparent that the basis of cell attachment to extracellular matrices centers on the molecular recognition, by specific amino acid sequences that reside along regions of extracellular matrix proteins, of complementary sequences that reside along regions of cell membrane receptors proteins or integrins. As an example Figure 5 depicts the profound effects that these binding sequences have not only on cell attachment but also on cell spreading. More specifically, it can be seen that fibroblasts spread according to specific sequence regions that are allowed to interact with the integrin receptors; the most spreading occurring only when a unique combination of both heparin and cell binding sequences interact with fibroblast cell receptors (91). It has now been shown that the ECM adhesive proteins are encoded with, what are considered to be, universal cell binding molecular recognition sequences, i.e., apparently only four binding sequences have been associated with these ECM protein self-assemblies for cell attachment as shown in Table I (76,92).

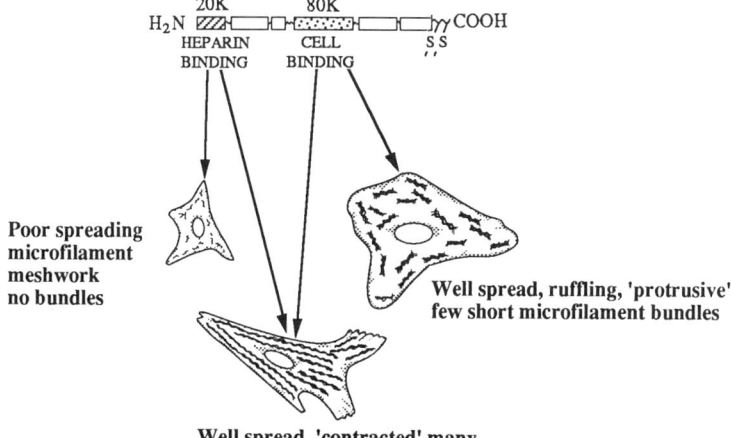

Figure 5. Affect of binding sequences on fibroblasts spreading (adapted from ref. 91)

Table I
Comparison of ECM cell binding sequences for various cell assembly systems

Cellular Assembly	Cellular Adhesive Protein	Cell Recognition Sequence
Platelet Adhesion (ref.92)	Fibronectin	ARG-GLY-ASP [RGD]
	Vitronectin	ARG-GLY-ASP [RGD]
	Fibrinogen	ARG-GLY-ASP [RGD]
	Von Willebrand Factor	ARG-GLY-ASP [RGD]
Endothelial Cells (ref.93,94)	Thrombin	ARG-GLY-ASP [RGD]
	Collagen IV	[TAGSCLRKFSTM]
Fibroblasts (ref.92,95)	Fibronectin	ARG-GLY-ASP [RGD]
	Serum Spreading Factor	ARG-GLY-ASP [RGD]?
Bacteria (ref.96)	Fibronectin/Fibrinogen	ARG-GLY-ASP [RGD]
Mast Cells (ref.97,98,99)	Laminin/Merosin	{ TYR-ILE-GLY-SER-ARG [YIGSR] ILE-LYS-VAL-ALA-VAL [IKVAV]
	Fibronectin	ARG-GLY-ASP [RGD]
Epithelial Cells (ref.100,101,102,103)	Laminin/Merosin	{ TYR-ILE-GLY-SER-ARG [YIGSR] ILE-LYS-VAL-ALA-VAL [IKVAV]
	Epinectin, Epibolin	{ TYR-ILE-GLY-SER-ARG [YIGSR]? ILE-LYS-VAL-ALA-VAL [IKVAV]?
	Epiligrin	Unknown
Neuronal Cells (ref.104,105)	Tenascin	Unknown
	Laminin/Merosin	ILE-LYS-VAL-ALA-VAL [IKVAV]?

As can be seen in Table I, there appear to be only four apparent generic binding sequences involved in cellular assembly that have been identified. They are the polypeptides Arg-Gly-Asp or RGD, Tyr-Ile-Gly-Ser-Arg or YIGSR, and Ile-Lys-Val-Ala-Val or IKVAV. The fact that different common binding sequences have been elucidated for these ECM adhesive protein systems suggest that different sequences are selected for different cell lines which infers that specific binding heuristics may exist.

ECM Adhesive Protein Nanostructure

The laminin family of proteins includes merosin (106,107), s-laminin (108), and many others. The first isolated and best known is Engelbreth-Holm-Swarm (EHS) murine tumor laminin (109,110). Laminins are large glycoproteins of about 900 KDa and are the primary proteins of the basement membrane (111). Laminin is composed of three chains (A,B1,B2) arranged in a cruciform shape as shown in Figure 6a based upon transmission electron microscopy (TEM) studies employing rotary shadowing techniques. The dimensions determined by TEM micrographs shown in Figure 6b are approximately 35 nm for the three short arms, and 75 nm for the long arm (110,112).

Several specific sites have been determined on the laminin molecule which contain minimal amino acid binding sequences. These sites serve as functional interaction points for both proteins and cells. The intersectional area of the cruciform and the three short arms contain sites for cell attachment (113), cell signalling of mitogenic activity, chemotaxis (114) and locomotion (115), binding sites for nidogen (entactin), laminin polymerization

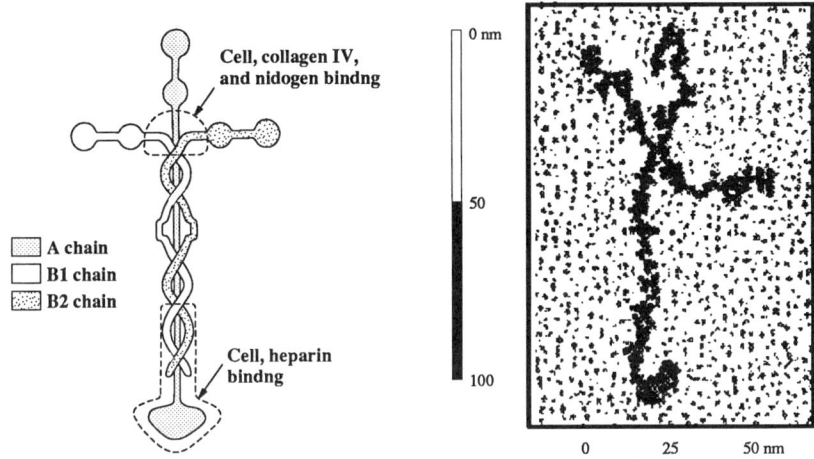

Figure 6a. Schematic representation of laminin (adapted from ref. 110)

Figure 6b. TEM of laminin (adapted from ref. 110)

(116,117,118), and collagen IV (119). The long arm of the "A" chain contains sites for cell attachment, cell signaling of neurite outgrowth (120) and cell metastasis (11-14), collagen IV (119), and heparin sulfate proteoglycan (121). The intrinsic self-assembly properties of laminin and the basement membrane (BM) have been demonstrated by incubating laminin, collagen IV, and heparin sulfate proteoglycan which resulted in a precipitate with portions having the organization and ultrastructure of BM (122). Figure 7a depicts the molecular organization and orientation of a laminin-cell adhesion system. Figure 7b illustrates how the macromolecular self-assembly of laminin is envisioned to occur.

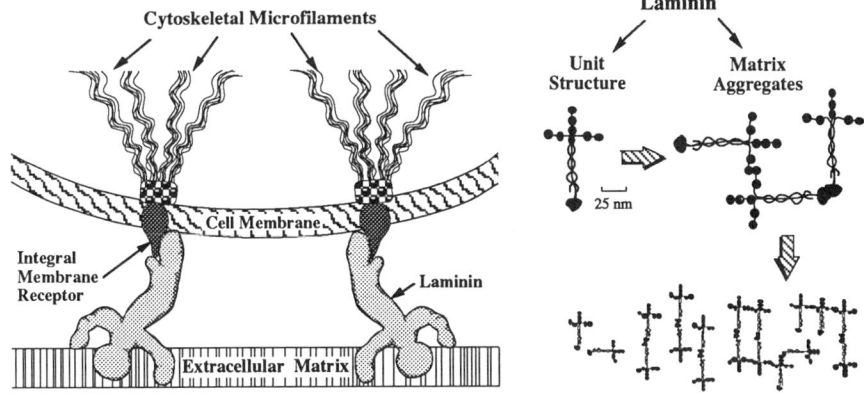

Figure 7a. Illustration of a laminin macromolecular self-assembly (adapted from ref. 123)

Figure 7b. Self-assembly of laminin (adapted from ref. 124)

Fibronectin is also a large glycoprotein of approximately 500 Kda which exist in two distinct forms; i.e., either its solubilized form as a dimer in blood where it plays a major role in the formation of cryoprecipitates, as well as, in its insoluble multimer form where it is incorporated into the ECM (125). Fibronectin is composed of two chains approximately 60 nm in length linked end-to-end at a 60° angle, forming a 'cotter pin" shaped molecule. A schematic representation is depicted in Figure 8a based upon TEM (Figure 8b) and related studies.

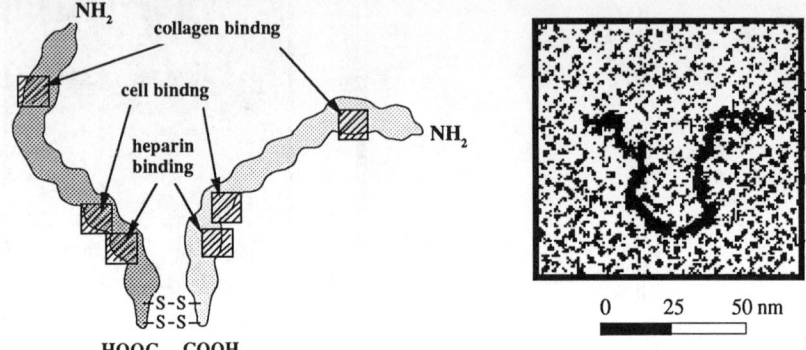

Figure 8a. Schematic representation of fibronectin (adapted from ref. 110)

Figure 8b. TEM of fibronectin (adapted from ref. 110)

Fibronectin contains sites for cell attachment, as well as, sites for fibrin and collagen. Figure 9a depicts the molecular organization and orientation of a fibronectin-cell adhesion system. Figure 9b illustrates how the molecular self-assembly of fibronectin (126) is envisioned to occur. A more detailed review on fibronectin structure and function is given elsewhere (127).

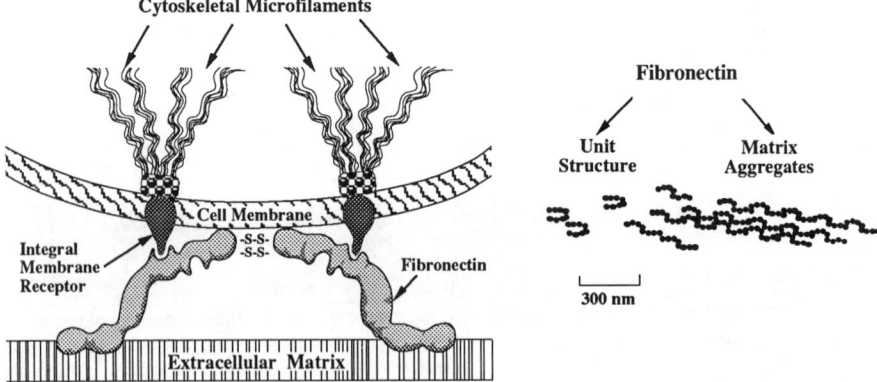

Figure 9a. Fibronectin cell assembly system (adapted from ref. 128)

Figure 9b. Fibronectin self-assembly (adapted from ref. 124)

Experimental

Two experimental approaches were taken in this study. The first set of experiments involved the study of direct cell adhesion to synthetic polymer surfaces using ECM proteins as linkers. Secondly, scanning tunneling microscopy studies were conducted on two ECM proteins, laminin and fibronectin. These proteins have been studied extensively with transmission electron microscopy and other related techniques as a means to better understand their structure-function role in binding a wide variety of cells.

Materials and Methods. MC/9, MC/9-2 and WEHI-3 cell lines were purchased from ATCC (Rockville, MD.). RBL-2H3 cells were kindly supplied by Dr. Juan Rivera affiliated with the NIAMS division of NIH. All cell lines were stored at -70°C until use if not used immediately for culture. Polystyrene culture plates and culture media fetal calf sera (FCS), Hepes buffered Dulbecco's modified Eagles media (H-DMEM) and laminin and fibronectin were purchased from Sigma Chemical. Both laminin and fibronectin were, upon receipt, divided into aliquots at 10 μgm/ml and stored at -70 °C until use. Mylar® film-covered biosensor prototypes, fabricated in our ASU bioengineering laboratories, were used to assess the immobilization capability of ECM proteins of MC/9 cells. For the STM studies, highly oriented pyrolytic graphite (HOPG-ZYB) was used as the conductive substrate (Union Carbide). Platinum-iridium tips coated with apiezon wax were fabricated in-house.

Procedures. The cellular adhesive proteins were applied as a solution (10 μgm/ml) of 20 μl drops to small areas on polystyrene culture plates. The drops were allowed to air dry for approximately 1 hour. All cell lines were spun down and resuspended in H-DMEM with 4% FCS at pH 7.3. A 20 μl drop of DI water was placed on the dried protein deposit immediately before 450 μl of cell suspensions were added. This was done so that the dried salts in the protein deposit did not effect the osmolarity, since small drops of cell suspensions were used. The cell drops were humidified with plastic wells of DI water placed in the culture plates. The plates were then covered and placed in the incubator at 38°C for 100 minutes. Plates were then rinsed of non-adherent cells by gently squirting fine jets of H-DMEM across the plates while holding them side ways. The plates were then tapped onto a paper towel to remove as much fluid as possible. The cell bound surface was not disturbed due to the depth of the culture plate. The plates were then allowed to air dry for 2-3 minutes after which they were rinsed with 100% ethanol and dried as before. Toluidine blue dye solution was then placed in drops over the cell bound surfaces for 30 seconds and was rinsed briefly with DI water. The plates were dried again and the adherent cells were visualized by light microscopy at 160x and 800x using a Universal Zeiss microscope.

Additional ECM adhesion experiments were conducted whereby laminin and fibronectin were used to adhere MC/9 cells, in situ, onto the plastic (Mylar®) protective surface of a solid state biosensor prototype (129). Laminin (100 μgm/ml) and fibronectin (100 μgm/ml) were applied to separate devices in minimal volumes to coat approximately one cm^2 of the Mylar® surface. These devices were then air dried for one hour at room temperature after which MC/9 cells (2×10^6 cells/ml) were placed directly onto the precoated Mylar® surfaces and incubated at 37° C for one hour. These surfaces were gently rinsed several times with H-DMEM prior to evaluation of cell immobilization with the optical microscope.

STM experiments were carried out at room temperature and at ambient pressure conditions using an Angstrom TAK 3.0® STM instrument (Tempe, Arizona). Fresh graphite surfaces were created before each experiment by carefully cleaving the exposed graphite surface layer with scotch tape. A small cylinder (5mm ID x 3mm H) was placed onto the graphite surface to create a liquid reservoir prior to tunneling. Approximately 100 μl of Tris buffer solution containing 15 μgm/ml of ECM protein was place into the microwell just prior to tunneling. The tunneling head was then carefully place into the desired position and followed by a manual advancement of the tip through the liquid and within close proximity to the surface. At this point the instrument was placed into an automatic approach mode until it reached a constant tunneling current of 1 nanoampere.

Results and Discussion
Direct Cell / ECM Protein Binding Studies. The results of the in vitro binding of both anchorage-dependent (RBL and WEHI) and suspension cell lines (MC/9 and MC/9-2) are summarized in Table II. As can be seen from polystyrene control experiments shown in Table II, both anchorage dependent and suspension cells behaved as expected, i.e., the anchorage-dependent cells did bind and spread to the polystyrene surface with the RBL line showing more attachment and spreading than the WEHI line while the MC/9 cells remaining in suspension.

Optical micrographs of ECM protein binding experiments using the RBL-2H3 cell line is shown in Figures 10a-10c. As demonstrated by the polystyrene control (Figure 10a), these cells have a natural adhesion and spreading ability with many surfaces. The prevalent spreading morphology for control cells is a 'flattened-like' appearance and having 1-4 filopodia (cell extension processes) per cell, termed 'protrusive' spreading.

RBL-2H3 cells in the presence of laminin (Figure 10b) and fibronectin (Figure 10c) displayed increased adhesion to the substrate, increased cell clumping and a drastically altered spreading type. In these cases filopodia were less common, virtually non-existent in the case of fibronectin and occurring rarely in the presence of laminin. Although the cells appear to have maintained a spherical shape from a two-dimensional perspective, they actually flattened out to a 'pancake-like' spreading conformation. The cell surfaces appeared venated with web-like patterns as a result of their flattened morphology and highly arranged cytoskeletons. Additionally, cell clumping was much more prevalent in the presence of ECM proteins probably due to cell-cell adhesion by resolubilized ECM protein resulting in less defined cell boundaries. Similar results were obtained using the WEHI-3 cell line, also a natural adhering cell line, for both control and fibronectin experiments. Laminin, however, induced a decrease in cell adhesion.

MC/9 cells, a suspension cell line, were adhered to culture plates coated with laminin or fibronectin. Fibronectin-bound cells primarily displayed both the 'protrusive' and 'pancake-like' spreading as primary spreading morphologies in approximately equal populations. This distribution was shifted using butyrate treated MC/9 cells (MC/9-2) in which most cells were of the 'pancake-like' morphology, presumably by an increased expression of membrane receptors. Fibronectin-bound cells resisted mild shear forces, however, laminin-bound cells detached as a cell-protein composite film when exposed to mild shear forces. This occurred in both the MC/9, MC/9-2 and possibly WEHI-3 cell lines.

Table II
Affect of ECM Proteins on Cell Attachment and Spreading

	POLYSTYRENE ADHERENCE	CONTROL SPREADING	LAMININ ADHERENCE	LAMININ SPREADING	FIBRONECTIN ADHERENCE	FIBRONECTIN SPREADING
RBL-2H3	+++	++	+++	+++	+++	+++
MC/9	--	--	+/-*	--	+++	++
MC/9-2	--	--	+/-*	--	+++	+++
WEHI-3	++	++	+	+	+++	+++

LEGEND

	Adherence	Spreading
--	None	No spreading cells spherical in shape
-	1-29%	Poor spreading in some cells (1 filopodia/cell) most cells spherical
+	30%-49%	Poor spreading in most cells (1 filopodia/cell) cells still somewhat spherical
++	50%-84%	Most cells spread, average 3 filopodia/cell some cells highly flattened and contracted
+++	85%-100%	Most cells spread and highly flattened and contracted

* *initially adhered but detached as a cell-protein film after prolonged exposure.*

The adhesive value of ECM proteins for applications, such as the engineering of cellular assemblies, depend on their binding affinity to cells, as well as, the desired substrate. In vivo, the specific high affinity sites in the ECM provide an optimum substrate for adhesive protein attachment. However, the nonspecific adsorption of ECM proteins to synthetic materials is a weak link in the design of hybrid microstructures using cellular assembly systems. For example, although MC/9 cells can be adhered to synthetic surfaces via laminin they tend to detach over time and more quickly under shear, at least for the synthetic materials used in this study. This detachment usually occurs as films of cells and protein. This phenomena may also explain why the WEHI-3 cell line appeared to adhere less with laminin than without it. In contrast, the RBL-2H3 line still tightly adhered to the laminin coated substrate even though the presence of laminin had obvious effects on cell adhesion, clumping and spreading. These differences are presumably due to variations in ECM binding receptors (i.e., in their total number and/or their binding site specificity) and/or the presence of proteins adsorbed to the synthetic surface. In order to resolve these issues, additional insight must be gained at the molecular level of binding specificity of cells and how adhesive protein structures assemble at interfaces.

An initial result of a practical application of laminin and fibronectin based molecular assembly systems is shown in Figures 11a and 11b. It can be seen in these figures that a highly dense, confluent assembly of MC/9 cells has been immobilized on the protective Mylar® surface which covers the active sensing elements of the whole cell thermopile biosensor. This is an optimum result as high cell assembly populations are required to obtain adequate cell signal transduction based upon the sensitivity of this thermopile (129) and the nominal metabolic capability of biological cells equivalent to approximately a picowatt per cell (130). However, even a mild rinsing of the thermopile surface containing these immobilized cell masses resulted in a significant loss of bound cells as shown in Figures 12a and 12b. This result is particularly evident in Figure 12a, the laminin-bound MC/9 cell assemblage, whereas fibronectin-bound cells were more shear resistant (Figure 12b). Note, that the dark, equally spaced lines are the antimony-bismuth micro-thermo-couples located on the opposite face of the transparent Mylar® film.

STM Studies. Both individual and molecular aggregates of laminin and fibronectin were seen in their respective tunneling experiments. However, individual laminin and fibronectin molecules were seen relatively less frequently and, thus, were less available to image. This was expected since there were no attempts made to control their self-association which is reported to occur at these room temperature conditions, although the concentrations used in these experiments (10 μgm/ml) were well below reported critical self-assembly levels of 0.1 mg/ml (117,118). However when individual molecules of laminin and fibronectin were located, high resolution STM images of their molecular structures were obtained in their hydrated states. This suggest that nonspecific, interfacial forces associated with these ECM proteins may predominate over forces induced by the tip during scanning.

Figure 13a shows a three dimensional STM image of laminin which portrays a remarkedly similar cruciform structure to those reported in TEM studies. In addition, the overall molecular dimensions are consistent with the dimensions of laminin measured from TEM. Figure 13b depicts a three dimensional STM image of fibronectin showing its characteristic 'cotter pin' like structure and molecular dimensions comparable to reported TEM studies. One noted difference between the results reported in this study and TEM results is the discrepancy in the absolute height and width dimensions, i.e., the STM results indicate that these structures are larger in diameter but lower in thickness relative to the reported TEM values. This result is consistent with other STM results of lower than expected molecular thicknesses when imaging uncoated biological structures (131). Since the molecular diameters of the individual chains appear to be significantly larger in the STM nanographs, it may suggests that the molecule may undergo deformation from local tip forces. This is an important issue to resolve for STM to realize its potential in characterizing biological structures.

Of particular note is the height data (shown as differences in color intensity with the lightest regions corresponding to portions of the structure having the highest structural domains) that corresponds to the intersections of the long and short arms which is coincident with one of the major cell binding regions as shown in Figure 6a. A closer inspection of Figure 13a also suggest that the long arm (A chain) may have folded back onto itself. This would result in the globular portion of the A chain to reside on or near this binding region thereby accounting for the increase in molecular dimension in this region. Similarly, the tallest regions (lightest areas) of the fibronectin molecule shown in Figure 13b (although not as distinct as those seen with laminin) generally correspond to its

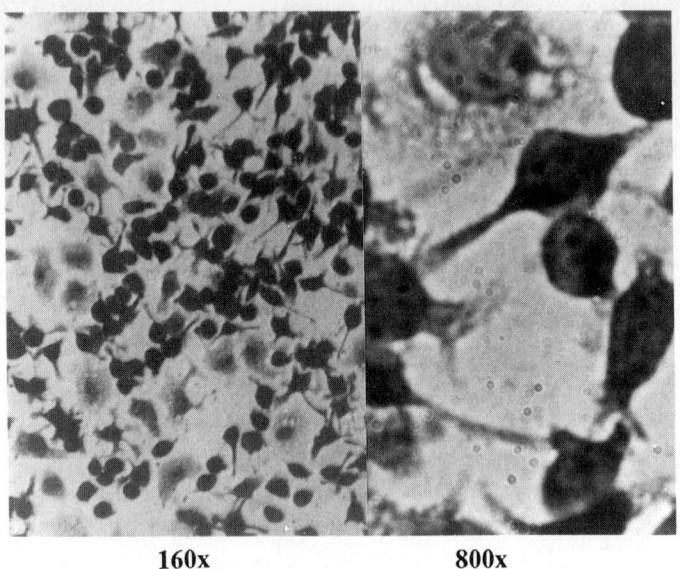

160x   800x

Figure 10a: RBL-2H3/Polystyrene (Control)

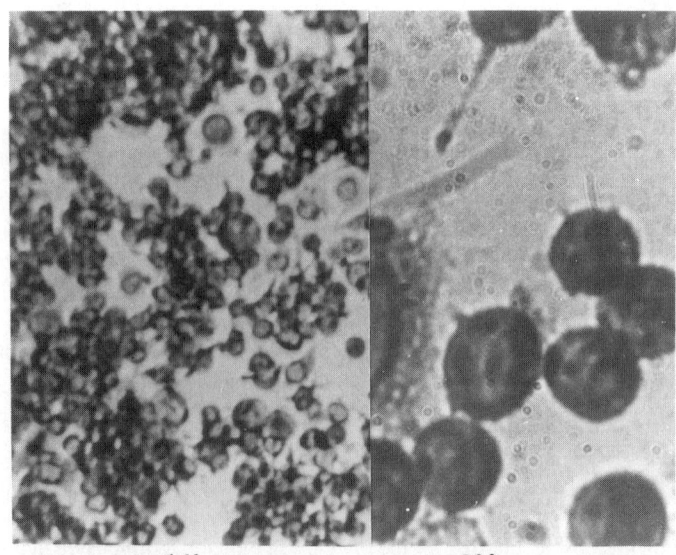

160x   800x

Figure 10b: RBL-2H3/Laminin/Merosin (10 μgm/ml)

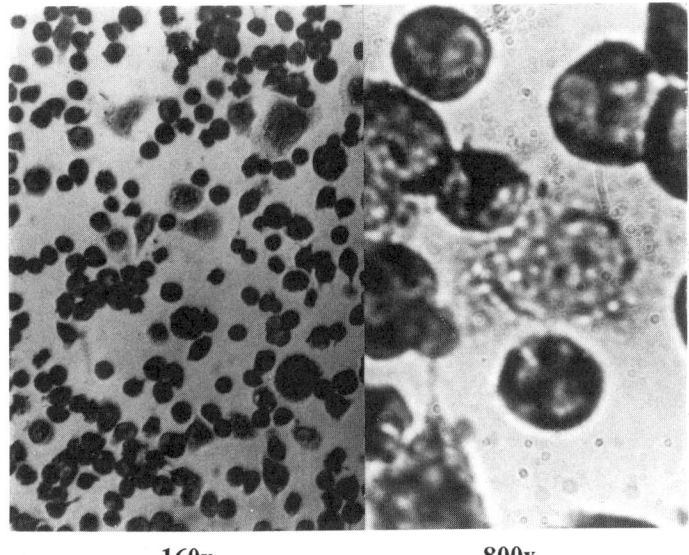

160x   800x

Figure 10c: RBL-2H3/Fibronectin (10 μgm/ml)

Figure 11a: MC/9 Laminin (100 μgm/ml) 225x

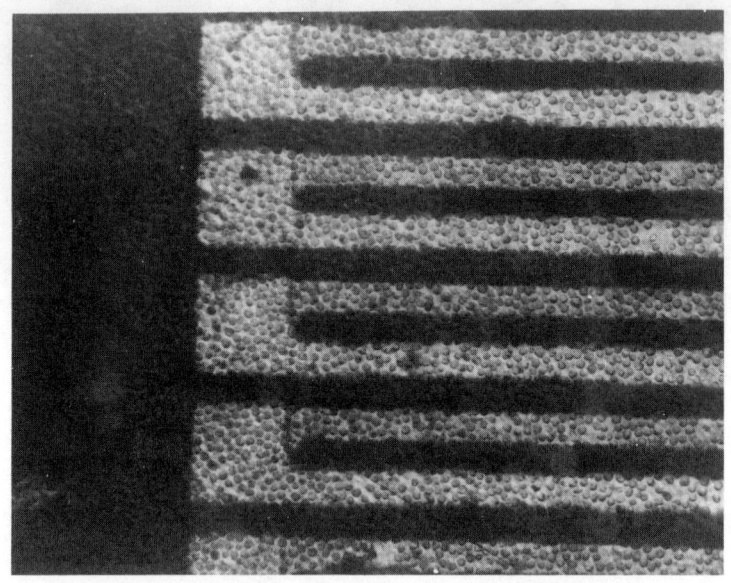

Figure 11b: MC/9 Fibronectin (100 µgm/ml) 225x

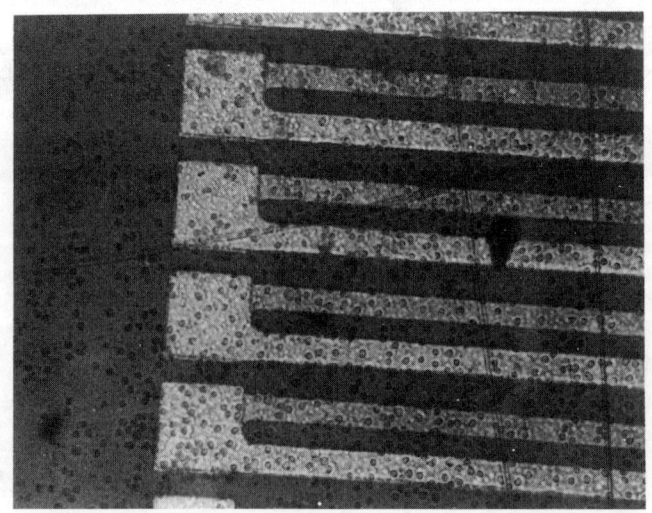

Figure 12a: MC/9 Laminin Rinse (100 µgm/ml) 225x

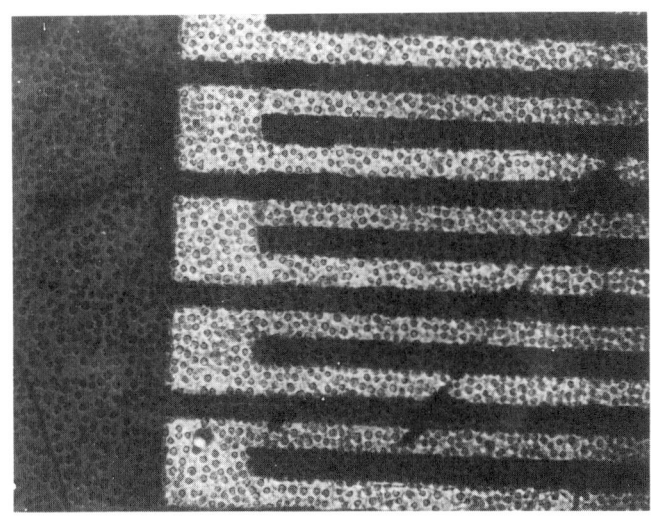

Figure 12b: MC/9 Fibronectin Rinse (100 μgm/ml) 225x

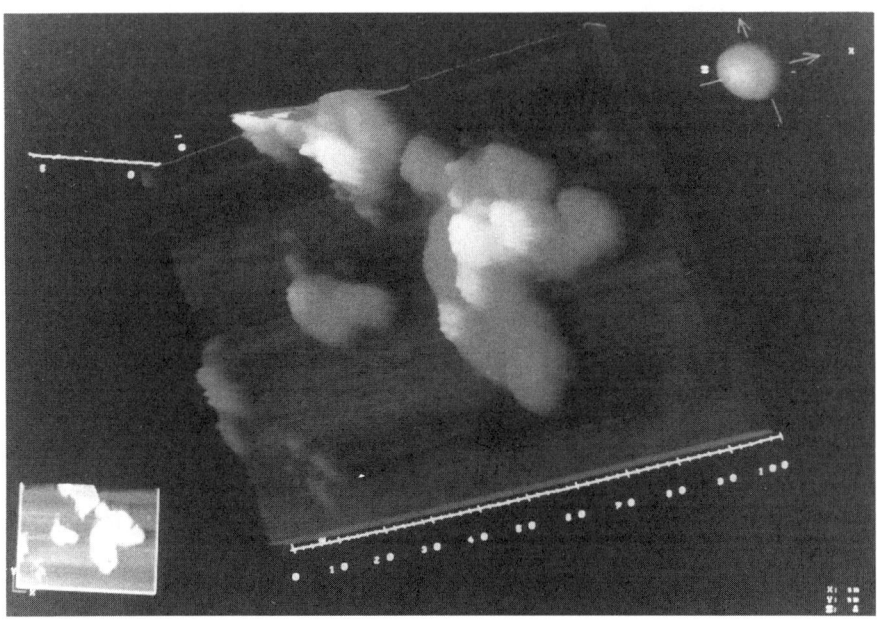

Figure 13a: Three dimensional STM image of uncoated, hydrated laminin molecule on HOPG graphite.

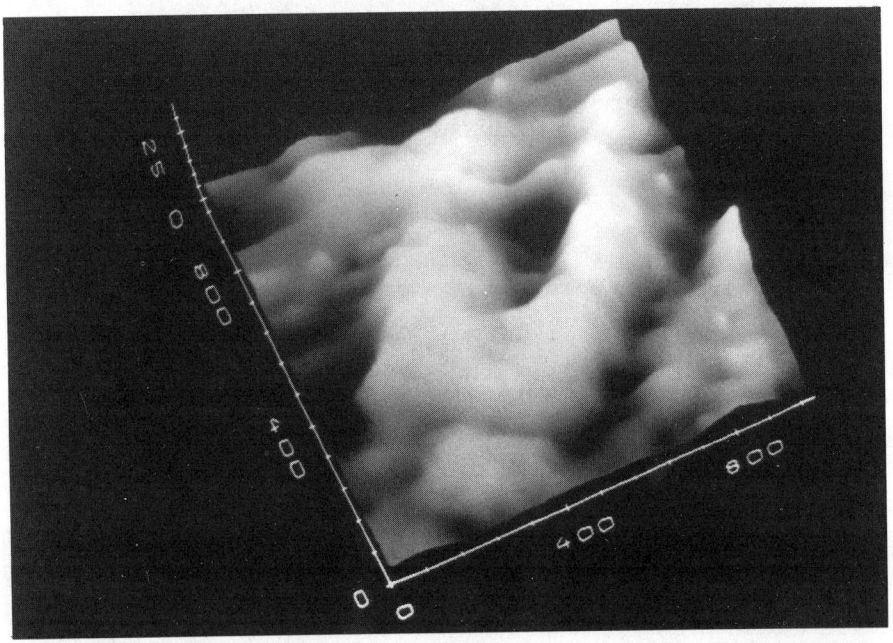

Figure 13b: Three dimensional STM image of uncoated, hydrated fibronectin molecule on HOPG graphite.

reported binding domains as shown in Figure 8a. Although these are very preliminary results, they can offer insight into the complex molecular organization involved in self-assembly processes.

Conclusions

The knowledge gained from biological self-assembly systems may serve as a basis for molecular device design using unique methods and novel materials. This study demonstrated that ECM proteins can serve as highly specific linkers that can allow the interfacing of synthetic microstructures with biological structures. A practical example of exploiting ECM proteins for this purpose was demonstrated with the use of the ECM proteins, fibronectin and laminin, for the fabrication of whole cell assemblies directly onto a biosensor surface. A preliminary result of this study indicates that fibronectin is capable of anchoring cells firmly to synthetic surfaces and thus appears to be a good candidate for the construction of cellular assembly systems. The same may not be true with laminin based upon its inability to maintain good adherence to Mylar® surfaces while maintaining good cell linkage. The different responses observed between these two proteins is unclear but may be due to variations in their structure-function heuristics that are not well understood at this time. This circumstance emphasizes the need to acquire a much more fundamental level of understanding of the molecular mechanisms associated with these and related systems. In this light, STM is a potentially valuable tool for gaining insight into complex biological systems at a molecular level despite the inherent problems encountered when imaging biological materials. Invaluable insight can be gained with STM as demonstrated in this study on fully hydrated protein structure presumably in their native conformational state. STM nanographs of laminin and fibronectin are consistent with TEM dimensions and additionally display important 3-D structural aspects which may prove to be functionally responsible for protein adsorption and cell binding properties.

In summary, a deeper understanding of cellular self-assembly processes can provide insight into the organization of various tissues in biological/physiological systems ranging from embryogenesis to wound healing, biofilm formation associated with many microorganism systems, as well as, the identification and utilization of molecular heuristics for innovation of molecular device design for bioengineering and biotechnology.

Acknowledgments

This work is supported in part by BRSG 2 S04 RR07112, Division of Research Resources, National Institutes of Health. The College of Engineering at ASU provided the support for the purchase of the STM instrument used in this study. Also, many fruitful discussions and helpful suggestions on STM were provided by my colleague Antonio Garcia. In addition, the support of the STM-IAP lab at ASU and the assistance of John Graham is appreciated. The graphical assistance provided by Sue Selkirk is also appreciated.

Literature Cited

1. Pizziconi, V.B., Extended Abstracts, AIChE National Spring Meeting, Houston, **1991**, p.91.
2. Clapper, D.L., Drexler, R.J., Guire, P.E., *Trans. Soc. Biomater.,* **1991**, *14*, p. 231.
3. Petit, D.K., Horbett, T.A., Hoffman, A.S. *Trans. Soc. Biomater.,* **1991**, *14*, p. 236.
4. Sirois, E., Doilon, C.J., *Trans. Soc. Biomater.,* **1991**, *14*, p. 237.
5. Drumheller, P.D., Desai, N.P., Hubbell, J.A., *Trans. Soc. Biomater.,* **1991**, *14*, p. 239.
6. Niu, S., Matsuda, T., Oka, T., *ASAIO Abstracts,* **1991**, *37*, p. 91.
7. Hall, E.H., In Biosensors, Open University Press, Milton Keynes, Buckingham, **1990**, pp. 3-52.
8. Massia, S.P., Hubbell, J.A., *Ann. N.Y. Acad. Sci.,* **1990**, *589*, pp. 261-270.

9. Grattarola, M., Perlo, G., Giannetti, G., Briozzo, In *Molecular Electronic Devices* Eds. Carter, F.L., Siatkowski, R.E., Wohltjen, *Proceedings of the 3rd International Symposium on Molecular Electronic Devices*, North Holland: New York, N.Y., **1988**, pp. 565-573.
10. Takemoto, Y., Matsuda, T., Kishimoto, T., Maekawa, M., Akutsu, T., *Trans. Am. Soc. Artif. Int. Organs*, **1989**, *35*, pp. 354-356.
11. Terranova, V.P., Williams, J.E., Liotta, L.A., Martin, G.R., *Science*, **1984**, *226*, pp. 982-984.
12. Iwamoto, Y., Robey, F.A., Graf, J., Sasaki, A., Kleinman, H.K., Yamada, Y., Martin, G.R., *Science*, **1987**, *238*, pp. 1132-1134.
13. Fridman, R., Giaccone, G., Kanemoto, T., Martin, G.R., Gazdar, A.F., Mulshin, J.L., *Proc. Natl. Acad. Sci. USA*, **1990**, *87*, pp. 6698-6702.
14. Kanemoto, T., Reich, R., Royce, L., Greatorex, D., Adler, S.H., Shiraishi, N., Martin, G.R., Yamada, Y., Kleinman, H.K., *Proc. Natl. Acad. Sci. USA*, **1990**, *87*, pp. 2279-2283.
15. Hansma, P.K., Elings, V.B., Marti, O., Bracker, C.E., *Science*, **1988**, *242*, pp. 209-216.
16. Quate, C.F., *Physics Today*, **1986**, August, pp. 26-33.
17. Sonnenfeild, R., Hansma, P.K., *Science*, 1986, *232*, pp. 211-213.
18. Gould, S.A.C., Drake, C.B., Prater, A.L., Weisenhorn, A.L., Manne, S., Hansma, H.G., Hansma, P.K., Massie, J., Longmire, M., Elings, V., Dixon Northern, B., Mukergee, B., Peterson, C.M., Albrecht, T.G., Quate, C.F., *J. Vac. Sci. Technol. A*, **1990**, *8*, pp. 369-373.
19. Lamb, D.R., *Electrical Conduction Mechanisms in Thin Insulating Films*, Spottiswoode Ballantyne, London, **1976**, p. 3.
20. Giaever, I., *Rev. Mod. Phys.*, **1974**, *4*, pp. 245-250.
21. Giaever, I., *Phys. Rev. Lett.*, **1960**, *5*, pp. 147-148.
22. Young, R., Ward, J., Scire, F., *Rev. Sci. Instrum.*, **1972**, *43*, pp. 999-1011.
23. Young, R., Ward, J., Scire, F., *Phys. Rev. Lett.*, **1971**, *27*, pp. 922-924.
24. Binnig, G.H., Rohrer, H., Gerber, Ch., Wiebel, E., *Phys. Rev. Lett.*, **1982**, *50*, pp. 57-61.
25. Binnig, G., Rohrer, H., *Helv. Phys. Acta*, **1982**, *55*, pp. 726-735.
26. Binnig, G., Rohrer, H., *Surface Sci.*, **1983**, *126*, pp. 236-244.
27. Golovchenko, J.A., *Science*, **1986**, *232*, pp. 48-53.
28. Welland, M.E., Taylor, M.E., In *Modern Microscopies: Techniques and Applications. Eds.*, Duke, P.J., and Michette, A.G., Plenum Press: New York, N.Y., **1990**, pp. 231-266.
29. Behm, R.J., Garcia, N., Rohrer, H., *Scanning Tunneling Microscopy and Related Methods;* Academic Publishers: Dordrecht, The Netherlands, **1990**.
30. Sarid, D., *Scanning Force Microscopy: With Applications to Electric, Magnetic, and Atomic Forces;* Oxford University Press: New York, N.Y., **1991**.
31. Binnig, G., Rohrer, H., In *Trends in Physics*, Eds. Janata, J., Panatoflicek, European Physical Society, Petit-Lancy: Switzerland, **1984**, pp. 38-46.
32. Yanagimoto, K.C., Fisher, K.A., Whitfield, S.L., Thomson, R.E., Gustafsson, M.G.L., Clarke, L., *Ultramicroscopy*, **1990**, *33*, pp.117-126.
33. Zasadzinski, J.A., Hansma, P.K., *Ann. N.Y. Acad. Sci.*, **1990**, *589*, pp. 476-491.
34. Clemmer, C., Beebe, Jr., T.P., *Science*, **1991**, *251*, pp. 640-642.
35. Blackford, B.L., Watanabe, M.O., Dahn, D.C., Jericho, M.H., *Ultramicroscopy*, 35 **1989**, *27*, pp. 427-432.
36. Feng, L., Andrade, J.D., Hu, C.Z., *Scanning Microscopy*, **1989**, *3(2)*, pp. 399-410.
37. Allison D.P., Thompson, J.R., Jacobson, K.B., Warmack, R.J., Ferrell, T.L., *Scanning Microscopy*, **1990**, *4(3)*, pp. 517-522.
38. Nagahara, L.A., Thundat, T., Oden, P.I., Lindsay, S.M., *Ultramicroscopy*, **1990**, *33*, pp. 107-116.
39. Coratger, R., Chahboun, A., Ajustron, F., Beauvillain, J., *Ultramicroscopy*, **1990**, *34*, pp. 141-147.

40. Arscott, P.G., Bloomfield, V.A., *Ultramicroscopy*, **1990**, *33*, pp.127-131.
41. Thundat, T., Nagahara, P., Oden, Linsday, S.M., *J. Vac. Sci. Technol. A,* **1990**, *8*, pp. 645-647.
42. Selci, S., Cricenti, A., Felici, A.C., Generosi, R., Gori, E., Djaczenko, W., Chiarotti, G., *J. Vac. Sci. Technol. A*, **1990**, *8*, pp. 642-644.
43. Salmeron, M., Beebe, T., Odriozola, J., Wilson, T., Olgletree, D.F., Siekhaus, W., *J. Vac. Sci. Technol. A*, **1990**, *8*, pp. 635-641.
44. Keller, R.W., Dunlap, D.D., Bustamante, C., Keller, D.J., Garcia, R.G., Gray, C., Maestre, M.F., *J. Vac. Sci. Technol. A*, **1990**, *8*, pp. 706-712 .
45. Bendixen, C., Besenbacher, F., Laegsgaard, E., Stensgaard, I., Thomsen, B.,Westergaard, O., *J Vac. Sci. Technol. A*, **1990**, *8*, pp. 703-705.
46. Lindsay, S.M., Barris, B., *J. Vac. Sci. Technol. A.*, **1988**, *6*, pp. 544-547,
47. Lindsay, S.M., Thundat, T., Nagahara, L., *J. Microsc.*, **1988**, 152, pp. 213-220.
48. Lindsay, S.M., Thundat, T., Nagahara, L., In *Biological and Artificial Intelligence Systems*, E. Clementi and S, Chin Eds., ESCOM Science Publishers B.V.: Leiden, **1988**, pp. 125-141.
49. Arscott P.G., Lee, G., Bloomfield, V.A., Evans, D.F., *Nature*, **1989**, *339*, pp. 484-486.
50. Beebe Jr., T.P., Wilson, T.E., Ogletree, D.F., Katz, J.E., Balhorn, R., Salmeron, M.B., Siekhaus, W.J., *Science*, **1989**, *243*, pp. 370-372.
51. Dunlap, D.D., Bustamante, C., *Nature*, **1989**, *342*, pp. 204-206.
52. Lee, G., Bloomfield, V.A., Evans, D.F., *Science*, **1988**, *244*, pp. 514-516.
53. Lindsay, S.M., Thundat, T., Nagahara, L., Knipping, U., Rill, R.L., *Science*, **1989**, *244*, pp. 475-477.
54. Amrein, M., Stasiak, A., Gross, H., Stoll, E., Travaglini, G., *Science*, **1988**, *240*, pp. 514-516.
55. Pohl, F.M., Jovin, T.M., *J.Mol.Biol.*, **1972**, *67*, pp. 375-396.
56. Miles, M.J., McMaster, T., Carr, H.J., Tatham, A.S., Shewry, P.R., Field, J.M., Belton, P.S., Jeenes, D., Hanley, B., Whittam, M., Cairns, P., Morris, V.J., Lambert, N., *J. Vac. Sci. Technol. A,* **1990**, *8*, pp. 698-702.
57. Edstrom, R.D., Meinke., M.H., *Ultramicroscopy*, **1990**, 33, pp. 99-106.
58. McMaster, T.J., Carr, H., Miles, M.J., Cairns, P., Morris, V. J., *J. Vac. Sci. Technol. A.*, **1990**, 8(1), pp. 648-651.
59. Voelker, M.A., Hameroff, S.R., He, J.D., Dereniak, E.L., McCuskey, R.S., Schneiker, C.W., Chvapil, T.A., Bell, T.S., Weiss, L.B., *J. Microscopy*, **1988**, *152*, pp. 557-566.
60. Snellman, H., Pelliniemi, L.J., Penttinen, R., Laiho, R., *J. Vac. Sci. Technol. A,* **1990**, *8*, pp. 692-694.
61. Emch, R., Clivaz, C., Taylor-Denes, C., Vaudaux, P., Descouts, P., *J. Vac. Sci. Technol. A*, **1990**, *8*, pp. 655-658.
62. Horber, J.K.H., Lang, C.A., Hansch, T.W., Heckl, W.M., Mohald, H., *Chem. Phys. Lett.*, **1988**, *145*, pp.151-158.
63. Masai, J., Sorin, T., Kondo, S., *J. Vac. Sci. Technol. A*, **1990**, *8*, pp. 713 -717.
64. Yovana, S.K., Kelley, M., Schneiker, C., Krasovich, M., McCuskey, R., Koruga, D., Hameroff, S., *FASEB J.,* **1989**, *3,* pp. 2184-2188.
65. Hameroff, S., Simic-Krstic, Y., Vernetti, L., Lee, Y.C., Sarid, D., Wiedmann, J., Elings, V., Kjoller, K., McCuskey, R., *J. Vac. Sci. Technol. A,* **1990**, *8*, pp. 687-691.
66. Fisher, K.A., Yanagimoto, K.C., Whitfield, S.L., Thomson, R.E., Gustafsson, M.G.L., Clarke, J., *Ultramicroscopy*, **1990**, *33,* pp. 127 -131.
67. Jericho, M.H., Blackford, B.L., Dahn, D.C., Frame, C., Maclean, D., *J. Vac. Sci. Technol. A,* **1990**, *8*, pp. 661-666.
68. Joseph, A.N., Zasadzinski, J., Schneir, J., Gurley, J., Elings, V., Hansma P.K., *Science*, **1988**, *239*, pp. 1031-1015.
69. Stemmer, A., Engel, A., *Ultramicroscopy*, **1990**, *34*, pp. 129-140 .
70. Baro, A.M., Miranda, R., Carrascosa, J.S., *IBM J. Res. Dev.,* **1986** , *30*, pp. 380-386.

71. Moller, R., Esslinger, A., Koslowski, B., *J. Vac. Sci. Technol. A*, **1990**, *84*, pp. 659 -660 .
72. McMaster, T.J., Carr, H., Miles, M.J., Cairns, P., Morris, V.J., *J. Vac. Sci.Technol. A*, **1990**, *8*, pp. 672-674.
73. Mizutani, W., Shigeno, M., Sakakibara, Y., Kajimura, K., Ono, M., Tanishima, S., Ohno, K., Toshima, N., *J. Vac. Sci. Technol. A*, **1990**, *8*, pp. 675-678.
74. Garcia, R., Garcia, N., *Chem. Phys. Lett.*, **1990**, *173(1)*, pp. 44-50.
75. Yuan, J-Y, Shao Z., *Ultramicroscopy*, **1990**, *34*, pp. 223-226.
76. Ruoslahti, E., Pierschbacher, M.D., *Science*, **1987**, *238*, pp. 491-497.
77. Grinnell, F., *International Review of Cytology*, **1978**, *53*, pp. 65-143.
78. Pethica, B.A., In, *Microbial Adhesion to Surfaces*, Eds., Berkeley, R.C.W. Lynch, J.M., Melling, J., Rutter, P.R., Vincent, B., Soc. Chem. Ind.: London, **1980**, pp. 19-45.
79. Evans, E., In *Physical Basis for Cell Adhesion*, Ed., Bongrand, P., *CRC Press*, Boca Raton, Fl., **1988**, pp. 91-123.
80. Jacobson, B.S., Branton, D., *Science*, **1976**, *195*, pp. 302-304.
81. Arkles, B.C., Miller, A.S., Brinigar, W.S., In *Silyated Surfaces*, Leyden, D.E. Collins, W.T., Eds., Gorden and Breach, Inc., **1980**, pp. 363-375.
82. Beck, K., Hunter, I., Engel, *J. FASEB*, **1990**, *4*, pp. 148-160.
83. Sanes, J.R., Engvall, E., Butkowski, R., Hunter, D.D., *J. Cell Biol.*, **1990**, *111*, pp.1685-1699.
84. Furthmayr, H., Yurchenco, A.S., Charonis, A.S., Tsilibary, E.C., In *Basement Membranes*, Shibata, S., Ed., Elsevier Science Publishers B.V.: Amsterdam, **1985**, pp. 169-179.
85. Yurchenco, P.D., Tsilibary, E.C., Charonis, A.S., Furthmayr, H., *J. Histochem. Cytochem.*, **1986**, *34*, pp. 93-102.
86. Sage, H., Vernon, R.B., Funk, S.E., Everitt, E.A., Angello, J. C., *J. Cell Biol.*, **1989**, *109*, pp. 341-356.
87. Elices, M.J., Urry, L.A., Hemler, M.E., *J. Cell Biol.*, **1991**, *112(1)*, pp. 169-181.
88. Hall, D.E., Reichardt, L.F., Crowley, E., Holley, B., Moezzi, H., Sonnenberg, A., Damsky, C.H., *J. Cell Biol.*, **1990**, *110*, pp. 2175-2184.
89. Gismondi, A., Morrone, S., Humpheries, M.J., Piccoli, M., Frati, L., Santoni, A., *J. Immunol.*, **1990**, *146*, pp. 384-392.
90. Singer, I.I., Kawka, D,W., Scott, S., Mumford, R.A., Lark, M.W., *J. Cell. Biol.*, **1987**, *104*, pp. 573-584.
91. Couchman, J.R., Woods, A., In *Interaction of Cells with Natural and Foreign Surfaces*, Eds. Crawford, N., Taylor, D.E.M., Plenum Press: New York, N.Y., **1984**, pp. 21.
92. Ruoslahti, E., Pierschbacher, M.D., *Cell*, **1986**, *44*, pp. 517-518.
93. Bar-Shavit, R., Sabbah, V., Lampugnani, M.G., Marchisio, P.C., Fenton II, J.W., Vlodavsky, I., Dejena, E., *J. Cell Biol.*, **1991**, *112(2)*, pp. 335-344.
94. Tsilibary, E.C., Reger, L.A., Vogel, A.M., Koliaakos, G.G., Anderson, S.S., Charonis, A.S., Alegre, J.N., Furcht, L.T., *J. Cell. Biol., 1990*, *111*, pp. 1583-1591.
95. Barnes, D.W., Silnutzer, *J. Biol. Chem.*, **1983**, *258(20)*, pp. 12548-12552.
96. Hook, M., Switalski, L.M., In *Fibronectin* , Ed. Mosher, D., *Biology of Extracellular Matrix: A Series*, Academic Press Inc.: San Diego, Ca., **1989**, pp. 47-121.
97. Thompson, H.L., Burbelo, P.D., Segui-Real, B., Yamada, Y., Metcalfe, D.D., *J. Immunol.*, **1989**, *143(7)*, pp. 2323-2327.
98. Thompson, H.L., Burbelo, P.D., Yamada, Y., Kleinman, H.K., Metcalfe D.D., *Immunology*, **1991**, 72, pp. 122-149.
99. Dastych, J., Costa, J.J., Thompson, H.L., Metcalfe, D.D., *Immunol.*, **1991**,73, pp.478-484.
100. Kleiman, H.K., Cannon, F.B., Laurie, G.W., Hassell, J.R., Aumailley, M., Terranova, V.P., DuBois-Dalcq, M., *J. Cell. Biochem.*, **1985**, *27*, pp. 317-325.
101. Enenestein, J., Furcht, L.T., *J. Invest. Dermatol.*, **1988**, *91(1)*, pp. 34-38.
102. Stenn, K.S., *Proc. Natl. Acad. Sci. USA*, **1981**, *78(11)*, pp. 6907-6911.

103. Carter, W.G., Ryan, M.C., Gahr, P.J., *Cell*, **1991**, *65*, pp. 599-610.
104. Ruegg, C.R., Chiquet-Ehrismann, R., Alkan, S.S., *Proc. Natl. Acad. Sci. USA*, **1989**, *86*, pp. 7437-7441.
105. Tashiro, K., Sephel, G.C., Weeks, B., Sasaki, M., Martin, G.R., Kleinman, H.K., Yamada, Y., Martin, G.R., *J. Biol. Chem.*, **1989**, *264*, pp. 16174-16182.
106. Ehrig, K., Leivo, I., Argraves, W.S., Ruoslahti, E., Engvall, E. *Proc. Natl. Acad. Sci. USA*, **1990**, *87*, pp. 3264-3268.
107. Leivo, I., Engvall, E., *Proc. Natl. Acad. Sci. USA*, **1988**, *85*, pp. 1544-1548.
108. Hunter, D.D., Shah, V., Merlie, J.P., Sanes, J.S., *Nature*, **1989**, *338*, pp. 229-234.
109. Hassell, J.R., Robey, P.G., Barrach, H.G., Wilczek, J., Rennard, S.I., Martin, G.R. *Proc. Natl. Acad. Sci. USA*, **1980**, *77*, pp. 4494-4498.
110. Engel, J., Odermatt, E., Engel, A., Madri, J., Furthmayer, H., Rohde, H., Timpl, R., *J. Mol. Biol.*, **1981**, *150*, pp. 97-120.
111. Timpl, R., Rhode, H., Robey, P.G., Rennard, S.I., Foidart, J.M., Martin, G.R., *J. Biol. Chem.*, **1979**, *254*, pp. 9933-9937.
112. Bruch, M., Landwehr, R., Engel, J. Eur. *J. Biochem.*, **1989**, *185*, pp. 271-279.
113. Aumailley, M., Nurcombes, V., Edgar, D., Paulsson, M., Timpl, R., *J. Biol. Chem.*, **1987**, *262(24)*, pp. 11532-11538.
114. Graf, J., Iwamoto, Y., Sasaki, M., Martin, G.R., Kleinman, H.K., Robey, F.R., Yamada, Y., *Cell*, **1987**, *48*, pp. 989-996.
115. Goodman, S.L., Risse, G., Von der Mark, K., *J. Cell Biol.*, **1990**, *109*, pp. 799-809.
116. Schittny, J.C., Yurencho, P.D., *J. Cell Biol.*, **1990**, *110*, pp. 825-832.
117. Yurchenco, P.D., Tsilibary, E.C., Charonis, A.S., Furthmayr, H., *J. Biol. Chem.*, **1985**, *260 (12)*, pp. 7636-7644.
118. Yurchenco, P.D., Cheng, Y., Schittny, J.C., *J. Biol. Chem.*, **1990**, *265 (7)*, pp. 3981-3991.
119. Charonis, A.S., Tsilibary, E.C., Yurencho, P.D., Furthmayr, H.J., *Cell Biol.*, **1985**, *100*, pp. 1848-1853.
120. Drago, J., Nurcombe, V.,and Bartlett, P.F., *Exp. Cell Res.*, **1991**, *192*, pp. 256-265.
121. Ott, U., Odermatt, E., Engel, J., Furthmayr, H., Timpl, R., *Eur. J. Biochem.*, **1982**, *123*, pp. 63-72.
122. Grant, D.S., Leblond, C.P., Kleinman, H.K, Inoue, S., Hassel, J.R., *J. Cell Biol.*, **1989**, *108*, pp. 1567-1574.
123. Abramhamson, D.R., Irwin, M.H., St.John, P.L., Accavitti, M.A., Heck, L.W., Couchman, J.R., *J. Cell Biol.*, **1989**, *109*, pp. 3477-3491.
124. Trelsted, R.L., *Ann. N.Y. Acad. Sci.*, **1990**, *580*, pp. 391-420.
125. Oh, E., Pierschbacher, M., Ruoslahti, E, *Proc. Natl. Acad. Sci. USA.*, **1981**, *278*, pp. 3218-3221.
126. Vucnto, M., *Eur. J. Biochem.*, **1980**, *105*, pp. 33-42.
127. *Fibronectin*, Mosher, D.F., Mecham, R.P. Eds., *Biology of Extracellular Matrix: A Series:* Academic Press, San Diego, Ca., **1989**.
128. Hynes, R.O., Yamada, K.M., *J. Cell. Biol.*, **1982**, *95*, pp. 369-377.
129. Guilbeau, E.J., Towe, B.C., Muehlbauer, M.J., *Trans. ASAIO*, **1987**, *10(3)*, pp. 329-335.
130. Monti, M., *Thermochim. Acta.*, **1990**, *172*, pp. 53-60.
131. Wang, Z., Hartmann, T., Baumiester, W., Guckenberger, R., *Proc. N.Y. Acad. Sci., USA*, **1990**, *87*, pp. 9343-9347.

RECEIVED October 14, 1991

POLYMERS AND LIQUID CRYSTALS

The use of polymers in macromolecular assemblies offers the possibility of constructing more robust systems than could be obtained with monomeric systems and opens new avenues to design polymeric molecules that exhibit a wide variety of functions when incorporated in macromolecular assemblies. As an example, polymeric liquid crystals are of considerable interest for constructing robust nonlinear optical sensors, optical storage devices, and optical displays. Block copolymers are known to assemble into specific morphologies that can create bulk materials with unusual mechanical properties or impart specific properties to the materials' surfaces. Although it is not yet possible to design a specific polymer for a specific function in a macromolecular assembly, the idea of molecular engineering for polymers has fascinated polymer chemists and materials scientists for some time. Active research programs are in progress to elucidate the fundamental principles that are necessary to establish such a discipline.

Chapter 22

Chiral Liquid-Crystalline Copolymers for Electrooptical Applications

E. Chiellini[1], G. Galli[1,2], A. S. Angeloni[3], M. Laus[3], and D. Caretti[3]

[1]Dipartimento di Chimica e Chimica Industriale, Università di Pisa, Via Risorgimento 35, Pisa 56126, Italy
[2]Department of Materials Science and Engineering, Cornell University, Ithaca, NY 14853
[3]Dipartimento di Chimica Industriale e dei Materiali, Università di Bologna, Viale Risorgimento 4, Bologna 40136, Italy

> The synthesis and liquid crystalline properties of four new series of copolymers consisting of variously substituted azobenzene mesogenic groups are presented. While three series contain chiral, optically active counits, the fourth series contains achiral dipolar chromophore counits. The incidence and stability of different chiral (nematic and smectic) mesophases are discussed with relevance to their potential ferroelectric and nonlinear optical properties. These samples may serve as models of liquid crystalline polymers to be used in electrooptical applications.

Liquid crystalline (LC) polymers present a unique combination of the characteristics peculiar to liquid crystals with those typical of polymers. While the former include molecular polarizability, self-assembly tendency, diversity of structures, and fast response to external electric or magnetic fields, the latter can feature variety of molecular architectures, dimensional stability, mechanical orientability, and ease of processability (*1*). The introduction of chirality in the molecular structure of LC materials induces the formation of chiral nematic (cholesteric) or chiral smectic supermolecular assemblies endowed with a macroscopic twist superposed on them (*2*).This consistently offers an additional valuable means of tuning the liquid crystal behavior and addressing specific responses of chiral LC polymers in optics and electrooptics. In this context, the ferroelectric and nonlinear optical properties of side chain LC polymers are currently the focus of intense research (*3*).

The *ferroelectric* properties of the chiral smectic C* mesophase are recognized in a number of thermotropic polymers (*4-9*). In this mesophase the optically active mesogens with high spontaneous polarization are assembled in a layered, helical superstructure which must then be untwisted into another superstructure with a resulting macroscopic electric polarization allowing switching bistability. While the switching times measured for ferroelctric LC polymers are in the range of milliseconds, it may be argued that the associated electroclinic effect (10) should provide faster response rates in an orthogonal smectic phase, such as the smectic A phase.

The second-order *nonlinear optical* properties of organic polymers are also well documented (9,11-15). For a material to exhibit significant second-order nonlinear optical responses it must have a noncentrosymmetric structure. Electric field poling of polymers containing a dipolar chromophore is a common method to introduce a polar axis into the LC medium, and the consequently reduced orientational averaging of molecular dipoles allows for a net second-order susceptibility to result. An alternative approach to originating noncentrosymmetry might be to introduce chirality and the other structural requirements of the ferroelectric smectic C* phase into a polymeric structure containing chromophores capable of producing a nonlinear response without poling in an external electric field (9,16).

However, the details of the crucial interplay of chemical, stereochemical and electronic factors in effecting the structures and properties of chiral supermolecular assemblies are still far from being elucidated (9,17,18).

Following our interest in chiral LC polymers (18), we have recently developed the synthesis and chemical modification of chiral LC side chain polymers based on different mesogenic units (19,20, Angeloni, A.S. et al. *Chirality*, in press). Among these, the azobenzene mesogenic core is characterized by sufficient chemical and thermal stability and

$$CH_3-\underset{\underset{CH_2}{|}}{C}-COO(CH_2)_6O-\bigcirc-N=N-\bigcirc-R \quad \mathbf{A}$$

$$CH_3-\underset{\underset{CH_2}{|}}{C}-COO(CH_2)_6O-\underset{\underset{R''}{}}{\bigcirc}-N=N-\bigcirc-R' \quad \mathbf{B}$$

Comonomer **A**		Comonomer **B**			Copolymer
R	No.	R'	R''	No.	Series
$C_2H_5CH^*(CH_3)CH_2O-$	1	$n\text{-}C_6H_{13}O-$	H	2	1/2
$C_2H_5CH^*(CH_3)CH_2O-$	1	$n\text{-}C_6H_{13}O-$	CH_3	3	1/3
$C_2H_5CH^*(CH_3)CH_2O-$	1	$n\text{-}C_{10}H_{21}O-$	H	4	1/4
NO_2	5	$n\text{-}C_{10}H_{21}O-$	H	4	4/5

π-electron conjugation and can be functionalized with various substituent groups in order to tune the structural and physical properties of the resulting LC polymers. Therefore, major objectives of this work were the design and synthesis of chiral LC side chain polymers (see general formula above) exhibiting a variety of mesophases according to the distinct tendencies of the azobenzene comonomers to impart either cholesteric or smectic properties. One further objective was to incorporate dipolar chromophore counits with potential nonlinear optical characteristics in LC polymers. We believe that these polymers may serve as models to address the structure-property correlation in polymeric materials for electrooptical applications.

Experimental

Synthesis of monomers. The synthesis of monomer **1** is described in detail as a typical example.

4-((S)-2-methylbutoxy)-4'-hydroxyazobenzene: A solution of 12.0 g (0.174 mol) of $NaNO_2$ in 30 mL of water was added dropwise to a solution of 25.8 g (0.144 mol) of 4-((S)-2-methylbutoxy)aniline (**6**, X = (S)-C_2H_5CH*(CH_3)CH_2O) in 150 mL of 3 M HCl at 0-5°C with vigorous stirring. After 1 h, the excess $NaNO_2$ was decomposed by addition of urea and the solution was slowly poured into a solution of 13.6 g (0.144 mol) of phenol (**7**, R" = H) in 150 mL of 2 M NaOH. After 10 min, the solution was acidified with HCl and the precipitated **8** was washed with water, dried and finally crystallized in cyclohexane (yield 47%): m.p. 85°C, $[\alpha]^{25}_D$ +10.4° ($CHCl_3$).

6-chlorohexyl methacrylate: A solution of 15.3 g (0.146 mol) of methacryloyl chloride in 50 mL of anhydrous 1,2-dichloroethane was added dropwise with vigorous stirring to a mixture of 20.0 g (0.146 mol) of 6-chlorohexanol, 29.6 g (0.293 mol) of triethylamine, and 0.5 g of 2,6-di-*tert*-butyl-4-methylphenol at ambient temperature under nitrogen atmosphere. The reaction mixture was then refluxed for 1 h, cooled down, washed with 1 M HCl, water and dried over Na_2SO_4. The product (**9**) was finally purified by distillation (yield 75%): b.p. 107°C/2mm.

4-((S)-2-methylbutoxy)-4'-(6-methacryloyloxyhexyloxy)azobenzene (**1**): A mixture of 5.4 g (0.019 mol) of 4-((S)-2-methylbutoxy)-4'-hydroxyazobenzene, 3.9 g (0.019 mol) of 6-chlorohexyl methacrylate, and 4.0 g (0.029 mol) of anhydrous K_2CO_3 in 60 mL of dry dimethyl sulfoxide was stirred at 80°C for 2 h, poured into 300 mL of cold water and washed with 1 M NaOH and water. The solid product (**1**) was crystallized twice in methanol (yield 82%): m.p. 82°C, $[\alpha]^{25}_D$ +6.9° (chloroform).

Copolymerization. In a typical copolymerization reaction (cf. (**1/2**)a copolymer) 0.542 g (1.20 mmol) of 4-((S)-2-methylbutoxy)-4'-(6-methacryloyloxyhexyloxy)azobenzene (**1**) and 0.361 g (0.75 mmol) of 4-hexyloxy-4'-(6-methacryloyloxyhexyloxy)azobenzene (**2**) were dissolved in 5 mL of dry benzene in the presence of 5 mg of AIBN. The solution was introduced under nitrogen into a glass vial that was sealed after repeated freeze-thaw pump cycles. After reacting 48 h at 60°C, the mixture was poured into 100 mL of methanol. The polymeric product was purified by several precipitations from chloroform solution into methanol (yield 90%).

Characterization. Optical rotatory power measurements were carried out with a Perkin Elmer 141 spectropolarimeter on chloroform solutions. Average molecular weights of the polymers were determined by size exclusion chromatography (SEC) of chloroform solutions using a 590 Waters chromatograph with a Shodex KF804 column. Polystyrene standards were employed for the universal calibration method.

Differential scanning calorimetry (DSC) was performed with a Perkin- Elmer DSC7 apparatus. The transition temperatures were taken as corresponding to the maximum of the enthalpic peaks obtained at a heating/cooling rate of 10 K/min. The mesophase textures were observed between crossed polarizers on a Reichert microscope equipped with a Mettler FP52 heating stage. X-ray diffraction data were recorded at the CHESS facilities at Cornell University. Oriented fibers, films and powders were examined using a Mettler FP82 hot stage mounted in the beam path to control sample temperatures. Fibers were formed by taking the polymer to the isotropic melt on a Fisher-Johns apparatus and drawing out the molten polymer with tweezers. Monochromated radiation of λ = 1.565 Å was used. The experimental setup and analytical procedures have been previously described (*21*).

Results and Discussion

Methacrylate monomers **1-5** were prepared following the synthetic route outlined in Scheme 1. Among them, only **2** presents a monotropic nematic phase (isotropic-nematic transition at 354 K) and **4** exhibits a smectic phase (smectic-isotropic transition at 363 K).

The polymers were obtained by free radical initiation (AIBN) at 60°C with polymerization yields greater than 85-90%. In the ^{1}H-NMR and ^{13}C-NMR spectra the relative intensities of the signals originating from the R, R', and R" substituents do not depend on the polymerization yield and correspond to those of the initial comonomer mixtures. Taking into account that the methacrylate moiety is well spaced apart from the mesogenic core, we may reasonably assume that the different monomers are characterized by a comparable reactivity and hence the corresponding copolymers have a random distribution of monomer counits. The copolymers incorporating the chiral comonomer **1** are optically active. There exists a linear relationship between optical rotatory power and weight fraction of **1** (X_1) of the different copolymer systems, the higher molar optical rotation being detected for poly(**1**) ($[\Phi]^{25}_D$ = +31° ($CHCl_3$)). It was, however, difficult to measure the optical rotatory power of the copolymers due to their strong absorption of the UV-visible light, particularly in copolymers with lower contents of counits from **1**. The chiroptical properties and the *trans-cis* photoinduced isomerization in solution of these polymers will be described in a forthcoming paper. The molecular weights of the polymers were determined by SEC and their number average molecular weight (M_n) ranges from 190,000 to 760,000 g/mol with first polydispersity index (M_w/M_n) between 1.9 and 2.6.

Scheme 1. Reaction pathway for the synthesis of methacrylates 1 - 5.

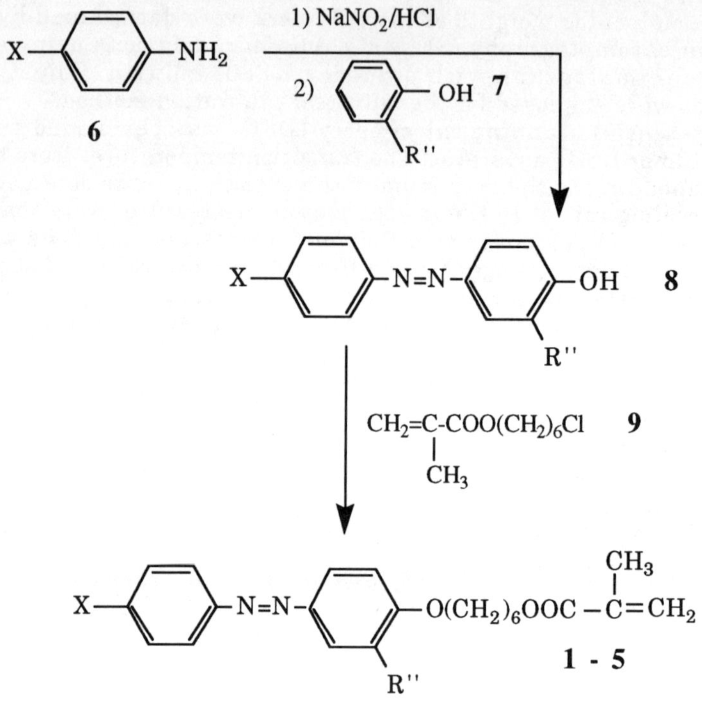

$$X-\underset{\mathbf{6}}{\underset{|}{\bigcirc}}-NH_2 \xrightarrow[\text{2)}]{\text{1) NaNO}_2/\text{HCl}} \underset{\mathbf{7}}{\underset{R''}{\bigcirc}-OH}$$

$$X-\underset{}{\bigcirc}-N=N-\underset{R''}{\bigcirc}-OH \quad \mathbf{8}$$

$$\underset{CH_3}{\underset{|}{CH_2=C}}-COO(CH_2)_6Cl \quad \mathbf{9}$$

$$X-\underset{}{\bigcirc}-N=N-\underset{R''}{\bigcirc}-O(CH_2)_6OOC-\underset{CH_3}{\underset{|}{C}}=CH_2 \quad \mathbf{1-5}$$

$X = (S)\text{-}C_2H_5CH^*(CH_3)CH_2O$, $n\text{-}C_6H_{13}O$, $n\text{-}C_{10}H_{21}O$, NO_2

$R'' = H$, CH_3

All of the polymers exhibit LC properties. There is no crystallinity detectable by DSC or X-ray scattering in the samples at temperatures below the onset of the thermotropic mesophase, which therefore extends from the glass transition temperature (T_g) to the isotropization temperature (T_i). In most cases we could not detect the glass transition by DSC and the lower temperature limit for the LC range of these copolymers remains to be identified better. Poly(**1**) and poly(**3**) have T_g values of 384 K and 364 K, respectively, and this suggests that in these systems the mesophase can easily be locked-in at room temperature.

The LC properties of **1/2** copolymers are summarized in Table I. For these samples there exists one smectic mesophase throughout all the composition range and the smectic-isotropic temperature increases linearly with increasing **2** content from that of poly(**1**) to that of poly(**2**). This reflects the tendency of counits from **1** to give rise to a less stable smectic mesophase, consistent with the presence of the branched chiral group, and suggests that the nature of the mesophase may be the same for all **1/2** copolymers. In agreement with this interpretation, the values of the smectic-isotropic entropy are very similar along the series.

Table I. LC properties of 1/2 copolymers containing azobenzene mesogens with R = (S)-2-methylbutoxy (1) and R' = n-hexyloxy (2) substituents

Sample	X_1[a]	M_n[b]	M_w/M_n[b]	T_i (K)	ΔS_i (J/mol·K)
poly(1)	1	760,000	1.9	388	18.7
(1/2)a	0.8	490,000	2.3	395	16.7
(1/2)b	0.6	390,000	2.6	402	18.8
(1/2)c	0.4	500,000	2.4	410	19.4
(1/2)d	0.2	540,000	2.2	417	20.4
poly(2)	0	440,000	2.4	423	18.8

[a] Weight fraction of monomer 1. [b] By SEC.

A similar trend is observable in the 1/4 copolymer series (Table II). All of these members exhibit smectic behavior and the isotropization temperature decreases with increasing X_1 (Figure 1). However, there is a parallel marked decrease in the smectic-isotropic entropy in going from poly(4) to poly(1) and the smectic mesophases in the copolymers may be different in character. Note also that for poly(4) and 1/4 copolymers the smectic-isotropic temperature is consistently higher than for poly(2) and respective 1/2 copolymers, this difference becoming less marked with increasing X_1. These results are in accordance with the general observation that longer alkyloxy substituents on mesogenic cores tend to favor the incidence of more stable and persistent smectic mesophases.

Table II. LC properties of 1/4 copolymers containing azobenzene mesogens with R = (S)-2-methylbutoxy (1) and R' = n-decyloxy (4) substituents

Sample	X_1[a]	M_n[b]	M_w/M_n[b]	T_i (K)	ΔS_i (J/mol·K)
poly(1)	1	760,000	1.9	388	18.7
(1/4)a	0.8	580,000	2.2	398	16.4
(1/4)b	0.6	540,000	2.2	410	19.4
(1/4)c	0.4	530,000	2.1	435	22.4
poly(4)	0	190,000	2.3	445	24.7

[a] Weight fraction of comonomer 1. [b] By SEC.

A more complex mesomorphic behavior is shown by 1/3 copolymers (Table III). Poly(3) presents one nematic mesophase of limited extension (4 K). Incorporation of counits from 3 in chiral 1/3 copolymers results in the occurrence of a cholesteric and a new smectic mesophase at intermediate compositions, and eventually the copolymers become purely smectic at 1 weight fractions greater than approximately 0.80.

Table III. LC properties of 1/3 copolymers
containing azobenzene mesogens with R = (S)-2-methylbutoxy (1)
and R' = n-hexyloxy, R" = methyl (3) substituents

Sample	X_1[a]	M_n[b]	M_w/M_n[b]	T_{SCh}[c] (K)	ΔS_{SCh}[c] (J/mol·K)	T_i (K)	ΔS_i (J/mol·K)
poly(1)	1	760,000	1.9	---	---	388	18.7
(1/3)a	0.8	360,000	2.5	---	---	380	13.3
(1/3)b	0.6	480,000	2.3	368	8.2	375	3.2
(1/3)c	0.4	470,000	2.3	367	5.6	370	3.4
(1/3)d	0.2	320,000	2.6	364	3.5	369	3.7
poly(3)	0	390,000	2.4	---	---	368	4.0

[a] Weight fraction of comonomer 1. [b] By SEC. [c] Smectic-cholesteric transition.

The introduction of the lateral methyl substituent in 3 reduces the aspect ratio of the azobenzene mesogen relative to 2, which greatly depresses its smectogenic tendency thus permitting the existence of two chiral mesophases in 1/3 copolymers ($0 < X_1 < 0.80$). In agreement with this mesophase behavior, the isotropization entropy (Figure 2) is low for the nematic phase ($X_1 = 0$) and lowest for the cholesteric mesophase whereas it is high for the smectic mesophase. Conversely, the smectic-cholesteric entropy increases regularly with increasing 1 content and, as a net result, the total entropy change from the glass to the isotropic liquid increases almost linearly with X_1.

4/5 copolymers present one smectic mesophase, whose isotropization temperature and entropy are not substantially altered by the incorporation of significant amounts (higher than 10 wt.%) of the nitrosubstituted comonomer 5 (Table IV). This polymer smectic structure can, therefore, easily accommodate different mesogenic units with strongly mismatching terminal substituents, including chromophores with pronounced dipolar character.

The structure of the mesophases was investigated using a high energy synchrotron source which allowed to perform real time X-ray scattering experiments. The phase transitions could, therefore, be followed dynamically as they were taking place in the same conditions as with DSC and polarizing microscopy. The X-ray observations are in agreement with the above results. Generally, the recorded diffraction patterns can be divided into two spectral regions (Figures 3 and 4). The small angle region signals (inner rings) are associated with the layered stacking of the molecules in the smectic phase, whereas the wide angle region diffractions (outer rings) occur from the lateral packing of the molecules within the smectic layers. In all cases a very diffuse outer signal is observed indicative of a liquid-like arrangement of the polymer side chains in a disordered smectic phase with an average intermolecular distance D ~ 4.4-4.5 Å. The cholesteric phase of the polymers displays a broad wide angle ring (D ~ 4.5-4.6 Å) only.

A typical diffraction pattern of the smectic phase of an unoriented

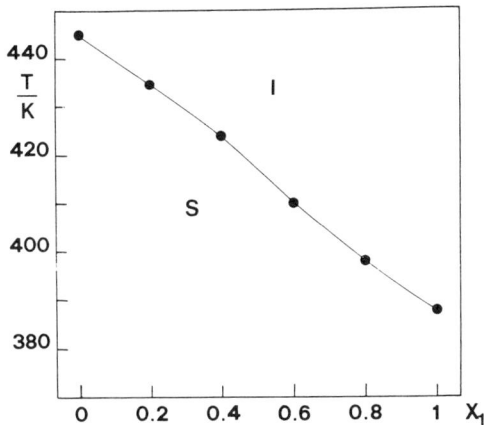

Figure 1. Smectic (S)-to-isotropic (I) transition temperatures for **1/4** copolymers as function of **1** weight fraction (X_1).

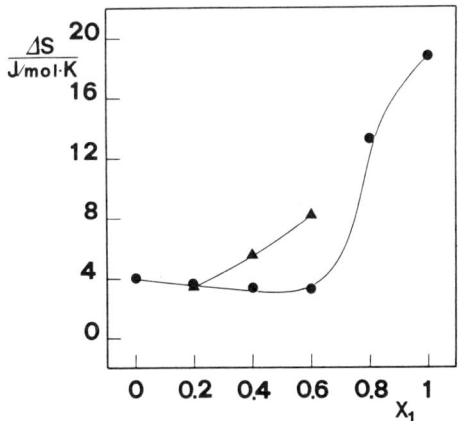

Figure 2. Phase transition entropies (ΔS) for **1/3** copolymers as function of **1** weight fraction (X_1): cholesteric (or smectic)-to-isotropic (●) and smectic-to-cholesteric (▲).

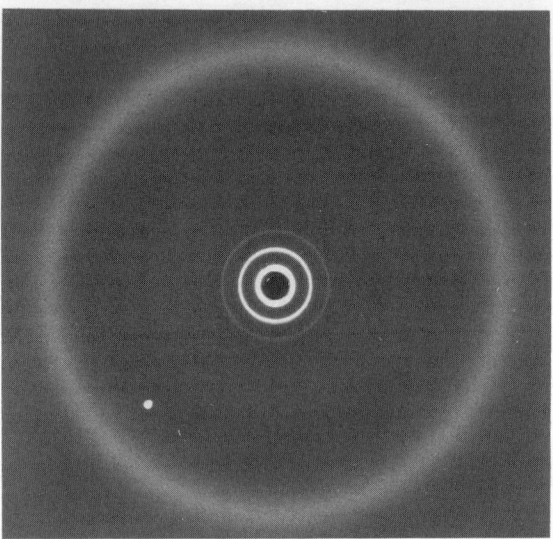

Figure 3. X-ray diagram of the smectic phase of poly(**4**) at 160°C.

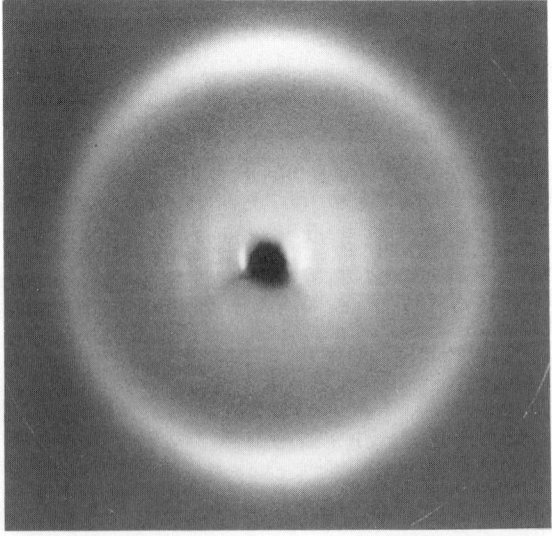

Figure 4. Fiber X-ray diagram of the smectic phase of poly(**1**) at room temperature (vertical fiber axis).

Table IV. LC properties of 4/5 copolymers
containing azobenzene mesogens with R = n-decyloxy (4)
and R' = nitro (5) substituents

Sample	X_4[a]	M_n[b]	M_w/M_n[b]	T_i (K)	ΔS_i (J/mol·K)
poly(4)	1	190,000	2.3	445	24.7
(4/5)a	0.98	450,000	2.4	445	23.5
(4/5)b	0.95	480,000	2.3	449	24.7
(4/5)c	0.90	240,000	2.6	449	22.9

[a] Weight fraction of comonomer 4. [b] By SEC.

polymer sample is given in Figure 3 for poly(4). There are three orders of reflection in the inner region with corresponding d spacings in the ratio 1:2:3 associated with the regular layer periodicity of the mesophase. This seems to be a common feature in thermotropic side chain polymers, in complete contrast to small molecule liquid crystals for which one small angle signal is observed and it is generally accepted that the projection of the electron density profile along the director can be described by the ideal model of a single sinusoidal modulation (22). The measured d spacing (d = 54 Å) is much longer than the calculated length of the side chains in their most extended conformation (L ~ 37 Å) and the mesophase must have a bilayer structure with partially interdigitated mesogenic units (S_{A2} or S_{C2}) (23). A typical fiber diffraction pattern is reported in Figure 4 for the smectic phase of poly(1). The six small angle Bragg arcs correspond to the three first orders of reflection on the layer plane. They are located on the equator, which shows that the smectic layers, and consequently the polymer backbones, are oriented parallel to the fiber axis. The detected d spacing of 44 Å is much longer than the calculated side chain length (L ~ 29 Å). The two broad crescents at wide angle are located on the meridian and suggest that the polymer side chains are orthogonal to the fiber axis and to the smectic layers of a S_{A2} phase. However, the existence of a small tilt angle in the present samples cannot be completely ruled out and investigations on annealed fibers and magnetically aligned samples are under way. A complete report will be given elsewhere.

Conclusions

A variety of chiral thermotropic copolymers can be prepared exhibiting a chiral smectic mesophase and, in some cases, a cholesteric mesophase. While the overall mesophase behavior is essentially dictated by the structure of the two parent comonomers, it can be somewhat accurately tuned by the proper adjustment of the relative amounts of the counits. We also anticipate that, by the synthetic scheme proposed, several chiral mesophases can be developed in copolymers containing strong dipolar chromophores to be tested in ferroelectric and nonlinear optical applications.

Acknowledgments

This work was carried out with partial financial support from the Ministero dell'Università e Ricerca Scientifica of Italy.

G.G. also thanks Italian CNR for a NATO Fellowship and the Cornell High Energy Synchrotron Source (CHESS) for use of the facility, as well as Prof. C. K. Ober, Cornell University, for helpful discussions.

Literature Cited

1. *Liquid Crystalline Polymers*, Weiss, R.A.; Ober, C.K., Eds.; ACS Symp.Ser.; American Chemical Society: Washington, DC, 1990; Vol.435.
2. (a) Chiellini, E.; Galli, G. In *Recent Advances in Mechanistic and Synthetic Aspects of Polymerization;* Fontanille, M.; Guyot, A., Eds.; Reidel: Dordrecht, NL, 1987; p.426.; (b) Chiellini, E.; Galli, G. In *Recent Advances in Liquid Crystalline Polymers;* Chapoy, L.L., Ed.; Applied Science: London, GB, 1985; p.15.
3. *Side Chain Liquid Crystalline Polymers;* McArdle, C.B., Ed.; Blackie: Glasgow, GB, 1989.
4. Uchida, S.; Morita, K.; Miyoshi, K.; Hashimoto, K.; Kawasaki, K. *Mol.Cryst.Liq.Cryst.* **1988**, *155*, 93.
5. Suzuki, T.; Okawa., T.; Ohnuma, T.; Sahon, Y. *Makromol.Chem., Rapid Commun.* **1988**, *9*, 755.
6. Vallerien, S.U.; Zentel, R.; Kremer, F.; Kapitza, H.; Fischer, E.W. *Makromol.Chem., Rapid Commun.* **1989**, *190*, 333.
7. Scherowsky, G.; Schliwa, A.; Springer, J.; Kuhnpast, K.; Trapp, W. *Liq.Cryst.* **1989**, *5*, 1281.
8. Dumon, M.; Nguyen, H.T.; Mauzac, M.; Destrade, C.; Achard, M.F.; Gasparoux, M. *Macromolecules* **1990**, *23*, 355.
9. Kapitza, H.; Zentel, R.; Twieg, R.J.; Nguyen, C.; Vallerien, S.U.; Kremer, F.; Willson, C.G. *Adv.Mater.* **1990**, *2*, 539.
10. Garaff, S.; Meyer, R.B. *Phys.Rev.Lett.* **1977**, *38*, 848.
11. LeBarny, P.; Ravaux, G.; Dubois, J.C.; Parneix, J.P.; Njeumo, R.; Legrand, G.; Levelut, A.M. *Proc. SPIE*, **1986**, *682*, 56.
12. Leslie, T.M.; Demartino, R.N.; Choe, E.W.; Khanarian, G.; Haas, D.; Nelson, G.; Stamatoff, J.B.; Stuetz, D.E.; Teng, C.-C.; Yoon, H.-N. *Mol. Cryst.Liq. Cryst.* **1987**, *153*, 451.
13. Eich, M.; Reck, B.; Yoon, D.Y.; Willson, C.G.; Bjorklund, G.C. *J.Appl.Phys.* **1990**, *66*, 3241.
14. Amano, M.; Kaino, T.; Yamamoto, F.; Takeuchi, Y. *Mol.Cryst. Liq.Cryst.* **1990**, *182A*, 81.
15. Yitzchaik, S.; Berkovic, G.; Krongauz, V. *Macromolecules* **1990**, *23*, 3539.
16. Taguchi, A.; Kajikawa, K.; Ouchi, Y.; Takezoe, H.; Fukuda, A. *Springer Proc.Phys.* **1989**, *36*, 250.
17. Chen, S.H.; Tsai, M.L. *Macromolecules* **1990**, *23*, 5055.
18. Chiellini, E.; Galli, G.; Carrozzino, S.; Gallot, B. *Macromolecules* **1990**, *23*, 2106 and references therein.
19. Angeloni, A.S.; Caretti, D.; Carlini, C.; Chiellini, E.; Galli, G.; Altomare, A.; Solaro, R.; Laus, M. *Liq.Cryst.* **1989**, *4*, 513.

20. Chiellini, E.; Galli, G.; Cioni, F. *Ferroelectrics* **1991**, *114*, 223.
21. Ober, C.K.; Delvin, A.; Bluhm, T.L. *J.Polym.Sci., Polym.Phys.Ed.* **1990**, *28*, 1047.
22. Davidson, P.; Levelut, A.M.; Achard, M.F.; Hardouin, F. *Liq.Cryst.* **1989**, *4*, 561.
23. Gray, G.W.; Goodby, J.W.G. *Smectic Liquid Crystals;* Leonard Hill: Glasgow, GB, 1984.

RECEIVED November 1, 1991

Chapter 23

Side-Chain Crystallinity and Thermal Transitions in Thermotropic Liquid-Crystalline Poly(γ-alkyl-α,L-glutamate)s

William H. Daly[1], Ioan I. Negulescu[1], Paul S. Russo[1], and Drew S. Poche[2]

[1]Department of Chemistry, Macromolecular Studies Group, Louisiana State University, Baton Rouge, LA 70803–1804
[2]Dow Louisiana, Plaquemine, LA 70765–0400

Thermal transitions in poly(γ-alkyl-α,L-glutamate)s with C16 (hexadecyl, PHLG) and C18 (stearyl, PSLG) side-chains were investigated by differential scanning calorimetry. Both PHLG and PSLG (Mw=40K-300K) were synthesized from the corresponding NCA. By annealing the polymers at appropriate temperatures it was possible to separate the melting of paraffinic side chain segments and a second transition that seems to be closely related to the formation of a liquid crystal. Depending on the thermal history of samples (repeated annealings and/or quenchings) the side chain segments melted and crystallized over a large temperature range suggesting the formation of crystallites of different dimensions and/or accommodating a different number of methylenic units. The thermal history also affected the transition to the LC state; the highest LC transition temperatures were observed for samples annealed above the melting domain of side chains.

Linear poly(γ-alkyl-α,L-glutamate)s, PALG-n (where n represents the number of carbon atoms of the paraffinic side chain segment)

$$(-NH-CH-C-)_x$$
$$\quad\quad\quad\;\; \underset{\displaystyle CH_2CH_2-C-O-(CH_2)_n-H}{|} $$

are unusual in the ranks of synthetic polymers because they exist in well defined, helical chain conformations and retain such structures even in solution (1). In contrast, the majority of synthetic macromolecules exhibit limited long range chain order on a statistical basis only. The peptide α-helix of PALG-n, which imparts rod-like character to the macromolecule, coupled with the aliphatic side chains emanating from the repeat units (Figure 1), creates a novel hydrophobic semi-rigid rod (2-5). Linear polypeptides are readily available in a wide range of molecular weights via anionic ring opening polymerization of N-carboxyamino acid anhydrides (NCA's) *(6)*. Recently, newly devised multi-functional initiators were used to synthesize PALG-18s with 3-9 rod-like arms *(7-10)*. Solution characterization and other observations *(9)* suggest that the stars have arms of uneven length, but that they were truly branched macromolecules. Both linear and star branched PALG-n exhibit interesting thermal and morphological properties in the bulk state *(10)*. They display side chain melting transitions at rather low temperatures if the alkyl side chain segment is sufficiently long, quite an unusual behavior for a nearly rigid rod-like polymer. Provided that the axial ratio exceeds a critical value *(11,12)* the melt may display liquid crystallinity; the paraffinic side chains evidently act as an efficient solvent or plasticizer for the backbone*(5)*. Their solubility in common organic solvents, a lack of polyelectrolyte effects, an easily accessible molten state and good molecular homogeneity make PALGs attractive model systems for the study of stiff polymers.

Polymeric glutamic acid esters are not just materials of academic interest. Their solubility in hydrocarbons suggests applications as thickening agents for mineral and vegetable oils, solidifiers for liquid fuels and toughening agents for waxes(13). Potential practical uses exploiting their novel liquid crystalline organization can be envisioned *(14)*. For example, Langmuir-Blodgett films of PALG-n can be prepared; the number of electrical or optical applications is greatly increased due to the improved long-range orientation order in the film plane *(15,16)*. Similarly, the availability of a helix-coil transition may prove useful in controlled release or other applications.

An unusual property of the PALG's is the coexistence of two types of structural units -- polyamide main chain backbones and hydrocarbon side chains-- that are normally incompatible. In effect, the main chain is immersed in a hydrocarbon environment; thus the solubility properties are more typical of a hydrocarbon than of a polypeptide. This duality of structure of the PALG-n macromolecule results in a splitting of the ordering behavior of the system as a whole. At low temperatures side chain crystallization is the dominant ordering force, but at temperatures where the side chains are molten, the mesogenic character of the polypeptide backbone controls the formation of a liquid crystalline phase. The objective of this paper is to analyze the complicated melting phenomenon that separates these regimes. This work is distinguished from previous studies*(2,5)* by extensive annealing steps and, especially, by the use of polymers synthesized from NCA monomers. PALG's prepared by transesterification of the side chain (e.g., from poly(γ-methyl-α,L-glutamate)) are really random copolymers containing some long alkyl side chains and some shorter side chains. In the bulk state, such structural defects could introduce extra free volume, altering the melting process and the melt itself.

Experimental

Synthesis of polymers. Both PALG-16 and PALG-18 were obtained by anionic polymerization of the corresponding NCA *(4,8)*. Only high molecular weight samples (MW=40K-300K) with a helicoidal conformation (as indicated by IR, e.g., C=O amide I stretching vibration at 1655cm^{-1}) are discussed in this study.
Measurements. IR spectra were recorded on a Perkin Elmer FTIR 1760X. Differential scanning calorimetry, DSC, measurements were performed with a Seiko DSC 220C calorimeter (heating/cooling rate = 10°C/min). Dynamic mechanical data were obtained using a Seiko DMS 200 Tension module (1 Hz) and thermomechanical data were measured using a Seiko TMA/SS analysis module.

Results And Discussion

The polypeptides examined have sufficient axial ratios to form lyotropic cholesteric liquid crystals at moderate concentrations, as shown in Figure 2. This feature they share with classic polypeptides such as poly(γ-benzyl-α,L-glutamate). A rather cursory inspection of this behavior has not yet revealed any surprises. But the thermotropic behavior in the absence of solvent is richly complex. A dry powder of PALG-n, fresh from precipitation and drying, is a material that once had the benefit of solvent to solubilize the side chains. During coagulation from a not-very-viscous reaction medium, it is possible that the side chains achieve a degree of crystalline order. As these materials are heated, a first order transition is observed at some temperature $T(1)$, representing the melting of the paraffinic side chains. Depending on the length of side chain, a second transition, also first order, may appear at temperature $T(2)$. It has been reported that the difference between these two temperatures diminishes as the number of carbon atoms in the side chains able to enter into paraffinic crystalline phase increases. For PALG-10 $T(1)=16^{\circ}C$, $T(2)=51^{\circ}C$ and for PALG-16 $T(1)=54^{\circ}C$, $T(2)=64^{\circ}C$ and only a $T(1)$ at $62^{\circ}C$ was reported for PALG-18 *(2)*. In contrast we have observed that there are two transitions in PALG-18, but the difference between the two is reduced to only $6^{\circ}C$ with $T(2)=68^{\circ}C$ *(4,8)*. This observation distiguishes PALG-18 prepared from monomer from that prepared by transesterification of poly(γ-methyl-α,L-glutamate); we found only single transitions in the latter sample. It was thought *(2)* that $T(2)$ represents the transition to a liquid crystalline state, but the nature of the new degrees of freedom that become excited to allow molecular flow are not known. In other words, the nature of intermolecular interactions between $T(1)$ and $T(2)$ is poorly understood. For copolypeptides made by transesterification, one possibility is that backbone-backbone intermolecular interactions between domains deficient in the long side chain might occur. Investigation of the thermal behavior of pure PALG-16 and PALG-18 is therefore in order.

Characterization of PALG-n Thermal Transitions by DSC

The thermal behavior of PALG-16 and PALG-18 was strongly dependent on the thermal history. There actually exists a large temperature domain in which the paraffinic parts of side chains can melt or crystallize. All samples were subjected to multiple heating/cooling cycles to establish a consistent thermal history. A minimum of one heating/cooling cycle was required before reproducible data for each cycle could be obtained.

For example, Figure 3A shows the thermal transition observed upon heating pristine PALG-16 obtained in powder form by precipitation with acetone from the reaction medium (dichloromethane); only one DSC peak (with a weak shoulder) is observed at $50.5^{\circ}C$. The cooling process revealed two different exotherms at $25^{\circ}C$ and $6^{\circ}C$ (Figure 4A). The onset of melting in the 2nd heating cycle occurs below zero, i.e., $-2^{\circ}C$ (Figure 3B). Two broad endotherms, at $11.9^{\circ}C$ and $38^{\circ}C$, characterize this DSC trace. The very broad "peak" at the higher temperature can be split, however, and shifted $10^{\circ}C$ or more by annealing at $43^{\circ}C$ (i.e., at a temperature higher than that of the highest transition temperature observed in the second heating thermogram). After cooling the annealed sample and reheating for the third time, multiple DSC peaks were observed at $14^{\circ}C$, $48^{\circ}C$ and $58^{\circ}C$ (Figure 3C). By repeated annealings, at temperatures closer and closer to but less than the highest temperature peak observed, higher melting samples can be prepared. The

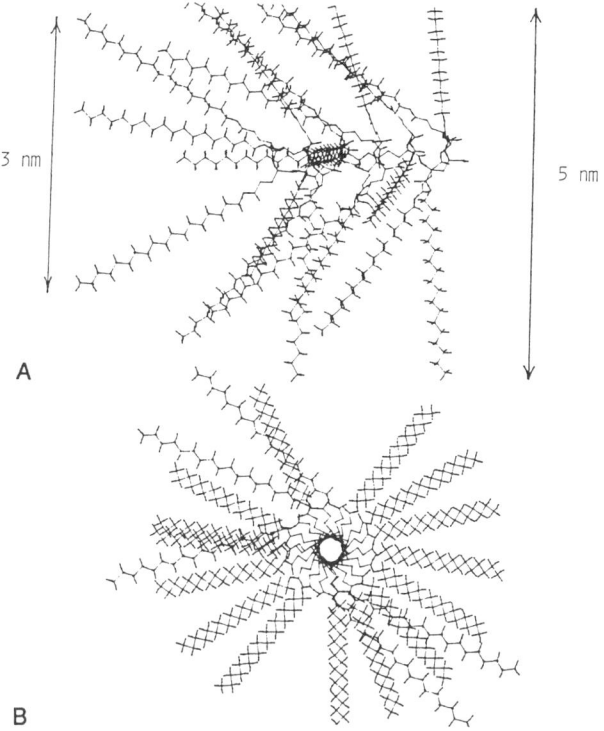

Figure 1. SYBIL molecular model for PALG-18 with a DP of 20. (A) Front view; (B) Side view.

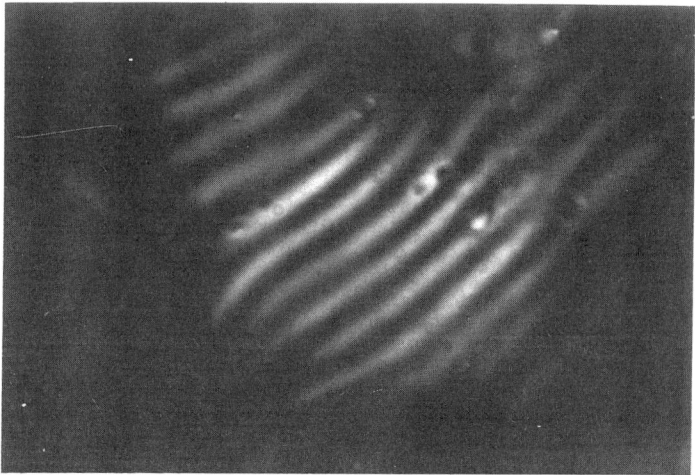

Figure 2. Cholesteric liquid crystal structure of 15% solution of PALG-18 MW=200K in toluene at 20°C. The objective was x10. The periodicity spacing is 34 μm.

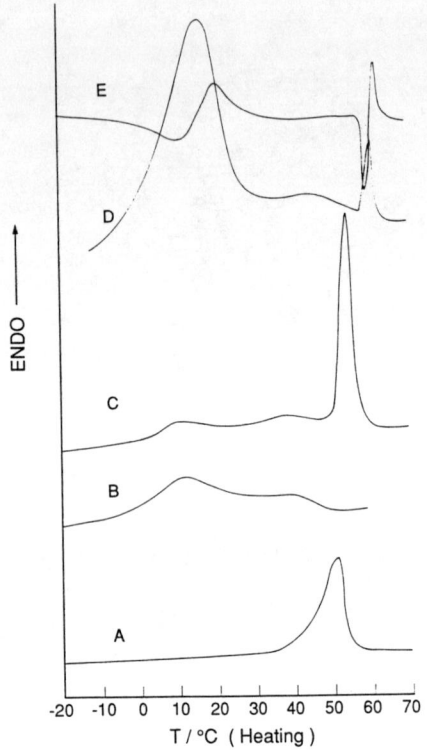

Figure 3. DSC traces of PALG-16 (heating). (A) Pristine sample (1st heating); (B) Reheated sample (2nd heating); (C) Sample annealed at 43°C for 60 min; (D) Sample annealed at 55°C for 60 min; (E) Derivative of (D).

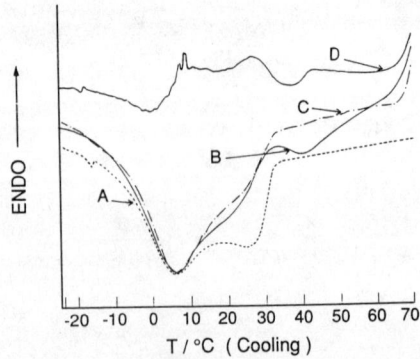

Figure 4. DSC traces of PALG-16 (cooling). (A) 1st Cooling from 100°C; (B) Cooling after annealing at 43°C (sample heated to 80°C); (C) Typical cooling trace from 100°C after reproducibility has been achieved in the heating cycles; (D) Derivative of curve C.

highest endothermic peak for this polypeptide was observed around 60°C. After repeated recycling and annealings a reproducible thermogram with only two transitions, at 13°C (major) and 60°C, was obtained (Figure 3D). The derivative curve (Figure 3E) shows that no other transition occurs in the region between these two temperatures. The reproducible cooling curves exhibited transitions at 40°C and 6.7°C (Figure 4C), but the derivative curve indicated that the second peak represents multiple transitions with maxima between 9°C and -1° (Figure 4D)

Similar phenomena were observed for PALG-18. The DSC thermogram of the pristine sample contained one major peak at 62°C with a shoulder at 67°C (Figure 5A). The first reheating resulted in a two-peak DSC trace with maxima at 60°C and 67.3°C (Figure 5B). Rapid cooling suppressed the appearance of the second maximum (Figure 5C). In contrast, annealing produced more ordered side chains melting at 68°C as the dominant component (Figure 5D). Transition temperatures as high as 72°C were recorded by annealing the sample just under the temperature of the highest endotherm observed (not shown). Annealing at lower temperatures, even for long periods of time, e.g., 600 min at 50°C, led to apparent mixture of side chain crystallites, e.g., thermograms with maxima at 45°C, 60°C (major peak) and 68°C were obtained. One difference between PALG-18 and PALG-16 is that only one crystallization exotherm at 43.6°C was observed for PALG-18 during cooling (Figure 6), regardless of the thermal history of the sample.

We interpret the reduction in T(1) (compared to pristine material) as a decrease in the number or order of methylene units entering into the side chain crystallites upon annealing. Presumably, the highly viscous nature of the material prevents chains from reaching a high degree of order, and the system must settle for those interactions that can be achieved in a finite time. Perhaps the side chains never again become as highly ordered as they were after precipitation and drying from solvent.; therefore, they exhibit lower and, sometimes, much broader melting transitions.

Annealing generally produces the opposite effect on T(2). Recall that this transition was not previously observed in PALG-18 *(2)*. If T(2) really represents the transition from a state in which the side chains are mobile to a true liquid in which the backbones also achieve mobility, then whatever forces prevent backbone motion at temperatures between T(1) and T(2) can be enhanced via judicious annealing. Perhaps the longer side chains work to prevent tight inter-backbone interactions, such that once the side chains melt, the whole system becomes immediately fluid. This is consistent with the appearance of only one exotherm upon cooling PALG-18. But the exact nature of the T(2) transition requires further study.

Preliminary Viscoelastic Measurements and Conclusion

The viscoelastic properties of PALG-18 films were probed using dynamic mechanical spectroscopy (DMS) and thermomechanical analysis (TMA). The temperature range was limited by the physical properties of the rather brittle film samples. Preliminary results of a DMS temperature sweep (at 1 Hz) correlate reasonably well with thermally stimulated current measurements to be described elsewhere*(17)*. The loss tangent of the DMS curve exhibited maxima at -100°C, -10°C and 20°C. After the onset of side chain melting at 43°C, the DMS sample failed. As already mentioned, T(2) has been identified by others*(2)* as liquid crystal transition, but there is the possibility that the liquid transition (leading, for example, to mechanical failure) occurs before the transition to an oriented state. However, as the axial ratio of the polymers is more than sufficient to support the crystalline liquid state, it is not clear what

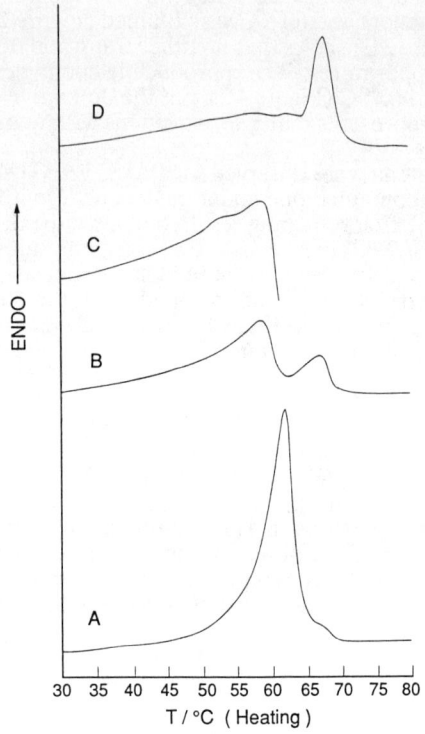

Figure 5. DSC traces of PALG-18 (heating). (A) Pristine sample (1st heating); (B) Reheated sample (2nd heating); (C) Heating of a quickly cooled sample; heating was interrupted (causing sharp downturn) once T(1) from run (B) was exceeded; (D) Sample annealed at 65°C for 60 min.

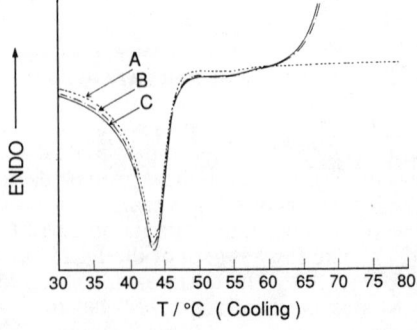

Figure 6. DSC traces of PALG-18 (cooling); (A) 1st Cooling from 110°C; (B) 2nd Cooling from 110°C; (C) Cooling of the sample annealed at 65°C and heated to 100°C.

prevents birefringence or other obvious signs of order from appearing until T(2) is exceeded. As these issues continue to be investigated, we anticipate a greater role for polymers of low molecular weight (that cannot form liquid crystals) and star-branched polymers (which might not do so without distortion of the symmetrical structure).

Acknowledgments

This work is supported by grants from Research Corporation to WHD and from NSF to PSR. We acknowledge the generosity of the Louisiana Educational Quality Support Fund for equipment acquisitions. We are grateful to Javier Nakamatsu for the synthesis of PALG-16.

Literature Cited

1. Block, H., *Poly(γ-benzyl-L-glutamate) and Other Glutamic Acid Containing Polymers;* Gordon and Breach Science Publishers: New York, N. Y., 1983, Chapter 5.
2. Watanabe, J.; Ono, H.; Uematsu, I.; Abe, A. *Macromolecules* **1985**, *18*, 2141.
3. Smith, J. C.; Woody, R. W. *Biopolymers* **1973**, *12*, 2657.
4. Daly, W.H.; Poche, D. S. *Polymer Preprints* **1989**, *30(1)*, 107.
5. Sakamoto, R.; Osawa, A. *Mol. Cryst. Liq. Cryst.* **1987**, *153*, 305.
6. Kricheldorf, H. R. α-*Aminoacid-N-Carboxy Anhydrides and Related Heterocycles. Syntheses, Properties, Peptide Synthesis, Polymerization;* Springer-Verlag, Berlin, 1987, Chapter 2.
7. Daly, W. H.; Russo, P. S.; Poche, D. S., Second Euro-American Conference on Functional Polymers and Biopolymers, Oxford, *Macromolecules Preprints,* **1989**, p. 25.
8. Poche, D. S. *Synthesis and Characterization of Linear and Star Branched Poly(γ-stearyl-L-glutamate),* PhD Dissertation, Louisiana State University, **1990**.
9. Poche, D. S.; Daly, W. H.; Russo P. S. *Polymer Preprints* **1990**, *31(2)*, 639.
10. Poche, D. S.; Daly, W. H.; Russo, P. S. *Polymer Preprints* **1990**, *31(2)*, 418.
11. Flory, P.; Ronca, G. *Mol. Cryst. Liq. Cryst.* **1979**, *54*, 289.
12. Flory, P. *Adv. Polymer Sci.* **1984**, *59*, 1.
13. Wasserman, D.; Garber, J.D.; Meigs, F.M., U. S. Patent 3,285,953, 1966.
14. Watanabe, J.; Goto, M.; Nagase, T. *Macromolecules* **1987**, *20*, 298.
15. Wegner, G.; Duda, G.; Bubeck, C.; Schouten, A. J., Ger. Offen. DE 3,724,542, 1989; *Chem. Abstr.* **1989**, *111*, 116449.
16. Ishii, T,; Sato, T., Eur. Pat. Appl. EP 232,113, 1987; *Chem. Abstr.* **1988**, *109*, 30223.
17. Neagu, E.; Neagu, E.; Daly, W. H.; Negulescu, I. I. *IEEE Trans. Electr. Insul.* accepted, **1991**.

RECEIVED January 3, 1992

Chapter 24

A Novel Route to Poly(ether ketone)–Polycondensate Block Copolymers

Robert J. Kumpf[1], Dittmar Nerger[2], Christopher Lantman[1], Harald Pielartzik[1], and Rolf Wehrmann[2]

[1] Mobay Corporation, Pittsburgh, PA 15205
[2] Bayer AG, Krefeld D–4150, Germany

Poly (etherketones) containing ester groups were synthesized and transesterified with an aromatic polycarbonate to give block copolymers. Block copolymer formation was proven using SEC, film studies, DSC, DMA, and TEM. The poly(etherketone) / polycarbonate block copolymers formed clear films whereas the analogous blends were cloudy and macrophase separated. The block copolymers exhibited only a single Tg as measured by DSC but further DMA and TEM studies showed the block copolymers to be microphase separated. The domains were ≈ 10 nm. The block copolymers showed synergistic physical properties superior to the analogous blends.

Block copolymers based on amorphous and semicrystalline engineering thermoplastics are of current scientific and industrial interest (*1,2*). These materials are candidates for applications which require very high performance . Block copolymers are especially attractive because they often exhibit synergistic physical properties - that is, the properties of the block copolymer are superior to the constituent homopolymers. For example, McGrath and coworkers found that the impact strength of a polysulfone / polycarbonate block copolymer was far superior to either homopolymer or the analogous physical blend (*3*).

Because of the ability of block copolymers to assemble into specific morphologies, they are predicted to modify surfaces and compatibilize immiscible blends (*4,5*). The results can be quite dramatic. Quirk and coworkers studied blends of polystyrene and polybutylene terephthalate. The binary blends were very brittle but became tough and ductile when compatibilized with 10 wt% of a polystyrene / polybutylene terephthalate AB diblock copolymer (*6*).

Block copolymers of engineering resins are classically prepared by synthesizing telechelic oligomers and then coupling these oligomers to form block copolymers (*7*). This route is scientifically appealing for its rigor; the final composition and microstructure of the block copolymer are well controlled. This approach is, however, often not industrially practical since it usually entails modifying a continuous production process, an untenable scenario.

Because of these commercial limitations, recent work has concentrated on devising practical routes to block copolymers of engineering resins. Such work has often focused on preparing block copolymers in processing equipment such as kneaders and twin screw extruders. Some examples are known. For instance, commercial polycarbonate / polyester "blends" are actually kinetically controlled block copolymers. Also, Mullins has used a very interesting ether cleavage / decarboxylation reaction to prepare polysulfone / polycarbonate block copolymers in a Haake mixer (8).

We are interested in this "reactive processing" approach to preparing block copolymers. In this paper, we describe a new synthetic route to block copolymers via transesterification. In comparison to conventional coupling reactions of telechelic oligomers, this route is applicable to a wide range of polycondensates and can be readily performed on larger scales. The one requirement: the condensation polymer must contain moieties that undergo cross reaction with aromatic esters. A schematic of the general approach is shown in Figure 1.

Specifically, in this paper we describe the synthesis and characterization of a poly(etherketone) which contains random ester groups and the subsequent block poly(aryl ether)-co-carbonates prepared therefrom.

Experimental

Materials. 4'-Hydroxyphenyl-4-hydroxy benzoate was prepared directly from hydroxybenzoic acid and hydroquinone and then purified as described by Kargas et. al. (9). Monomer grade bisphenol A (Mobay) and difluorobenzophenone (Cipsy Chemical) were used as received. Potassium carbonate (Aldrich) was dried in vacuum then stored in a drybox. N-Methyl-2-pyrolidinone (NMP) was either vacuum distilled from P_2O_5 or used as received (Aldrich-HPLC grade). Toluene and chlorobenzene (Aldrich) were used as received. Commercial grade polycarbonate was obtained additive free in-house (designated APEC HT).

Polymer Characterization. Molecular weights were measured by size exclusion chromotography (SEC) in tetrahydrofuran using polystyrene standards. Glass transition temperatures (Tg's) were determined by dynamic scanning calorimetry (DSC) using a Perkin-Elmer DSC 7 and a heating rate of 20°C / min. All Tg values were taken from second heating scans to ensure a consistent thermal history. Polymer films were prepared by casting $MeCl_2$ solutions at constant thickness with a drawdown bar. Dynamic mechanical spectra were obtained on a Reometrics RDA II at 1 Hz and a heating rate of 2.5°C / min. Infrared spectra were recorded on a Nicolet FTIR spectrophotometer. Samples for transmission electron microscopy were prepared as follows: solution cast films were dried at 80 oC under vacuum; imbeddded in a crosslinked polyurethane; microtomed and then stained with RuO_4.

Synthesis of Ester-Containing Poly(etherketone). Difluorobenzophenone (21.82 g: 0.1 mol), bisphenol A (21.73 g; 0.0952 mol), 4'-hydroxy phenyl - 4-hydroxybenzoate (1.104 g; 0.0048 mol.), and K_2CO_3 (15.2 g; 0.11 mol) were combined with 100 ml of NMP and 65 ml toluene in a 250 ml 3-neck flask (all glassware was previously flame dried). A mechanical stirrer, Dean-Stark trap (12 mL volume) with an N_2 outlet and a Claisen adapter with N_2 inlet and thermocouple were attached and the entire system was purged with N_2 for 10 minutes. The solution was heated at 155°C for 8 hours. During this time toluene / H_2O collected in the Dean-Stark trap. After 8 hours 20 ml of toluene / H_2O was allowed to drain from the trap and the temperature was raised to 180°C. After 5 additional hours the toluene was drained and the temperature increased to 187°C for 2 hours. During this time the yellow/green solution became very viscous. The solution was cooled to room temperature, diluted with 100 ml NMP, then poured into a large excess of MeOH to precipitate a fibrous white polymer which was collected and redissolved in $MeCl_2$. The $MeCl_2$ solution was washed with 10% HCl, H_2O, then poured into MeOH to precipitate the polymer which was dried in vacuum at 80°C for 8 hours. A typical reaction gave a yield of 97% - 99% polymer with an inherent viscosity (NMP, 30°C) of 0.37 - 0.50 dL/g.

Synthesis of Polycarbonate/Poly (etherketone) Block Copolymer. A 50wt% / 50wt% (polycarbonate / poly(etherketone) block copolymer was synthesized by combining ester-containing poly(etherketone) (8.75 g), polycarbonate (8.75 g), and 30.0 ml chlorobenzene in a 100 ml resin kettle equipped with mechanical stirrer, N_2 inlet and short path condenser. The mixture was degassed 3 times, placed under a continuous N_2 flow and then heated to 275°C. The chlorobenzene quickly distilled off to give a polymer melt which was stirred for a total of 2.5 hours at 275°C. The block copolymer was cooled to room temperature, dissolved in $MeCl_2$, precipitated into MeOH then dried in vacuo at 80°C for 8 hours. Block copolymers of other compositions were prepared by changing the relative amounts of ester-containing poly(etherketone) and polycarbonate.

Alternately, the block copolymer was formed by melt blending the ester-containing polyester ketone and the polycarbonate for 5 minutes at 250°C in a Haake mixer.

Results and Discussion

Poly (etherketones) which contain ester groups were prepared as shown in Figure 2. Difluorobenzophenone was reacted with a mixture of bisphenol A and 4'-hydroxyphenyl-4-hydroxy benzoate (ester bisphenol) in the presence of potassium carbonate in an aprotic solvent such as NMP. High polymer was formed through a nucleophilic substitution stepwise polymerization. Obviously side reactions involving the aromatic ester group were a concern. Accordingly the reaction was studied in some depth to prove that the ester bisphenol was incorporated into the polymer backbone. Two side reactions of concern were endblocking of the polymer chain by the ester bisphenol (because of the decreased reactivity of the carbonyl end of the molecule) and hydrolysis of the ester bond.

Ester hydrolysis was easily ruled out using FTIR spectroscopy. Figure 3 shows the infrared spectrum for an ester PEK (16.7 mole% ester bisphenol / 83.3 mole% BPA). The absorption band due to the ester carbonyl stretch is clearly visible at 1735 cm^{-1}. There is, however, no evidence of bands due to the benzoic acid group which would result (upon acid workup) from hydrolysis of the ester bisphenol.

Endblocking was ruled out by saponification experiments (8) in which the ester bond was selectively cleaved and the decrease in molecular weight was measured; poly(etherketone) homopolymer (no ester groups) was essentially unchanged in molecular weight whereas ester-containing polymers decreased in proportion to the amount of ester bisphenol (see Table I). This proves that the ester bisphenol is indeed incorporated into the polymer backbone. The saponification experiment is also important because it shows that the average block length of the poly(etherketone) segment is controlled by the ratio of BPA to ester bisphenol. A relative increase in the amount of ester bisphenol leads to a decrease in the number of poly(etherketone) repeat units between ester groups.

Table I: Saponification study

BPA: Ester bisphenol	Mw	Mw*	% change
homopolymer	44600	46100	+ 3%
20:1	47000	25400	- 45%
10:1	34200	14000	- 60%

Molecular weight by SEC; Mw - before saponification; Mw* - after saponification

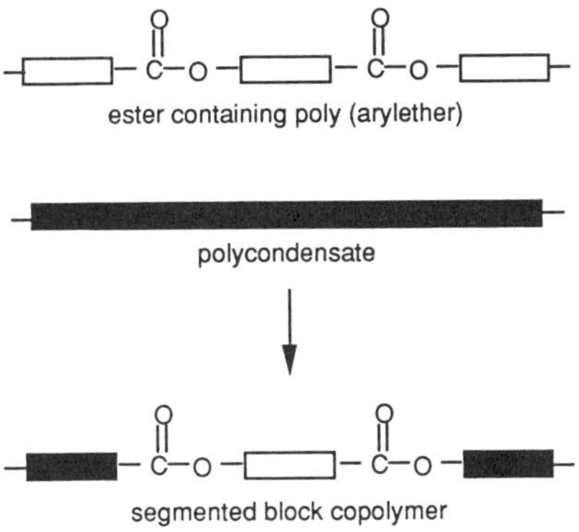

Figure 1: General approach to preparing block copolymers

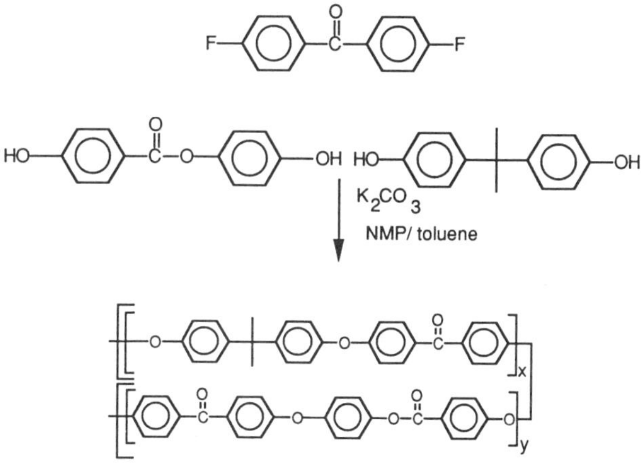

Typical results for y = 4.8 mole %:
Yield = 97%
Mw = 47,000
PD = 2.1
Inherent viscosity = 0.50

Figure 2: Synthesis of poly (etherketones) containing ester groups

Synthesis of Poly (etherketone) / Polycarbonate Block Copolymer. To reiterate, our general approach to preparing block copolymers of engineering resins is as follows. Poly(arylethers) which contain ester groups along the polymer backbone are reacted with condensation polymers. If the condensation polymers can undergo cross reactions with aromatic ester groups, a block copolymer will result as transesterification reactions couple poly(arylether) segments to the polycondensate. The number of ester groups in the backbone and the overall composition (hence the ratio of ester groups / labile groups) will control the block copolymer microstructure.

As an initial model study we examined the transesterification of an ester PEK and a polycarbonate. Devaux et al. have - with daunting thoroughness - investigated all the possible reactions between polybutylene terephthalate and BPA polycarbonate (a good model for our reaction). They conclude that alcoholysis by phenol does not occur and that the dominant process is direct transesterification (10,11).

The synthesis of a poly(etherketone) / polycarbonate block copolymer is shown in Figure 4 (note that only one of the possible linking groups is shown). APEC HT was deliberately chosen because its Tg is higher than BPA based polycarbonate and thus thermal analysis could be used as a characterization tool (BPA polycarbonate and BPA poly(etherketone) have similar Tg's). Initial reactions were conducted in a resin flask using chlorobenzene as a mixing aid. The chlorobenzene was quickly distilled off to give a polymer melt which was stirred at 275 °C for 3 hours.

The product of this reaction was characterized using numerous techniques. Our first task was to prove that a block copolymer had indeed formed and that it had formed through the tranesterification process described above rather than some other mechanism. Accordingly we will discuss the characterization of these materials in some depth.

Characterization of Poly (etherketone) / Polycarbonate Block Copolymer. There are three possible products from this reaction: a simple blend, a microphase separated block copolymer, or a homogeneous block copolymer. To distinguish between the possibilities we used size exclusion chromotography (SEC), film studies, dynamic scanning calorimetry (DSC), dynamic mechanical analysis (DMA), and transmission electron microscopy (TEM).

When block copolymers are prepared using telechelic oligomers SEC can be used to distinguish a blend of oligomers from a true block copolymer. Unfortunately this is not the case with block copolymers prepared via transesterification - the reactants are themselves high polymers. SEC curves of APEC, ester PEK, and the resulting block copolymer (80 wt% polycarbonate / 20 wt% poly(etherketone)) are reproduced in Figure 5. The results are consistent but not conclusive. The molecular weight of the block copolymer is intermediate between the constituent homopolymers and the polydispersity is not broadend. The results do prove, however, that the product of the transesterification is a high molecular weight polymer and reaction paths that would lead to degradation of molecular weight do not dominate.

A simple method to distinguish a true amorphous block copolymer from an amorphous blend is to cast films from a common solvent. Immiscible blends will assemble into cloudy, macrophase separated films (assuming dissimilar refractive indices) whereas a film of a block copolymer will be clear. Films of block copolymers and blends were cast from MeCl$_2$. The results are collected in table II. As expected, physical blends of APEC with ester PEK's and with BPA poly(etherketone) gave cloudy, macrophase separated films. The products of melt reactions between APEC and BPA poly(etherketone) (no ester groups) also gave cloudy films. Only the products of the transesterification of ester PEK's and APEC gave clear films.

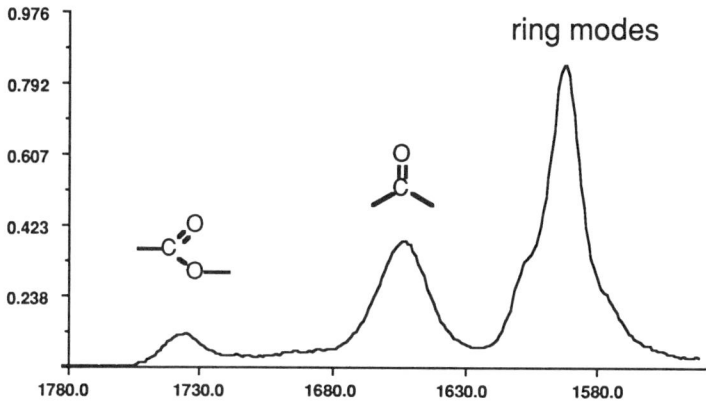

Figure 3: FTIR spectrum of ester-containing poly(etherketone) (carbonyl region)

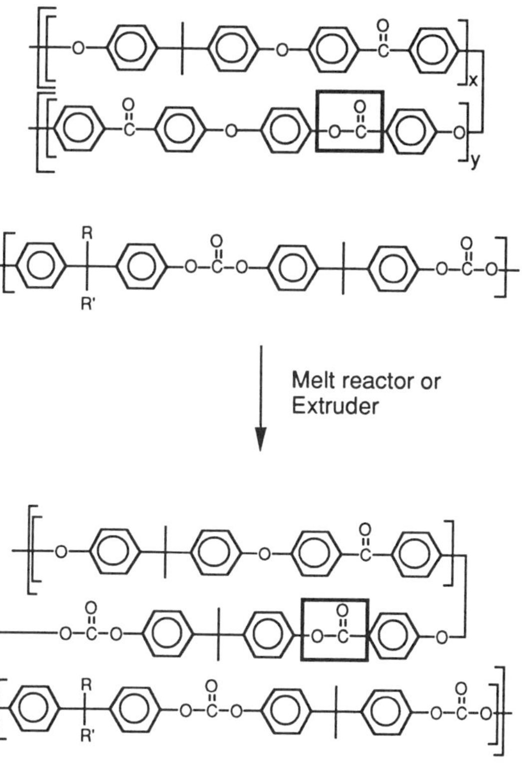

Figure 4: Synthesis of poly(etherketone) / polycarbonate block copolymer

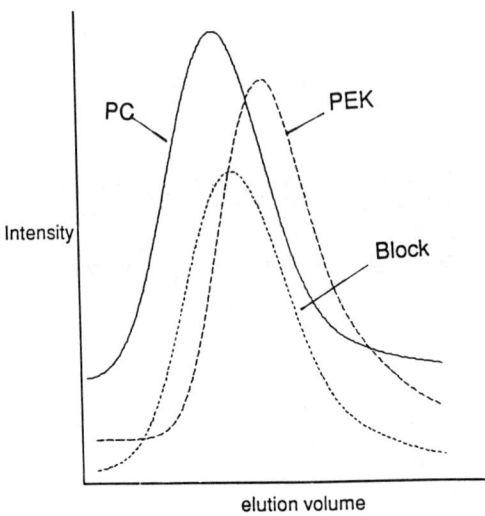

Figure 5: SEC curves for polycarbonate (bottom), ester-containing poly (ether ketone) (middle), and resulting block copolymer (top)

Table II: Composition and morphology of films cast from MeCl$_2$

Film	Description	[1]Composition	Film Morphology
1	solution blend	80/20	cloudy, discontinuous
2	solution blend	80/20[2]	cloudy, discontinuous
3	melt blend	80/20[2]	cloudy, discontinuous
4	block copolymer	80/20	transparent

[1] wt% polycarbonate / wt% ester poly(etherketone); [2] poly(etherketone) homopolymer

Two conclusions can be drawn from this experiment. One, the product of the transesterification reaction is a true block copolymer rather than a simple blend. Two, the block copolymer forms via reaction of the ester groups of the ester PEK with the carbonate groups in the APEC rather than from an ether cleavage reaction. If the latter were the case the control reaction between a poly(etherketone) homopolymer and APEC (film 3) would also have given a transparent film.

To determine the morphology of these block copolymers we performed a combination of physical experiments. Since the constituent homopolymers are immsicible, the block copolymers should strive to phase separate. First, DSC experiments were performed on precipitated powder samples of the block copolymers. These materials exhibit a single Tg intermediate between the two homopolymers (see Figure 6 and 7). These DSC results suggest that the block copolymers are homogeneous when precipitated as powders.

Of course detection of microphase separation is a function of sample history and the sensitivity of the particular analytical technique employed. When the block copolymer is isolated from solution by a less rapid method, such as solution casting, the system has more time to self assemble into its preferred microphase separated state. To study this self-assembly process in more detail we performed DMA on solution cast films (which were dried at 80 °C for 12 hours). This technique is known to be especially sensitive. The DMA plot of a 50/50 (wt% poly(etherketone) / wt% polycarbonate) block copolymer is shown in Figure 8. There are clearly two relaxations in the tan δ spectrum ; this indicates that microphase separation is occurring. DSC experiments were then repeated on these same films. During the first scan a single Tg was again observed. After annealing at 250 °C for 15 minutes this transition broadened and shifted to higher temperatures; again this is indicative of microphase separation.

TEM experiments confirmed microphase separation and quantified the size scale of the microphases. Figure 9 reproduces micrographs of a 50/50 blend and the corresponding 50/50 block copolymer. The blend is comprised of macrophases of poly(etherketone) (diam $\geq$ 100 nm) dispersed in a polycarbonate matrix. The block copolymer differs substantially in morphology; microphases of PC (diam. $\approx$ 10 nm) are dispersed in a poly(etherketone) continuous phase.

Hence we observe that the self-assembly process is governed by both thermodynamic and kinetic forces in this system. The inherent chemical immiscibility of the different blocks drives the system to microphase separation but this process is kinetically limited. When rapidly precipitated (as powders) these materials do not have sufficient time to self-assemble but when films are prepared more slowly by solution casting, the self-assembly process leads to a more microphase separated morphology.

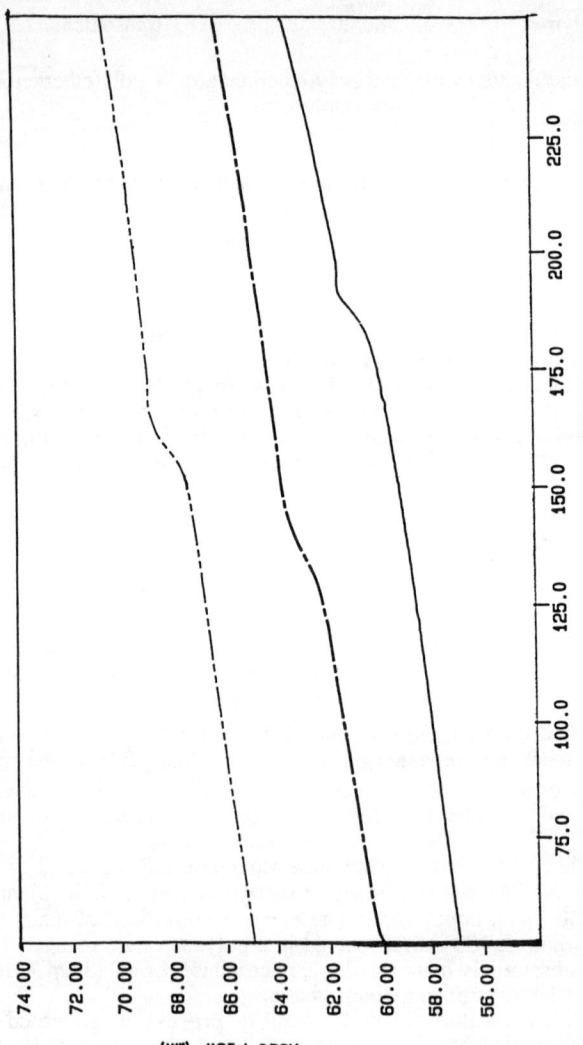

Figure 6: DSC plot of polycarbonate (bottom), ester-containing poly (ether ketone) (middle), and resulting 50/50 block copolymer (top) (all precipitated powders)

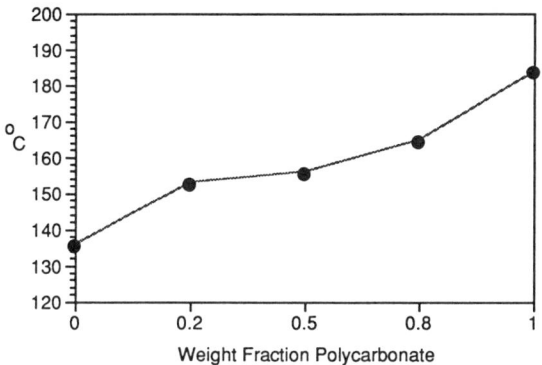

Figure 7: Tg of block copolymers vs composition

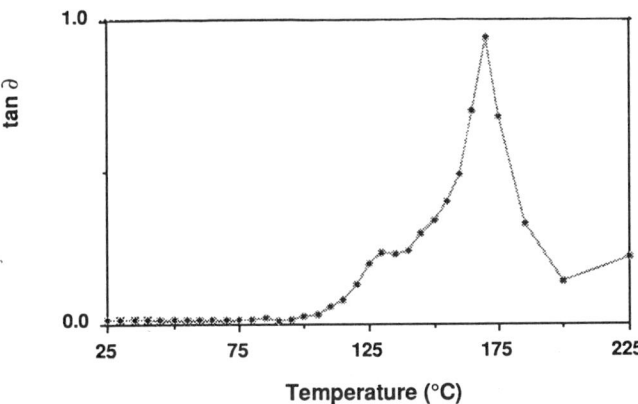

Figure 8: DMA plot of a 50/50 (wt% poly(etherketone) / wt% polycarbonate) block copolymer (solution cast film)

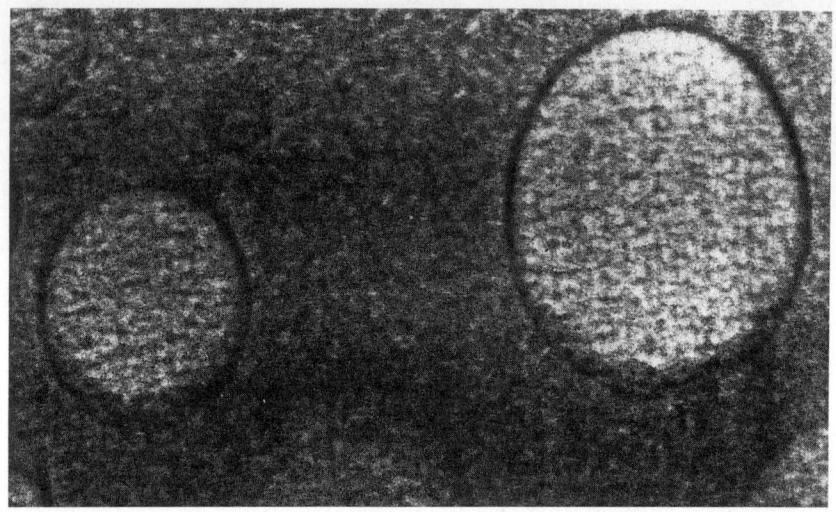

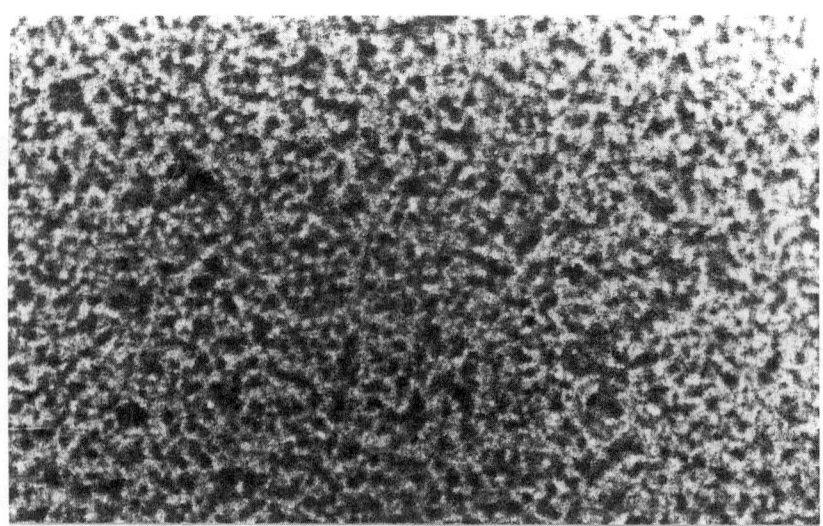

Figure 9: Electron micrographs of a 50/50 blend (top) and a 50/50 block copolymer(bottom)

Detection of this microphase separation is also dependent on the sensitivity of the technique. Our studies indicate that DMA is more sensitive than DSC. Similar differences between films and powders were found by McGrath and coworkers with block copolymers of BPA poly (ethersulfone) and BPA polycarbonate (*12*)

The size scale of microphase separation in block copolymers is controlled by many factors but is primarily influenced by the average block length of the segments. The average domain size should be on the order of one to two times the average block length. In the present system, the average length of each block can be calculated from the stoichiometry. The average DP of the polycarbonate segments in a 50/50 block copolymer was found to be 30. From this value and the average steplength of polycarbonate (10Å), the RMS end-to-end distance of an average polycarbonate block was calculated to be 55Å. This correlates nicely to the domain size observed by TEM (100Å). A study is underway to examine the influence of structure and block length on the domain size - and subsequently on the physical properties - of poly(arylether) / polycarbonate block copolymers.

Taken together the results of the SEC, film, DSC, DMA, & TEM experiments are compelling. The ester groups along the backbone of the poly(etherketone) transesterify with the carbonate groups in the polycarbonate. The result is a true block copolymer that (at these particular block lengths) is microphase separated on a scale readily discerned by DMA and TEM.

Physical Properties

As mentioned earlier block copolymers sometimes exhibit synergistic physical properties. Studies on the physical properties of these block copolymers have been quite encouraging. The storage modulus (as measured by DMA) of a block copolymer (consisting of 80% polycarbonate and 20% poly(etherketone) increased 33% as compared to pure polycarbonate. As compared to blends of the same composition, injection molded block copolymers showed increased modulus and impact strength and superior flame properties. The dependence of these various physical properties on morphology (macrophase separated vs microphase separated) is quite interesting and will be studied in detail.

Reactive Processing

Because of the great potential of block copolymers derived from engineering thermoplastics, interest in new, industrially feasible synthetic routes to block copolymers remains high. One approach is to do chemistry in processing equipment rather than in chemical reactors. Since transesterification reactions are known to be very fast, we were interested in studying the formation of block copolymers from ester PEK's in processing equipment. Preliminary experiments showed that the block copolymers described above could easily be prepared in a Haake mixer and a twin screw extruder. Further reactive processing experiments are underway. The results of these experiments will be reported elsewhere.

Conclusion

Poly (etherketones) containing ester groups were prepared and reacted with polycarbonates to give poly(etherketone) / polycarbonate block copolymers. Although the constituent homopolymers are immiscible and simple blends form cloudy, macrophase separated films, the the block copolymers which result from transesterification form transparent films. The rapidly precipitated powders had a single Tg (as measured by DSC) and appeared to be homogeneous block copolymers. When less severe kinetic constraints are imposed during sample preparation (solution cast films) the self-assembly process has enough time to form microphases which can be detected by DMA and TEM. The observed domain size (approximately 10 nm) can be explained in terms of the average block length. Detection of this microphase separation is also dependent on the sensitivity of the technique. Our studies indicate that DMA and TEM are more sensitive than DSC. The block copolymers showed interesting physical properties. Preliminary experiments indicated that these block copolymers could be prepared in processing equipment such as mixers and twin screw extruders.

Acknowledgments

The authors would like to thank David Derikart for his help with the preparation of the block copolymers and Alex Karbach for the TEM photomicrographs.

Literature Cited

1) Jensen, B.J.; Hergenrother, P.M.; Bass, R.G. *Polym Prep.* **1990**, *31(1)*, 618.
2) Hedrick, J.L.; Labadie, J.W. *J.Polym.Sci. Chem.* **1990**, *28*, 2255.
3) McGrath, J.E.; Robeson, L.M.; Matzner, M.; Barclay, R. *J.Polym. Sci. Symp.* **1977**, *60*, 29
4) Paul, D.R.; Locke, C.E.; Vinson, G.E. *Polym. Eng. Sci.* **1973**, *13*, 202.
5) Balazs, A.C.; Siemasko, C.P.; Lantman, C.W. *J. Chem. Phys.* A.C. Balazs, C.P. Siemasko, C.W. Lantman, *J. Chem. Phys.* **1991**, *94(2)*, 1653.
6) Yoshida, M.; Ma, J.; White. J.; Quirk, R. *Polym. Eng. Sci.* **1990**, *30*, 30.
7) Noshay, A.; McGrath, J.E. "Block Copolymers: Overview and Critical Survey" Academic Press, New York, **1977**.
8) Mullins, M. EP 0 353 478 A1 (assigned to DOW)
9) Kargas, S.L.; Menzies, R.H.; Wang, D.; Jones, F.N. *Polym. Prep.* **1989**, *30(2)*, 462.
10) Devaux, J.; Godard, P.; Mercier, *J. Polym. Sci. Chem.* **1982**, *20*,1875, 1895, & 1901.
11) Devaux, J.; Godard, P.; Mercier, J.P., Touillaux, R.; Dereppe, J.M. *J. Polym. Sci. Chem.* **1982**, *20*, 1881.
12) Wnuk A.; McGrath, J.E; Ward,T.C. *Government Report 14072.7-C*, **1980**.

RECEIVED December 10, 1991

INDEXES

Author Index

Advincula, R. C., 10
Anderson, Valerie C., 154
Angeloni, A. S., 280
Balazs, Anna C., 1
Blumstein, A., 10
Brinkhuis, R. H. G., 49
Cabrerizo-Vilchez, M. A., 135
Caretti, D., 280
Chen, James K., 227
Chiellini, E., 280
Chu, Yen-Ho, 227
Coleman, L. B., 83
Cornelio-Clark, Paula A., 113
Daly, William H., 292
Dixon-Northern, B., 242
Duran, R. S., 10,20,31
Ford, W., 20
Gálvez-Ruiz, M. J., 135
Galli, G., 280
Gardella, Jr., Joseph A., 113
Gramain, P., 20
Grinstaff, Mark W., 218
Henry, R. A., 94
Hoover, J. M., 94,104
Huruguen, J. P., 171
Johnson, Jr., Robert W., 113
Koren, Roni, 83
Kumpf, Robert J., 300
Lantman, Christopher, 300
Laus, M., 280
Leidner, Charles R., 202
Lindsay, G. A., 94
Liu, Min D., 202
Longo, M. L., 242
Nadler, M. P., 94
Nee, S. F., 94
Negulescu, Ioan I., 292
Nerger, Dittmar, 300
Page, Darren L., 256
Patterson, Dale H., 202
Petit, C., 171
Pielartzik, Harald, 300
Pileni, M. P., 171
Pizziconi, Vincent B., 256
Poche, Drew S., 292
Prime, Kevin L., 227
Prouty, Muriel S., 184
Rabolt, John F., 104
Reichert, W. M., 122
Rikukawa, M., 64
Ringsdorf, H., 20
Roberts, M. J., 10
Royappa, A. T., 76
Rubner, M. F., 64,76
Russo, Paul S., 292
Saperstein, David D., 104
Scheper, William M., 202
Schmid, Walther, 227
Schouten, A. J., 49
Schuster, A., 20
Seltzer, M. D., 94
Seto, Christopher T., 227
Skoulios, A., 20
Spaltenstein, Andreas, 227
Stenger-Smith, J. D., 83,94
Stroeve, Pieter, 1,83,104
Suslick, Kenneth S., 218
Thibodeaux, A. F., 20
Thompson, David H., 154
Wehrmann, Rolf, 300
Whitesides, George M., 227
Woodward, J. T., 242
Zasadzinski, J. A. N., 242
Zemb, T., 171
Zerkowski, Jonathan A., 227
Zhang, X., 10
Zhao, Shulei, 122
Zhou, Huanchun, 31

Affiliation Index

Arizona State University, 256
Bayer AG, 300
CEN Saclay, 171
Cornell University,
Dow Louisiana, 292
Duke University, 122
Harvard University, 227
IBM, 104
Institut Charles Sadron, 20
Johannes Gutenburg Universität, 20
Louisiana State University, 292
Massachusetts Institute of Technology, 64,76
Mobay Corporation, 300
Naval Weapons Center, 83,94,104
Oklahoma State University, 20
Oregon Graduate Institute of Science
 and Technology, 154
Purdue University, 202
State University of New York
 at Buffalo, 113
Università di Bologna, 280
Università di Pisa, 280
Université Pierre et Marie Curie, 171
University of California—Davis, 1,83,104
University of California—Santa
 Barbara, 242
University of the District
 of Columbia, 184
University of Florida, 10,20,31
University of Granada, 135
University of Groningen, 49
University of Illinois, 218
University of Lowell, 10
University of Pittsburgh, 1

Subject Index

A

Accordion polymer samples, 95f,96
Acetylenic compounds, liquid-crystalline,
 behavior at air–water interface, 10–18
Aerosol OT reverse micelles containing
 cytochrome c
 distance between particle determination,
  179,180t
 effect of cytochrome c on conductivity,
  175,176f
 effect of cytochrome c on critical
  exponents, 177,178t
 effect of temperature on conductivity,
  175,176f,177
 effect of temperature on permittivity
  vs. frequency, 177,178f
 effect of temperature on static
  permittivity, 175,176f,177
 experimental procedure, 174–175
 formation of spheroidal aggregates, 172
 particle distance for geometric
  model, 181
 percolation threshold for Aerosol
  OT–water–isooctane solution, 172–174
 radii from X-ray scattering, 179,180t

Aerosol OT reverse micelles containing
 cytochrome c—Continued
 scattered intensity determination, 177,179,180
 scattering behavior, 179,180f
 specific interface determination, 179,181
Affinity gel electrophoresis, advantages
 for study of receptor–ligand
 interactions, 227–228
Affinity polymers, molecular recognition
 in gels, 227–231
Affinity surfaces, Langmuir–Blodgett,
 targeted binding of avidin to biotin-doped
 films at tip of optical fiber sensor, 122–132
Air–aqueous solution interface, effect of
 temperature and subphase pH on
 lecithin–bile acid monolayers, 136–152
Air-filled microbubbles, description, 219
Air–water interface
 behavior of liquid-crystalline
  acetylenic compounds, 10–18
 behavior of side-chain liquid-
  crystalline polymer blends, 20–30
Aluminum chlorophthalocyanine
 tetrasulfonate, role in photoinduced
 morphological changes in plasmalogen
 liposomes using visible light, 156–167

INDEX

Aqueous solution–air interface, effect of temperature and subphase pH on lecithin–bile acid monolayers, 136–152
Atomic force microscopy
 applications, 242–243
 Langmuir–Blodgett films, 247–252
 phospholipid bilayers, 251–254
 principles of operation, 245
Average barrier speed, relationship to mean molecular area and polymerization rate, 38,40
Avidin, targeted binding to biotin-doped Langmuir–Blodgett films, 122–132

B

Bacterial photosynthetic reaction center, example of nature's structuring to elicit specific redox reactions, 202
Bilayer membranes, reduction of phospholipid quinones, kinetics and mechanism, 202–217
Bilayers, surfactant, scanning probe microscopy, 242–255
Bile acid–lecithin monolayers, effect of temperature and subphase pH at air–aqueous interface, 136–152
Bimolecular model, ligand–receptor binding, 127
Biological cellular assembly, STM, 259,260f
Biological membranes, main structural pattern and nonbilayer structures, 171
Biological self-assembly, interest in understanding cellular self-assembly processes, 256
Biological structures, STM, 258–259
Biomacromolecular assemblies, functions, 3
Biotin-doped Langmuir–Blodgett films, targeted binding of avidin, 122–134
Block copolymers
 amorphous and semicrystalline engineering thermoplastics, 300
 engineering resins, reactive processing synthetic approach, 301,303f

C

Carbonic anhydrase B, characteristics, 228
Cell adhesion
 multistep paradigm, 259,260f
 schematic representation of spatial arrangement and order, 260,261f
Cell binding sequences, universal molecular recognition sequences, 261f,262t
Cellular adhesive proteins, applications, 256
Cellular self-assemblies, molecular bioengineering, use of STM, 256–278
Chemically selective layers, potential applications in sensing, 122
Chenodeoxycholic acid–lecithin monolayers, effect of temperature and subphase pH at air–aqueous solution interface, 136–143
Chiral liquid-crystalline copolymers for electrooptical applications
 characterization measurement procedure, 283
 copolymerization procedure, 282
 fiber X-ray diagram of smectic phase of (S)-2-methylbutoxy polymer, 286,288f,289
 liquid-crystalline properties of various copolymers, 284–286,289t
 monomer synthetic procedure, 282
 phase transition entropies for (S)-2-methylbutoxy–N-(hexyloxy)methyl copolymers, 286,287f
 properties of polymers, 283
 smectic to isotropic transition temperatures vs. weight fraction for (S)-2-methylbutoxy–N-decyloxy copolymers, 285,287f
 structures, 281–282
 synthetic reaction pathway for methacrylates, 283–284
 X-ray diagram of smectic phase of N-decyloxy polymer, 286,288f,289
Chiral smectic C* mesophase of thermotropic polymers, 280–281
Chiral liquid-crystalline copolymers for electrooptical applications, 280–291
Cholic acid–lecithin monolayers, effect of temperature and subphase pH at air–aqueous solution interface, 136–137,146,149–152

Chromophoric polymer, *See* Fluorinated main-chain chromophoric polymer
Conducting polymers, Langmuir–Blodgett films of polyion complexes, 76–82
Crystalline ultrathin film design and growth, research interest, 2
α-Cyanocinnamate chromophores, future research, 102
Cyanuric acid, idealized structure of complex with melamine, 234
Cytochrome *c*, effect on percolation process in Aerosol OT reverse micelles, 171–181

D

Deoxycholic acid–lecithin monolayers, effect of temperature and subphase pH at air–aqueous solution interface, 136–137,143–148
Deoxy sickle cell hemoglobin, 184–201
  comparison of condensation in macromolecular and simple systems, 195,196*f*
  effect of temperature on osmotic pressure, 193*f*,194
  experimental apparatus, 192–194
  gelation inhibitor tests, 197,200
  gels, 188
  phase transitions in various phosphate buffers and media, 194–199
  polymerization mechanism, 186,188
  polymer packing and alignment in stressed gels, 195,196*f*
  structure, 186,187*f*
  ultracentrifugation test of solubility, 194–195
Diethylbarbituric acid, X-ray crystal structure of complex, 235
Diphenylmelamine, X-ray crystal structure of complex with diethylbarbituric acid, 235
Dip stick type evanescent fiber optic sensor, 123,124*f*
Docusate sodium reverse micelles containing cytochrome *c*, percolation process and structural study, 171–180

E

Electrically conducting polymers, 76
Electroactive multilayer thin films, factors affecting properties, 64
Electrooptical applications, chiral liquid-crystalline copolymers, 280–289
Encapsulation of pharmaceuticals in liposomes, 154–155
Erythrocytes, affinity polymer inhibition of influenza-induced agglutination, 229–230,231*f*
Extracellular matrix, 256,260,261*f*,262*t*
Extracellular matrix adhesive protein, 262–264*f*,265,266*t*,267,268–269*f*
  STM studies, 267,271–272*f*,273

F

Fibronectin, 263–264*f*,267,270–271*f*,272*f*,273
Fluorescence intensity of fluorescently labeled avidin, 126
Fluorinated main-chain chromophoric optically nonlinear polymers, 94–103
  characteristics, 96,97*t*
  ellipsometric characterization, 99*t*
  experimental procedure, 96
  film thickness measurement by null ellipsometry, 100*f*
  FTIR spectroscopy, 83–92,101
  future research, 102
  information obtained from ellipsometric measurements, 100
  pressure–area isotherms of Langmuir–Blodgett films, 96,97*f*,98
  second harmonic generation, 101*t*,102
  UV–visible spectra, 98–99*f*
Fluorinated main-chain chromophoric polymer
  effect of aging on Langmuir layer, 87,88*f*
  effect of temperature on compression–expansion curves, 87,88*f*
  equilibrium isotherm for Langmuir layer, 84–85,86*f*
  experimental procedure, 84
  Fourier transform IR spectroscopy of Langmuir–Blodgett films, 87,89–92
  IR band assignment, 90*t*

INDEX

Fluorinated main-chain chromophoric polymer—*Continued*
  IR reflection–adsorption spectra, 87–92
  IR transmission spectra, 90–92
  molecular structure, 84f
  stability of surface pressure, 85,86f,87
Fourier transform IR studies, fluorinated main-chain chromophoric polymer, 83–92
Freeze–fracture replication STM effect of freezing rate on results, 246,247,248f,249f

G

Gelation inhibitors, description, 189
Gelation of deoxy sickle cell hemoglobin, mechanism, 188
Gels of affinity polymers, molecular recognition, 228–230

H

Hemoglobin, deoxy sickle cell, polymerization and phase transitions, 184–201
4-Hexadecylaniline
  effect of environmental conditions on isotherm, 36–37
  isotherm on sulfuric acid, 32,33f,34
  isotherm on water, 32,33f,34
  pictorial representation of conformation at air–water interface, 34,35f
  structures, 32,33f
hubM$_3$–R(CA)$_2$ complex
  ^{1}H-NMR spectra of titration with R(CA)$_2$, 235,237f
  schematic representation of self-assembly, 235,236f
Hysteresis technique, description, 21

I

Influenza-induced agglutination of erythrocytes, affinity-polymer inhibition, 229–230,231f
Initial mean molecular area, 37–38
Isotactic poly(methyl methacrylate)
  monolayer behavior, 49–55
  thin film behavior, 56–62

K

Kinetics, reduction of phospholipid quinones in bilayer membranes, 202–217

L

Laminins
  binding studies, 267,269–270f
  STM studies, 267,271f
  schematic representation, 262,263f
  TEM, 262,263f
Langmuir–Blodgett affinity surfaces, targeted binding of avidin to biotin-doped films at tip of optical fiber sensor, 122–132
Langmuir–Blodgett deposition, formation of macromolecular assemblies, 2–3
Langmuir–Blodgett film(s)
  atomic force microscopy, 247,250–251f,252
  multilayers, static SIMS of sampling depth, 113–120
  polyion complexes of conducting polymers, 77–82
  stilbazolium chloride polyethers, 105–109
  See also Langmuir–Blodgett thin films
Langmuir–Blodgett multilayers
  fluorinated main-chain chromophoric optically nonlinear polymers, 94–102
  static SIMS analysis, 113–121
Langmuir–Blodgett polymerization of 2-pentadecylaniline, 31–48
  comparison to bulk polymerization, 40–41
  effect of applied surface pressure, 41–48
  effect of blend, 44,45f
  effect of spread volume on mean molecular area and average barrier speed, 38,39f,40
  kinetics, 46,47f,48
  polymerization rate determination, 38–39
  procedure, 37
  surface pressure, mean molecular area, and average barrier speed vs. reaction time, 37,39f
Langmuir–Blodgett technique
  advantages, 31,76,94–95
  studies of liquid-crystalline acetylenic compound behavior at air–water interface, 10–18

Langmuir–Blodgett thin films of
poly(3-hexylthiophene) and
3-octadecanoylpyrrole, 64–75
  conductivities of doped films, 71,73t
  effect of doping with FeCl$_3$ on
    absorption spectra, 71,73,74f
  effect of temperature on conductivity, 73,74f
  electrical properties, 71,73t,74f
  Fourier transform IR spectra, 66,68f,69
  polarized absorption spectra, 71,72f
  structural differences from pure
    3-octadecanoylpyrrole system, 71,72f
  surface pressure–area isotherm, 66,67f
  X-ray diffraction pattern, 69,70f,71
Langmuir layer studies, fluorinated
  main-chain chromophoric polymer, 83–92
Lecithin–bile acid monolayers, effect of
  temperature and subphase pH at
  air–aqueous solution interface, 136–152
Ligand–receptor binding, bimolecular
  model, 127
Linear poly(γ-alkyl-α-L-glutamate)s, 292–293
Lipidic particles, structural
  arrangements, 171
Liposomes, plasmalogen, photoinduced
  morphological changes using visible
  light, 154–170
Liquid crystal(s), self-organization
  ability, 10
Liquid-crystalline acetylenic compounds at
  air–water interface, 10–19
Liquid-crystalline polymers
  advantages, 20
  characteristics, 280
  chiral copolymers for electrooptical
    applications, 280–291
  poly(γ-alkyl-α,L-glutamate)s, side-chain
    crystallinity and thermal
    transitions, 292–299
  properties, 5
  side-chain, behavior at air–water
    interface, 20–30

M

Macromolecular assemblies, 1–5
  advantages of Langmuir–Blodgett
    fabrication, 94–95
  elucidation using STM, 256–273

Main-chain chromophoric, optically nonlinear
  polymers, fluorinated, 83–103
Mean molecular area, 38,40
Mechanism, reduction of phospholipid
  quinones in bilayer membranes, 202–217
Melamine, idealized structure of complex
  with cyanuric acid, 234
Membranes, bilayer, reduction of phospholipid quinones, kinetics and
  mechanism, 202–217
Metamorphic mosaic model of
  biomembranes, description, 171
Micelles, reverse, docusate sodium
  containing cytochrome c, percolation
  process and structural study, 171–180
Microencapsulation, applications, 218
Microspheres
  compositions, 218–219
  proteinaceous, See Proteinaceous
    microspheres
Mixed monolayers of lecithin and bile acids
  at air–aqueous solution interface, 135–152
  effect of pH on molecular area vs. lecithin
    mole fraction for chenodeoxycholic acid
    monolayers, 137,140–141f,142t
  effect of pH on molecular area vs.
    lecithin mole fraction for cholic
    acid monolayers, 146,151f,152
  effect of pH on molecular area vs. lecithin
    mole fraction for deoxycholic acid
    monolayers, 143,146,147–148f
  effect of temperature on molecular area
    vs. lecithin mole fraction for
    chenodeoxycholic acid monolayers,
    142,143t
  effect of temperature on molecular area
    vs. lecithin mole fraction for
    cholic acid monolayers, 146,151f,152
  effect of temperature on molecular area
    vs. lecithin mole fraction for deoxycholic
    acid monolayers, 143,146,147–148f
  experimental procedure, 136–137
  forces controlling interaction between
    amphiphilic molecules, 136
  surface pressure–area isotherms for
    chenodeoxycholic acid monolayers,
    137,138–139f
  surface pressure–area isotherms for
    cholic acid monolayers,
    146,149–150f

Mixed monolayers of lecithin and bile acids at air–aqueous solution interface—*Continued*
  surface pressure–area isotherms for deoxycholic acid monolayers, 143,144–145f
Molecular bioengineering of cellular self-assemblies in molecular device design, elucidation of macromolecular assemblies, 256–273
Molecular device–system design in bioengineering and biotechnology, 256–278
Molecular recognition in gels of affinity polymers, 228–230,231f
Molecular recognition in monolayers and solids, self-assembly, 230–237
Monolayer(s)
  mixed, lecithin and bile acids at air–aqueous solution interface, 135–152
  molecular recognition, 230–237
  surfactant, scanning probe microscopy, 242–255
Monolayer behavior of isotactic poly(methyl methacrylate), 49–63
  comparison to syndiotactic monolayer behavior, 49–51,51f
  crystallization kinetics, 52,53f,54
  crystal structure, 50
  effect of molecular weight, 54,55f
  effect of temperature on crystallization, 52
  isobaric stabilization experiment, 52,53f
  pressure–area isotherms, 50,51f
  schematic representation of crystallization process, 52,53f,54
  structural nature of monolayer transition, 50,51f,52
Morphological changes, photoinduced, in plasmalogen liposomes using visible light, 154–170
Multicomponent Langmuir–Blodgett thin films, poly(3-hexylthiophene) and 3-octadeca-noylpyrrole, preparation, 64–74

N

Nonaqueous liquid filled microcapsules, 219
Noncovalent macromolecules, self-assembly, 234–237f

Nonlinear optical polymers, noncentrosymmetric arrangement of polarizable species, 94–95
Number of water droplets, definition, 174

O

3-Octadecanoylpyrrole thin films
  Fourier transform IR spectra, 66,68f,69
  structural differences from multilayer films, 71,72f
  structure of pyrrole, 65,67f
  surface pressure–area isotherm, 66,67f
  synthesis of pyrrole, 65
  X-ray diffraction pattern, 69,70f,71
Operating principles of scanning probe microscopes, 243–245
Optical computing and communications technologies, requirements for advancements, 94
Optical fiber sensor, targeted binding of avidin to biotin-doped Langmuir–Blodgett films, 122–132
Optically nonlinear polymers, fluorinated main-chain chromophoric, 83–103
Organic materials
  damage threshold to laser irradiation, 1
  with large nonlinear susceptibilities, organization into films, 104
Organic polymers, second-order nonlinear optical properties, 281
Osmotic stress method, 189,191–192,193f,196f

P

2-Pentadecylaniline
  effect of environmental conditions on isotherm, 36–37
  effect of pH on surface pressure onset points, 34,35f,36
  isotherm on sulfuric acid, 32,33f,34
  isotherm on water, 32,33f,34
  Langmuir–Blodgett polymerization, 37–41
  pictorial representation of conformation at air–water interface, 34,35f
  structures, 32,33f
  twist energy, 36

Percolation process, Aerosol OT reverse micelles containing cytochrome c, 171–181
Percolation threshold, Aerosol OT–water–isooctane solution, 172–174
pH, subphase, effect on mixed monolayers of lecithin and bile acids at the air–aqueous solution interface, 135–152
Phase transitions of deoxy sickle cell hemoglobin, 193–200
Phospholipid bilayers, atomic force microscopy, 252–254
Phospholipid quinones, reduction in bilayer membranes, 202–217
  effect of BH_4^- addition on percent quinone reduced, 206,208,209f
  effect of $S_2O_4^{2-}$ addition on UV-visible spectra, 206,207f
  effect of $S_2O_4^{2-}$ concentration on pseudo-first-order rate constants, 210,213f,214
  effect of $S_2O_4^{2-}$ on kinetics, 210,213f
  experimental materials, 204
  identification of quinone equilibrium, 214,216
  liposome preparation procedure, 204
  mechanism of $S_2O_4^{2-}$–quinone reaction, 214,215–216f
  NMR spectra, 208,210,211f
  phospholipid structures, 204–205
  representation of quinone head group, 210,212f
  size exclusion chromatograms for sonicated solutions, 206,207f
  spectrophotometric experimental procedure, 204
  stopped-flow experimental procedure, 204,206
Photodynamic therapy of diseased sites, factors affecting effectiveness, 155
Photoinduced morphological changes in plasmalogen liposomes using visible light, 154–170
  comparison to previous studies, 163–164
  dark-subtracted apparent glucose release kinetics, 161,164f
  DSC results, 158,159t
  effect of irradiation in presence of oxygen, 166–167
  effect of NaCl on photoinduced glucose release, 161,162f
Photoinduced morphological changes in plasmalogen liposomes using visible light—*Continued*
  effect of sensitizer on photoinduced glucose release, 159,160f,161
  effect of sodium azide on photoinduced glucose release, 161,162f
  electron micrographs before and after irradiation, 163,165f
  experimental procedure, 156
  factors affecting gel-to-liquid-crystalline transition, 158,159t
  penetration efficiency of $^1O_2$, 166
  photoinduced glucose release, 157
  plasmalogen liposome preparation, 157
  products isolated in photolyzed samples, 167
  structures of plasmalogen and photosensitizer, 156–157
  TLC, 157
  TLC analysis of photolytic products, 161,163t
  UV–visible absorption spectra, 163,164f
Phototriggered delivery systems, 155
Plasmalogen liposomes, photoinduced morphological changes using visible light, 154–170
Polyacetylenes, hyperpolarizability, 10
Poly(γ-alkyl-α-L-glutamate)s, *See* Thermotropic liquid-crystalline poly(γ-alkyl-α-L-glutamate)s
Polycondensate–poly(ether ketone) block copolymers, *See* Poly(ether ketone)–polycondensate block copolymers
Polyether(s), stilbazolium chloride, Langmuir–Blodgett films, structural studies, 104–111
Poly(ether ketone)–polycondensate block copolymers, 300–311
  block copolymer synthetic procedure, 302,304,305f
  characterization of block copolymer, 304,306–311
  composition and morphology of films cast from $MeCl_2$, 304,306t
  DSC of polymer and block copolymer, 306,308f
  dynamic mechanical analysis, 306,309f
  ester-containing poly(ether ketone) synthetic procedure, 301–302,303f

INDEX

Poly(ether ketone)–polycondensate block copolymers—*Continued*
  experimental materials, 301
  Fourier transform IR spectrum of ester-containing poly(ether ketone), 302,305*f*
  glass transition temperature vs. composition, 306,309*f*
  physical properties, 311
  polymer characterization procedure, 301
  reactive processing, 311
  saponification study results, 302*t*
  SEC of polymers and block copolymers, 304,307*f*
  TEM, 306,310*f*
Poly(3-hexylthiophene) thin films, 65,66,67*f*
Polyion complexes of conducting polymers, Langmuir–Blodgett films, 76–82
Polymer(s)
  conducting, Langmuir–Blodgett films of polyion complexes, 76–82
  fluorinated main-chain chromophoric optically nonlinear, 83–103
  use in ultrathin films, 2
Polymeric chromophores, future optoelectronic applications, 83
Polymeric glutamic acid esters, 293
Polymeric matrices and vesicles, use as delivery vehicles for controlled release of drugs, 154
Polymerization of deoxy sickle cell hemoglobin, mechanism, 186,188,190*f*
Polymerization rate, relationship to barrier speed and mean molecular area, 38,40
Poly(methyl methacrylate), isotactic
  monolayer behavior, 49–55
  thin film behavior, 56–62
Poly(octadecyl methacrylate), effect on properties of liquid-crystalline polymers, 21–30
Poly(thiophene-3-acetic acid), 77
Proteinaceous microspheres, 218–226
  effect of acoustic power on microcapsule and microbubble formation, 219,222*f*
  effect of disulfide bond formation on formation, 223
  effect of radical traps on microcapsule formation, 223,224*f*
  mechanism of sonochemical synthesis, 219,223,224*f*

Proteinaceous microspheres—*Continued*
  morphology, 219,220–221*f*
  particle distribution, 219,221*f*
  scanning electron micrograph, 219,220*f*
  stability, 219,223,224*f*
Protein adsorption to man-made surfaces, use of self-assembled monolayers as substrates for studying mechanisms, 231

Q

Q cycle in respiratory energy transduction, schematic representation, 202,203*f*
Quadrupole mass filter secondary ion mass spectrometry, advantages, 114

R

Red blood cell, effect of deoxy sickle cell hemoglobin, 185
Reduction of phospholipid quinones in bilayer membranes, kinetics and mechanism, 202–216
Relative fluorescence intensity, measurement procedure, 126–127
Reverse micelles, docusate sodium containing cytochrome *c*, percolation process and structural study, 171–180

S

Sampling depth of Langmuir–Blodgett film multilayers studied using static secondary ion mass spectrometry, 113–121
  analysis of arachidic acid salt films on germanium, 117,119*t*
  analysis of fatty acid(s) on germanium, 117,119*t*
  analysis of fatty acid(s) on silver, 116*t*
  analysis of fatty acid salt films on germanium, 117,120*t*
  barium adsorption onto germanium surface, 117,118*f*
  experimental procedure, 115–116
Scanning probe microscopy
  applications, 257
  principles of operation, 243–245,257,258

Scanning probe microscopy—*Continued*
  STM of biological structures, 258–259
  surfactant bilayers and monolayers, 246–254
Scanning tunneling microscopy
  biological cellular assembly, 259,260*f*
  biological structures, 258–259
  description, 257
  elucidation of macromolecular assemblies, 256–278
  fibronectin, 267,272*f*,273
  laminin, 267,271*f*
  principles of operation, 242–245,257,258
  schematic representation of tip hardware, 258*f*
Scattered intensity, definition, 177,179
Secondary ion mass spectrometry, static, analysis of sampling depth of Langmuir–Blodgett film multilayers, 113–120
Self-assembled monolayers, substrates for studying mechanisms of adsorption of proteins to man-made surfaces, 231,232–233*f*
Self-assembled structures, 230–231
Self-assemblies
  cellular, STM for molecular bioengineering, 256–278
  description, 230
  formation of macromolecular assemblies, 2
  molecular recognition in monolayers and solids, 230–237
  noncovalent macromolecules, 234–237*f*
  use as synthetic strategy, 231
Semisynthetic plasmalogen, photoinduced morphological changes using visible light, 156–157
Sickle cell anemia
  approaches to therapy, 189,190*t*
  hemoglobin, polymerization and phase transitions, 184–201
  role of sickle cell hemoglobin molecule in pathophysiology, 184–185
Side-chain crystallinity, thermotropic liquid-crystalline poly(γ-alkyl-α-L-glutamate)s, 297,298*f*
Side-chain liquid-crystalline polymer blends at air–water interface, 20–30
  deposition of polymer, 28,29*f*
  deposition of poly(octadecyl methacrylate), 29

Side-chain liquid-crystalline polymer blends at air–water interface—*Continued*
  deposition of poly(octadecyl methacrylate) blends, 29
  effect of temperature on polymer stability, 26
  effect of temperature on poly(octadecyl methacrylate) blend stability, 26–27,28*f*
  effect of temperature on poly(octadecyl methacrylate) stability, 26,28*f*
  experimental materials, 21
  future research, 30
  isotherm of poly(octadecyl methacrylate), 23,24*f*
  isotherm of pure polymer, 22*f*,23
  isotherm of stearic acid blend, 23,24*f*
  isotherm procedure, 21–22
  mean molecular area vs. concentration for immiscible polymer blend, 23–24,26*f*
  mean molecular area vs. poly(octadecyl methacrylate) concentration, 23,25*f*
  mean molecular area vs. stearic acid concentration, 23,25*f*
  stability of polymer, 26,27*f*
  stability of poly(octadecyl methacrylate) blend, 26,27*f*
Single monolayers, preparation technique, 2
Solids, molecular recognition, 230–237
Specific interface, definition, 179,181
Stability of monolayer film, 21
Star-branched poly(γ-alkyl-α-L-glutamate)s, 292–295
Static secondary ion mass spectrometry, analysis of sampling depth of Langmuir–Blodgett film multilayers, 113–120
Stearic acid
  effect of environmental conditions on isotherm, 36–37
  effect on properties of liquid-crystalline polymers, 21–26
  isotherm on sulfuric acid, 32,33*f*,34
  isotherm on water, 32,33*f*,34
  structure, 21*f*,32,33*f*
Stilbazolium chloride polyethers, Langmuir–Blodgett films, structural studies, 104–111

INDEX

Structural study in docusate sodium reverse micelles containing cytochrome c, 171–180
Subphase pH, effect on mixed monolayers of lecithin and bile acids at the air–aqueaus solution interface, 135–152
Surface pressure, effect on Langmuir–Blodgett polymerization of 2-pentadecylaniline, 31–48
  effect on apparent activation energy, 43
  effect on barrier speed vs. reaction time, 41,42f,43t
  effect on conformation, 44,46
  effect on kinetics, 46,47f,48
  effect on mean molecular area vs. reaction time, 41,42f,43t
  effect on polymerization rate, 44,45f
Surfactant bilayers and monolayers, scanning probe microscopy, 242–255

T

Targeted binding of avidin to biotin-doped Langmuir–Blodgett films at tip of optical fiber sensor, 122–134
  avidin binding experimental procedure, 126
  best fit values of fluorescence intensity, 129,130f
  bimolecular model of ligand–receptor binding, 127
  deposition of biotinylated monolayers, 125–126,128f
  detection limit, 131,132t
  effect of surface density on binding, 129,131
  fiber transfer technique, 125–126,128f
  fitted parameters, 129,130t
  fusion of fibers, 123,124f,125
  relative fluorescent intensity determination procedure, 126–127
  sensor fabrication procedure, 123,124f,125
  sensor sensitivity, 131,132t
  time course of binding, 127,128f,129
Temperature, effect on mixed monolayers of lecithin and bile acids at the air–aqueaus solution interface, 135–152
Therapeutic strategies for deoxy sickle cell hemoglobin modifications, designs, 189,190t
Thermal transitions, thermotropic liquid-crystalline poly(γ-alkyl-α-L-glutamate)s, 294–297
Thermotropic liquid-crystalline poly(γ-alkyl-α-L-glutamate)s
  cholesteric liquid-crystalline structure, 294,295f
  DSC traces of thermal transition on cooling, 294,296f,297
  DSC traces of thermal transition on heating, 294,296f,297
  effect of side-chain crystallinity on thermal transitions on cooling and heating, 297,298f
  measurement procedure, 293
  polymer synthetic procedure, 293
  thermal behavior, 294
  viscoelastic property measurements, 297,299
Thin-film-layer behavior of isotactic poly(methyl methacrylate), 49–63
  lateral orientation parameter vs. molecular weight and concentration conditions, 56,57f
  orientation of transferred monolayer, 62
  polarized transmission spectra of isotropic amorphous films, 59,61f,62
  schematic representation of flow-induced orientation process vs. aspect ratio of crystallites, 56,58–59,60f
  schematic representation of use of Langmuir–Blodgett layers as surface crystallization nuclei, 59,60f
Third-order susceptibility, influencing factors, 12
Time of flight SIMS, advantages, 114
Total internal reflection fluorescence spectroscopy and microscopy, 123
Transmission electron microscopes, limitations, 243
Triggered release targeted drug delivery methods, research and development efforts, 154
Tunneling current, 243–258

U

Ultrathin films, 2
Universal molecular recognition cell
  binding sequences, comparison for
  various cell assembly systems, 261,262f

V

Visible light used to induce morphological
  changes in plasmalogen liposomes,
  154–167

Production: Peggy D. Smith
Indexing: Deborah H. Steiner
Acquisition: Anne Wilson
Cover design: Amy Meyer Phifer

Printed and bound by Maple Press, York, PA